Rice, Agriculture, and the Food Supply in Premodern Japan

A large number of studies on the agricultural history of Japan have focused on the public administration of land and production, and rice, the principal source of revenue, has received much attention. However, while this cereal has clearly played a decisive role in the public economy of the Japanese State, it can be argued that it has not had a predominant place in agricultural production. Far from confining its scope to a study of rice growing for tax purposes, this volume looks at the subsistence economy in the plant kingdom as a whole.

This book examines the history of agriculture in premodern Japan from the 8th to the 17th century, dealing with the agricultural techniques and food supply of rice, wheat, millet and other grains. Drawing extensively on material from history, literature, archaeology, ethnography and botany, it analyses each of the farming operations from sowing to harvesting, and the customs pertaining to consumption. It also challenges the widespread theory that rice cultivation has been the basis of "Japaneseness" for two millennia and the foundation of Japanese civilization by focusing on the dietary diversity of Japan that is still the basis of Japanese nutrition today. Furthermore, the book will play a role in the current dialogue on the future of sustainable agricultural production from the viewpoints of ecology, biodiversity, environment, dietary culture and food security throughout the world as traditional techniques such as crop rotation are explored in connection with the safeguarding of the minerals in the soil.

Surveying agricultural techniques across the centuries and highlighting the biodiversity and polycultural traditions of Japan, this book will appeal to students and scholars of Japanese history, the history of science and technology, medieval history, cultural anthropology and agriculture.

Charlotte von Verschuer is a Professor at the École Pratique des Hautes Études (EPHE), Paris, France. Her research interests focus on agriculture, economics, foreign relations and material culture. Her books include *Across the Perilous Sea: Japanese Trade with China and Korea from the Seventh to the Sixteenth Centuries* (2006).

Wendy Cobcroft is a translator of premodern Japanese history and literature.

Needham Research Institute Series
Series Editor: Christopher Cullen

Joseph Needham's 'Science and Civilisation' series began publication in the 1950s. At first it was seen as a piece of brilliant but isolated pioneering. However, at the beginning of the twenty-first century, it became clear that Needham's work had succeeded in creating a vibrant new intellectual field in the West. The books in this series cover topics that broadly relate to the practice of science, technology and medicine in East Asia, including China, Japan, Korea and Vietnam. The emphasis is on traditional forms of knowledge and practice, but without excluding modern studies that connect their topics with their historical and cultural context.

Celestial Lancets
A history and rationale of acupuncture and moxa
Lu Gwei-Djen and Joseph Needham
With a new introduction by Vivienne Lo

A Chinese Physician
Wang Ji and the Stone Mountain medical case histories
Joanna Grant

Chinese Mathematical Astrology
Reaching out to the stars
Ho Peng Yoke

Medieval Chinese Medicine
The Dunhuang medical manuscripts
Edited by Vivienne Lo and Christopher Cullen

Chinese Medicine in Early Communist China, 1945–1963
Medicine of revolution
Kim Taylor

Explorations in Daoism
Medicine and alchemy in literature
Ho Peng Yoke

Tibetan Medicine in the Contemporary World
Global politics of medical knowledge and practice
Laurent Pordié

The Evolution of Chinese Medicine
Northern Song dynasty, 960–1127
Asaf Goldschmidt

Speaking of Epidemics in Chinese Medicine
Disease and the geographic imagination in Late Imperial China
Marta E. Hanson

Reviving Ancient Chinese Mathematics
Mathematics, history and politics in the work of Wu Wen-Tsun
Jiri Hudecek

Rice, Agriculture, and the Food Supply in Premodern Japan
Charlotte von Verschuer
Translated and edited by Wendy Cobcroft

Rice, Agriculture, and the Food Supply in Premodern Japan

五穀文化

Charlotte von Verschuer

Translated and edited by
Wendy Cobcroft

LONDON AND NEW YORK

First published 2016
by Routledge
2 Park Square, Milton Park, Abingdon, Oxon OX14 4RN

and by Routledge
711 Third Avenue, New York, NY 10017

First issued in paperback 2017

Routledge is an imprint of the Taylor & Francis Group, an informa business

Rice, Agriculture, and the Food Supply in Premodern Japan is the revised and updated edition of Charlotte von Verschuer, *Le riz dans la culture de Heian: Mythe et réalité*, Paris: Collège de France, Institut des Hautes Études Japonaises, 2003.

Published with the assistance of the Research Centre of East Asia Cultures CRCAO, Paris.
Translation rights by kind permission of the Institut des Hautes Études Japonaises, Collège de France, Paris.

British Library Cataloguing in Publication Data
A catalogue record for this book is available from the British Library

Library of Congress Cataloging in Publication Data
Names: Verschuer, Charlotte von. | Cobcroft, Wendy.
Title: Rice, agriculture, and the food supply in premodern Japan / Charlotte von Verschuer with Wendy Cobcroft.
Description: Milton Park, Abingdon, Oxon : Routledge, 2016. | Series: Needham Research Institute series | Includes bibliographical references and index.
Identifiers: LCCN 2015011680| ISBN 9781138885219 (hardback) |
ISBN 9781315715605 (ebook)
Subjects: LCSH: Agriculture–Japan–History. | Rice–Japan–History. | Food supply –Japan–History. | Subsistence farming–Japan–History. | Agrobiodiversity–Japan –History. | Food security–Japan–History. | Japan–History–To 1185. | Japan–History –1185–1600. | Japan–Economic conditions. | Japan–Environmental conditions.
Classification: LCC S471.J3 V47 2016 | DDC 338.10952–dc23
LC record available at http://lccn.loc.gov/2015011680

ISBN 13: 978-1-138-09929-6 (pbk)
ISBN 13: 978-1-138-88521-9 (hbk)

Typeset in Times New Roman
by Taylor & Francis Books

Contents

List of illustrations

Figures

Maps

Tables

Preface

The Japanese learn from their earliest years that irrigated rice cultivation has played a dominant role in their history. Having been born outside this context, I wanted to find out when this "culture based on rice" appeared and how this cereal came to be a fundamental element of Japanese civilization. This book therefore grew out of a desire to go back to the sources of Japan's rice-growing culture, and to examine the existence of a very widespread theory according to which rice cultivation has constituted the basis of what it means to be Japanese for two millennia. If one is to believe the supporters of this theory, who are as numerous among the Japanese as they are in non-Japanese scholarly circles, rice growing is the foundation of Japanese civilization, the rice field symbolizing the land and the society of this country.

For several years I worked my way through the historical and archaeological documents of early medieval Japan, looking for elements that would show the dominant place of rice in Japanese culture. But I found nothing that led in this direction. If many present-day scholars speak of a cultural primacy of rice, this is perhaps because of the wealth of information on rice found everywhere in the historical sources. However, these are concerned mainly with government and the life of the élites, so much so that the assessments of historians tend to focus on public accounting at the expense of the subsistence economy that is far less in evidence in the texts. The public administration is the subject of many publications. I will here focus my investigations on the origins of Japanese rice cultivation in the historical sources. To do this, I had to change my standpoint and approach. I tried to put myself in the place, no longer of the government, but of the farmers, namely the majority of the population. It became clear that rice was neither a staple food nor a fundamental element of the culture of early medieval Japan. In my view, the subsistence economy and the traditions of that time are characterized rather by a cereal polyculture (including rice and millets). The early medieval written sources speak of the "five grains", a notion that symbolizes the Japanese idea of polyculture. Rather than rice alone, these diverse "five grains" played an essential role in the annual rituals and celebrations of early medieval Japan. My first aim, which was to go back to the origins of the rice-growing culture, led me to re-examine the role of rice in Japanese society. My research has led, in short, to a deconstruction of the myth of rice in early medieval Japan. On the other hand, it has revealed the full array of the country's biodiversity and the rich variety of its agricultural practices.

Acknowledgements

In the course of this work I have drawn on viewpoints from history, literature, archaeology, ethnology and botany. I have thus benefited from the judicious advice of many colleagues and friends. Francine Hérail, Marie Maurin and François Sigaut kindly read various chapters of the original French text and made many helpful suggestions. Jacqueline Pigeot and Michel Vieillard-Baron checked the French translations of many poems. The seminars of François Sigaut on agricultural techniques and of Jean-Marie Pesez on medieval archaeology gave me a much better understanding of the specificity of the Japanese context. Perrine Mane helped me to analyse the implements using the Japanese agricultural iconography that I showed her. Bruno Smolarz and Yoshio Abe were kind enough to contribute useful comments at my seminar on rice cultivation. Takeshi Watabe introduced me to the iconography of Chinese agriculture and – with Yin Shaoting – to the techniques of swidden farming (shifting cultivation) in the province of Yunnan. Sam Quoc Chan helped me to conduct a survey with the farmers of Dali, in Yunnan. Noriko Maeyama accompanied me around Japan during my research into modern swidden fields and early medieval toponyms and made a number of valuable comments. Wilhem Grootaers drew my attention to the importance of toponyms and dialects. Ishii Susumu gave me useful advice about the value of toponyms. Georges Métailié was kind enough to share his knowledge of traditional Chinese botany. Abe Gihei kept me regularly informed about the latest archaeological discoveries. Kōno Michiaki generously shared his knowledge of the agricultural techniques of early medieval Japan and provided me with a great deal of information on the subject. Kimura Shigemitsu told me a lot about the non-irrigated crops of late medieval Japan and acquainted me with the farming traditions of the village Akiyama-gō in Nagano. Toshio Araki supported me in some of my hypotheses that depart from the views of certain Japanese historians. Ishigami Eiichi often gave me advice and helped me to question his colleagues, thereby enabling me to refine my thinking on specific points. Suzuki Yasutami helped me with the philological interpretation of some early medieval texts. Jane Cobbi and Ernst Lokowandt drew my attention to the material and economic aspects, when I first went into this new field of research many years ago. Everyone I have mentioned welcomed me warmly and greatly contributed to the progress of my work by their encouragement, the valuable information they provided and the advice they gave me. I am most grateful to them all. This English edition would not have been possible without my collaborator Wendy Cobcroft. Thanks to her knowledge of Japanese history and her meticulous attention to every detail of the text, this English version has become a revised and updated edition of the original French publication.

Abbreviations

BZ	Dainihon bukkyō zensho
DNK	Dainihon komonjo
ES	Engishiki
F	Fudoki
GR	Gunsho ruiju
Hi	Heian ibun
IJ	Iroha jiruishō
K	Kojiki
Kjg	Kōtaijingū gishikichō
M	Man'yōshū
MJ	Nihon Montoku tennō jitsuroku
N	Nihon shoki
Ni	Nara ibun
NK	Nihon kōki
NKBT	Nihon koten bungaku taikei
NR	Nihon kiryaku
RGi	Ryō no gige
RK	Ruiju kokushi
RMS	Ruiju myōgishō
RR	Ritsuryō
RSh	Ryō no shūge
RSK	Ruiju sandai kyaku
ShJ	Shinsen jikyō
SJ	Nihon sandai jitsuroku
SKT	Shinpen kokka taikan
SN	Shoku Nihongi
SNK	Shoku Nihon kōki
SNKBZ	Shinpen Nihon koten bungaku zenshū
SNKS	Shinchō Nihon koten shūsei
SY	Seiji yōryaku
SZKT	Shintei zōho kokushi taikei
TM	Tōdaiji monjo
WR	Wamyō ruijushō
WSZ	Wakansansaizue
ZGR	Zoku gunsho ruiju

Introduction

Up to the 20th century, Japan's economy was based on agriculture, which provided for the needs of its inhabitants. The open countryside that has inspired poets over the centuries was characterized by rice fields, but rice was reserved for upper-class tables. For centuries, work in the irrigated and dry fields regulated the lives of a predominantly farming population. The agricultural history of Japan has been the subject of numerous studies, most of them focusing on the public administration of land and production. Rice, the principal source of revenue, has received most attention. However, while this cereal has clearly played a decisive role in the public economy of the Japanese State, it has not had a predominant place in agricultural production. Agriculture is no more synonymous with rice growing in Japan than in other Asian countries. The present work proposes an approach to agriculture that encompasses not only irrigated and dry cultivation, but also the fruits of plant gathering in uncultivated areas. Far from confining its scope to a study of rice growing for tax purposes, it looks at the subsistence economy in the plant kingdom as a whole. Leaving aside for the moment the public economy and its administration, we will attempt to paint a picture of the private (household) economy of Japan from the 8th to the 17th century.

Chapter 1 focuses on rice cultivation, especially the annual cycle of farm work from sowing in the seedbed to harvesting and storage. It also deals with the other permanent crops – cereals, vegetables and tree fruits – these too being part of a regular cycle of farming tasks. It shows the production processes and techniques mainly from the farmers' point of view as they can be observed through the administrative texts and early medieval poetry. The chapter ends with a summary of the technical innovations of the Edo period (1603–1867).

Chapter 2 describes the methods of itinerant forest agriculture, namely swidden farming or shifting cultivation and agri-sylviculture. Given the dearth of historical sources, we have borrowed the conceptual framework from ethnology and have based our documentation on early medieval Japanese lexicography, poetry and toponymy. The practice of swidden farming, which has now been largely abandoned, appears to have been very widespread both in premodern Japan and in other civilizations.

Chapter 3 gives a catalogue of the wild plants harvested by the Japanese of the medieval period, with a great diversity of fruits, nuts, vegetables, herbaceous plants and tubers, as well as wild graminae. The information drawn from the regulations relating to taxes and the kitchens of the imperial court has been supplemented by archaeological and ethnographical data.

Chapter 4 is an attempt to evaluate the production of the various agricultural sectors in the plant kingdom. We have carried out estimates of the proportion of rice and other foods in order to get some idea of the diet of the premodern Japanese. We have therefore looked at the data from the point of view of consumption and, in order to do this, we have compared the administrative texts with data from the prehistoric and late medieval periods.

Chapter 5 deals with cultural factors. It analyses the notion of grains held by the Japanese of the early medieval period and shows that rice did not occupy a dominant place in the mythology, the historiography, or the rituals and celebrations of the imperial court.

Each chapter includes a historical review of the debate on a specific problem. The questions raised relate to various topics: the Neolithic origin of swidden farming in Japan; the coexistence of the gathering of wild plants and agriculture; the role of rice in the dietary history; the theory of a so-called "rice-growing culture" that is specifically Japanese.

The book ends with a catalogue of the edible and industrial plants known from early medieval Japan (see the Appendix). To draw it up, we consulted the most reliable authors as regards the identification and names of plants, but the reconstruction of the ancient Japanese flora remains a subject for future research.

In order to situate our study in the natural setting of the Japanese archipelago, we have included an overview of the geographical environment and historical evolution of the agrarian space in Japan (see Map I.1).

The natural environment

The Japanese archipelago has several climates and two major vegetation zones. The far north-east of the country consists of a subarctic zone with a conifer forest. The north-eastern half of Honshū is characterized by a cool-temperate climate and a forest of broad-leaved deciduous trees that includes fagaceae of the species beech *buna* and oak *konara*. The north-eastern region is called by the Japanese "Beech/Fagus zone *buna-tai*" or "oak forest *nara-bayashi*". The south-west, taking in Honshū, Shikoku and most of Kyūshū, corresponds to a warm-temperate zone with a forest of shiny leaved trees, i.e. a laurel-forest *shōyō jurin*, thickly wooded with oaks of the species cyclobalanopsis *kashi* and castanopsis *shii*. However, the primary forest has receded owing to reafforestation with useful conifers. A process begun at the end of the Neolithic, reafforestation, especially with pines *matsu*, would have been

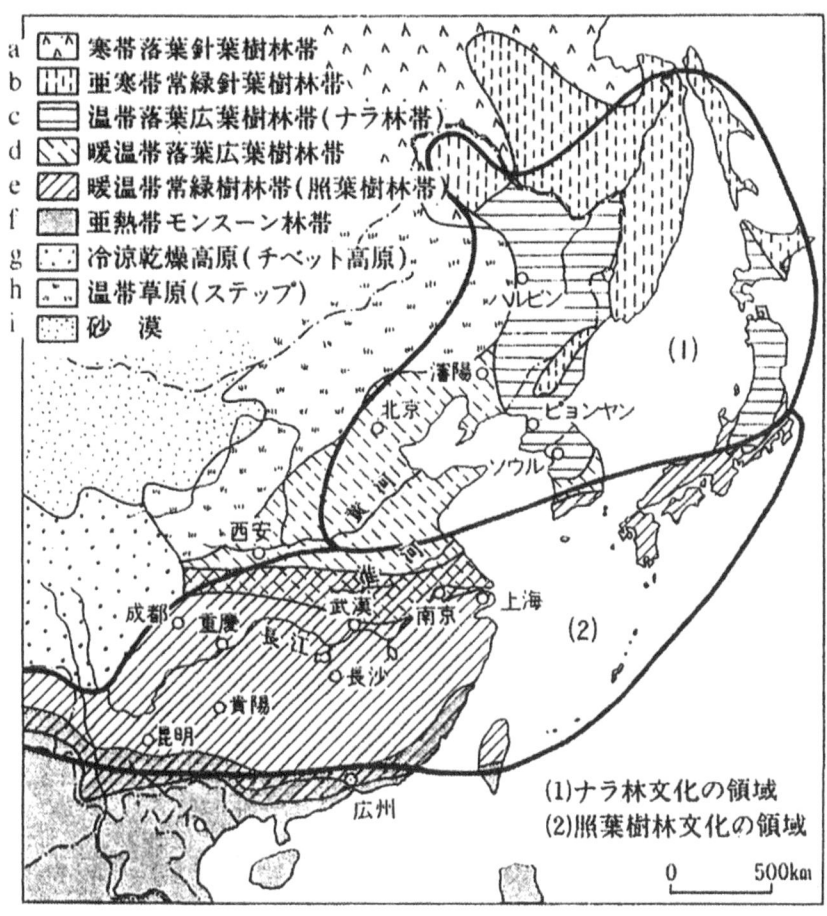

Map I.1 Vegetation zones of East Asia
Notes: (1) Beech/Fagus zone; (2) laurel-forest zone (a) Arctic zone, forest of conifers and deciduous trees; (b) subarctic zone, conifer forest; (c) cool-temperate zone, forests of broad-leaved deciduous trees, Fagus zone; (d) warm-temperate zone, forest of broad-leaved deciduous trees; (e) warm-temperate zone, forest of evergreen trees, laurel-forest zone; (f) subtropical zone, monsoon zone; (g) Tibetan plateau; (h) steppe; (i) desert.
Sources: Sasaki Kōmei, *Nihon bunka no kiso o saguru*, NHK Books, 1994, based on Zhongguo zhipi bianji weiyuanhui, ed., *Zhongguo zhipi*, Beijing Kexue Chubanshe, 1983, p. 53.

accelerated with the progress of civilization from the 5th century AD. In the 1970s, the planting of resinous trees, such as the cryptomeria *sugi* and cypress *hinoki*, was once again encouraged by the government, so much so that the forest landscape changed markedly up to very recent times. From an agricultural point of view, Japan has benefited from a dual influence from the continent. Dry crops arrived from the northern Fagus zone, while irrigated rice growing made its appearance from the southern laurel-forest zone.

The introduction of irrigated rice cultivation

Traditionally (from the mid-20th century), the introduction of irrigated rice growing into the Japanese archipelago was dated to the beginning of the Yayoi period, which in its turn was dated from 300 BC–300 AD. It was thought that rice growing spread very rapidly, in a few centuries, up to the far northeast of the island of Honshū. Since the distribution and management of irrigation systems required a social organization, the Yayoi period would have witnessed the birth of an agricultural society based on rice cultivation. This notion formed part of everyone's general knowledge and was included in Japanese primary school textbooks up to the beginning of the 21st century. However, this notion has recently been re-examined and redefined, as it is no longer supported by material evidence.

It is known today that the beginnings of irrigated rice cultivation in Japan date from the early 1st millennium BC, that this agricultural technique first arrived in the north of the island of Kyūshū and that it followed a long journey of several centuries before being adopted in other regions. Moreover, the discoveries relating to rice growing have led scholars to question the dating of the Yayoi period itself. Depending on whose work one consults, it begins around 400, 500, 780 or 950 BC. In the early 2000s, scholars tended to date the Yayoi period from the early 1st millennium BC to the late 3rd century AD as regards the introduction of irrigated rice. As of 2014, however, the dating for the beginning of Yayoi has been reassessed, so that it now coincides with the start of agriculture in general (beginning with dry crops before irrigated rice), namely between the mid-2nd and the mid-1st millennium BC (see Table A.3 in the Appendix).

Four archaeological sites have prompted theoretical speculations by those in favour of a very rapid expansion of irrigated rice cultivation in the archipelago:

- Itazuke in Fukuoka, discovered in 1978, dating from 1000–500 BC, or the end of the final Neolithic (according to the former dating);
- Nabatake in Fukuoka, discovered in 1979, dating from the middle of the final Neolithic (depending on the scholar consulted, between 800 and 400 BC);
- Tareyanagi in Aomori, dating from the middle phase of Yayoi (the first centuries around our era), discovered in 1981;
- Sunazawa in Aomori, dating from the end of the early phase of Yayoi (2nd century BC), discovered in 1987 (all according to the former dating).

While the four sites mentioned do indeed prove that irrigated rice cultivation in the south (Kyūshū) began in the early 1st millennium, these sites that are supposed to document the spread of rice growing from Kyūshū (Fukuoka) towards the far north of Honshū (Aomori) in three or four centuries remain isolated. Moreover, the spread towards the north-east (Aomori) was not overland, but via the Sea of Japan. Subsequent archaeological discoveries have shown other geographical gaps between the regions of the Inland Sea and the Aomori sites and, what is more, these new discoveries attest to the presence of rice, though seldom of irrigation. The rice discovered in the form of marks on pottery, charred grains, pollens or phytoliths does not make it possible to distinguish between irrigated and dry crops; and, apart from the two sites in Aomori, archaeologists have found few traces of hydraulic systems (irrigation canals or dykes) in remains before the 1st century AD outside Kyūshū.

On the other hand, the biologist Satō Yōichirō has revealed the genetic DNA structure of two types of rice in samples taken from sites of the Yayoi period: temperate Oryza sativa L. var. japonica (*ontai japonica*) and the tropical japonica variety (*nettai japonica*). The former, irrigated, is originally from eastern China; the latter is still being widely grown on swidden fields in many parts of South-East Asia. Temperate rice, irrigated, gradually made its way into the Japanese archipelago, after tropical dry rice which, according to Satō Yōichirō, was still grown until the Heian period (784–1185).

There remains the problem of the origin of this rice. In the present state of knowledge, it is thought that the centre of rice domestication is likely to have been in the region of the Middle and Lower Yangzi in China. However, this process has been the subject of ongoing research since the discovery of the Hemudu site in Zhejiang in the 1970s. In fact, it took several thousand years from the beginning of wild rice harvesting in the late Pleistocene until the emergence of fully domesticated rice in the early–middle Holocene. Researchers also agree that, even at this time, the selection pressure was apparently not intensive and that rice played only a minor role in the subsistence economy (Li Liu and Xingcan Chen, 2012). The Chinese archaeological sites for domesticated rice include both indica and japonica varieties and provide evidence of irrigated as well as tropical dry rice (see Map I.2). It is therefore possible that not only irrigated rice, but also tropical rice reached Japan from the Lower Yangzi region.

Transmission may have been directly via the East China Sea, or via southern Korea, albeit with a significant time difference. Irrigated rice would only have reached Kyūshū in southern Japan several thousand years after its beginnings in China (see Map I.3), that is, between 1000 and 500 BC.

Irrigation techniques subsequently spread across the Japanese archipelago (except Hokkaidō) during the Yayoi period (500 BC–300 AD), but with regional differences. The process of adopting such techniques imported from the Chinese mainland was probably not as rapid and uniform as might have been thought until recently. Dry or wet (non-irrigated) rice cultivation is attested in

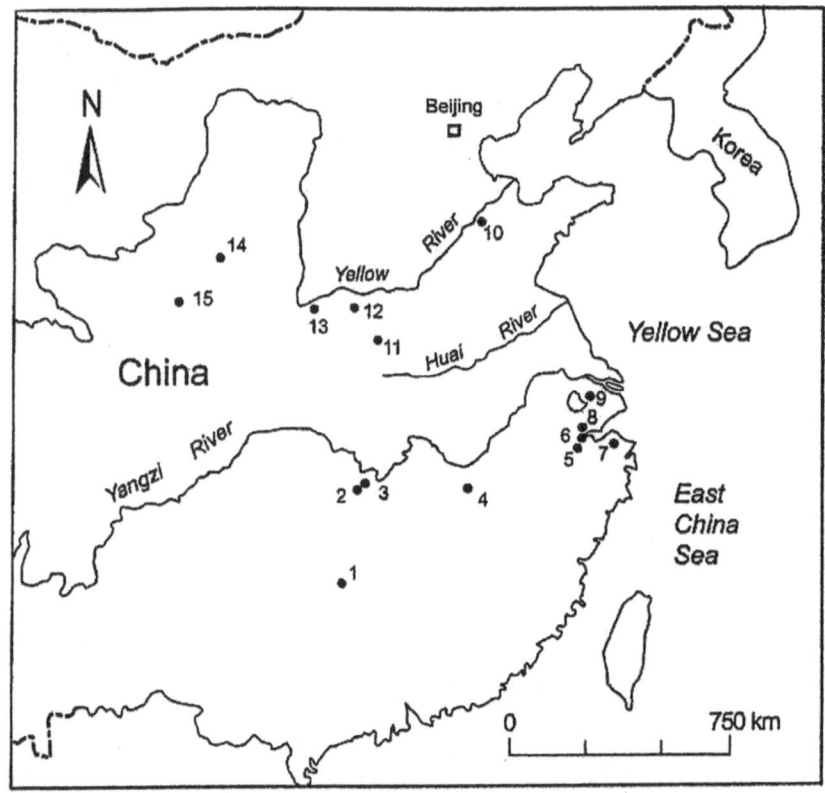

Map I.2 Archaeological sites associated with rice remains in China
Notes: Sites excavated between 1971 and 2009, including wild and domesticated rice remains, and dated to between 12000/11000 BP (or 10000/9000 BC) and 5000 BP (or 3000 BC). Numbered dots: 1. Yuchanyan; 2. Pengtoushan; 3. Bashidang; 4. Xianren-dong-Diaotonghuan; 5. Shangshan; 6. Kuahuqiao; 7. Hemudu; 8. Luojiaojiao; 9. Caoxieshan; 10. Yuezhuang; 11. Jiahu; 12. Huizui; 13. Nanjiaokou; 14. Qingyang; 15. Xishanping.
Source: Li Liu and Xingcan Chen, *The Archaeology of China: From the Late Paleolithic to the Early Bronze Age*, Cambridge: Cambridge University Press, 2012, p. 77.

the archipelago long before the introduction of irrigation techniques. Some regions may have adopted the new irrigation method for a time, without changing their system of rice production in the long term. In any event, it seems that in the Yayoi period rice was not always irrigated and that the two types, temperate (irrigated) and tropical (which may be dry or wet, and in this case rain fed), coexisted.

Several regions would have seen the emergence of agricultural societies based on rice cultivation. However, archaeological circles in Japan have in recent times distanced themselves from the traditional view of a Japanese archipelago that converted *fully* and *all at once* to a rice-growing society at the dawn of Yayoi culture (Satō Yōichirō, 1996, 2001, 2002; Negita Yoshio, 2000; Hirose Kazuo, 1997; Tamada Yoshihide, 2009; Fujio Shin'ichirō, 2011; Shitara Hiromi, 2014).

The indigenous "Japanese" population in the south of the archipelago, notably in Kyūshū, was joined by groups from the continent, mainly from the Korean peninsula, during the Yayoi period. They were farmers who practised, among other things, irrigated rice growing. The repeated waves of immigration, which no doubt continued beyond Yayoi and up to the 5th–6th century, resulted in the intermixing of peoples and the spread of irrigation techniques side by side with other non-irrigated shifting and permanent crops from Kyūshū towards the central areas.

Local élites were formed; it is likely that they took on the management of the hydraulic systems and imposed rice growing in their respective regions. These

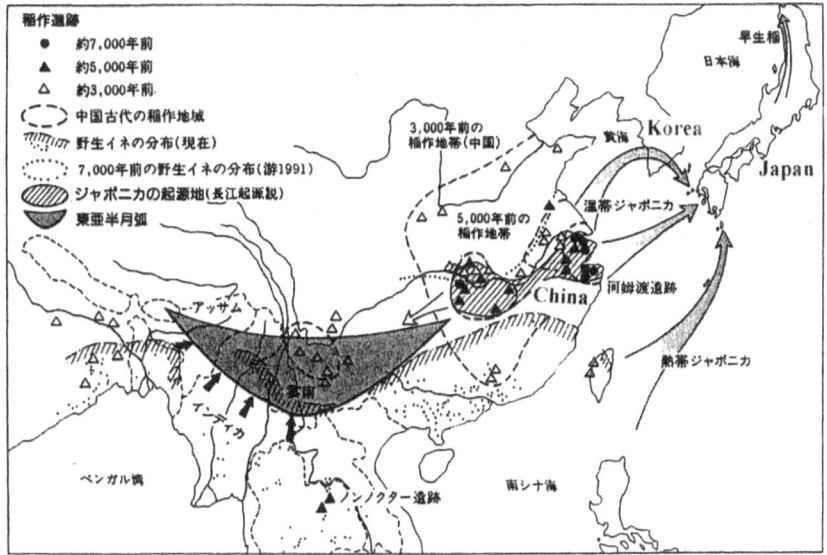

Map I.3 Transmission of irrigated rice cultivation to Japan

Notes: Rice-growing sites on the Asian continent: Black dot: site dating from 7000 BP or 5000 BC; Black triangle: site dating from 5000 BP or 3000 BC; White triangle: site dating from 3000 BP or 1000 BC; Dotted line: zone of irrigated rice cultivation in China; Striped line: zone of self-sown rice; Striped grill: probable centre of the emergence of rice cultivation in the Middle and Lower Yangzi; at the eastern extremity, the site of Hemudu at Yuyao, province of Zhejiang; Grey crescent: second centre of rice cultivation in Yunnan-Assam; Arrows: directions of the spread. The dating of Chinese sites is provided by Chinese archaeologists and has been reproduced by their Japanese counterparts.

Source: Sasaki Kōmei, *Nihon bunka no kiso o saguru*, NHK Books, 1994, p. 121; map based on data from Satō Yōichirō, *Ine no kita michi* (1992).

élites founded on rice cultivation may have gradually increased in number towards the centre of the archipelago. Then in the 7th century AD, the Yamato clan gained power. Starting from the seat of its imperial court in central Honshū, this clan imposed rice growing on the population as an obligatory system of production and in the administrative code decreed rice to be the foundation of the fiscal and land system. Henceforth, irrigated rice was the basis of the public economy of the Japanese State and remained the value standard over the centuries and up to the modern period. It might be thought that the rice-growing phenomenon reached all the territory controlled by the Yamato clan, but in our opinion this is not so. Irrigated rice cultivation was widely practised on the plains and plateaux of the archipelago, which cover about one-quarter of the land surface. Moreover, rice was largely reserved for fiscal purposes, with most of its production used to feed the élites. In our view, rice was not the main staple food of the rural populations, either during Yayoi or during the following centuries and throughout the medieval period. Furthermore, irrigated rice arrived from the continent long after the other cereals and vegetables and has coexisted with these dry crops up to the present day (Miyamoto Kazuo, 2000; Terasawa Kaoru, 2000; Harada Nobuo, 2006; Fujio Shin'ichirō, 2011; Barnes, 2015).

The introduction of dry grains

When irrigated rice cultivation reached northern Kyūshū between 1000 and 500 BC, dry rice and other graminae had already been known there for one or two millennia. Archaeologists have discovered traces of buckwheat *soba* in fourteen Neolithic sites. Buckwheat appeared with barnyard millet *hie* in Hokkaidō from the early Neolithic (4000–3000 BC) and spread to northern Honshū during the mid-Neolithic (3000–2000 BC), together with common millet *kibi* and foxtail millet *awa*. These plants are of continental origin, though for barnyard millet archaeologists have also discovered a self-sown indigenous variety. It may be noted that all these cereals spread from the north-east to the south-west. They have in fact followed an opposite path from that of irrigated rice within the archipelago. Map I.1 shows the influence of the southern laurel-forest zone, the centre of irrigated rice; but there has also been an undeniable impact of the northern Fagus zone through which dry cereals reached Japan (Map I.4).

Other edible plants have been present across the country since the early Neolithic, such as beans *ryokutō* and tubers. Dry rice was known in western Honshū and Kyūshū long before irrigated rice, i.e. from the mid-Neolithic. Given the geographical location of the Neolithic dry-rice sites in southern Japan, and the fact that tropical dry rice is attested in China (Lower Yangzi) and South-East Asia, it might be thought to have come from the laurel-forest zone. However, the possibility of another transmission route via the southern islands cannot be ruled out. A rice transmission route via the Pacific has been under consideration for a long time based on the theories of Yanagita Kunio (1875–1962), but it is not supported by any evidence (Map I.3). The existence of this route may be confirmed in the future by the results of research into the

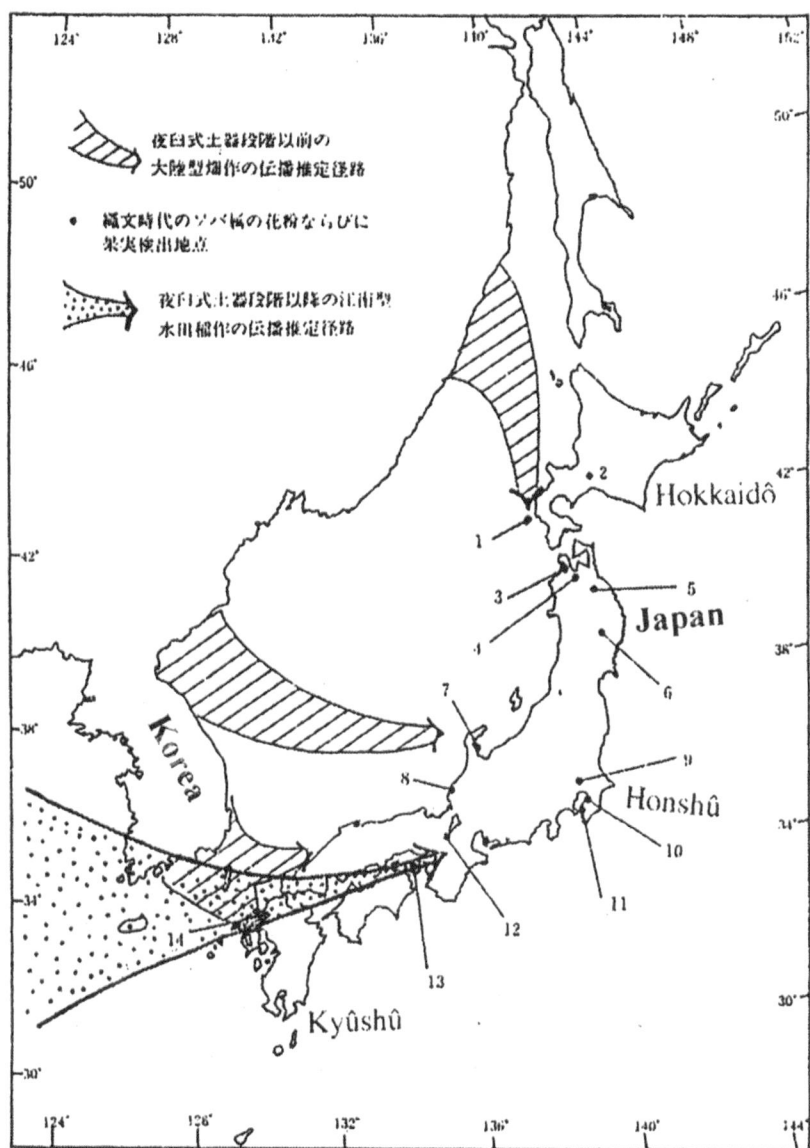

Map I.4 Transmission of cereals to Japan from the continent
Notes: Striped arrows: transmission of dry cereals from the Beech/Fagus zone; up to the late Neolithic (up to 1000 BC). Dotted arrow: transmission of irrigated rice cultivation from the laurel-forest zone; during the final Neolithic (1000–500 BC). Numbered dots: archaeological sites of the Japanese Neolithic where remains of buckwheat and/or nuts and seeds have been discovered.
Source: Yasuda Yoshinori, *Sekaishi no naka no Jōmon bunka*, Yūzankaku (1987: 214).

origin of rice, taro and/or yams. All these spontaneous plants later gave rise to pre-domestic practices including systematic collecting and self-sowing, as well as to domestication, that is, the selection and manipulation of varieties. In any event, Japan had grain acquisition techniques long before the introduction of irrigated rice growing (Yasuda Yoshinori, 1987; Miyamoto Kazuo, 2000; Satō Yōichirō, 2000; Yamaguchi Hirofumi, 2011).

The earliest forms of dry cropping are dated to the late or final Neolithic or to the early Yayoi period. The dates are the subject of a debate that will be summarized at the end of Chapter 2. But we may assume that the early forms of farming, after pre-domestic acquisition and domestication, evolved into swidden and permanent farming of dry (non-irrigated) cereals and leguminous plants, and that irrigated rice growing, far from supplanting non-irrigated agriculture, coexisted with it from that time.

Since rice fields need level surfaces, rice was first grown on lowlands, as terracing is required to prepare slopes. This is why the coastal plains and intra-mountainous basins were ideally suited to and probably reserved for irrigated rice cultivation. Yet this space represents only a very small part of the land area of the archipelago (Map I.5). In fact, between two-thirds and three-quarters of the country is covered by forests and mountains, which are the preferred locations for dry swidden or permanent agriculture and the harvesting of wild plants. The topography of Japan is thus characterized by a duality of landscape: rice-growing plains and wooded mountains (Pelletier, 1994).

The forest environment

With a forest cover of 66.5%, second only to that of Finland with 68.7%, Japan is characterized by a high rainfall (1,800 mm or twice the world average) favourable to the regrowth of forests, and these forests are situated almost entirely in the mountains. The wooded mountain is the defining element of the Japanese landscape. Mountains cover 61% of the nation's territory and hills (up to 300 m above sea level) cover 11%. Three-quarters of the country therefore consists of mountainous terrain. This proportion was perhaps close to 80% in early medieval Japan, if one excludes the areas of northern Honshū and Hokkaidō, which were not part of the country until modern times. Today, Japan has only 14% of lowland plains and 11% of moderately elevated flatlands (20–100 m above sea level).

Japan's terrain is characterized by its omnipresent slopes. Moderate or steep slopes (8–30° gradient) account for two-thirds of the total land area of the archipelago and very steep slopes (more than 30°) another one-tenth. One-quarter of the land area is on slopes of less than 8°, only half of which is actually flat (less than 3°). In other words, even the slightest reliefs stand out clearly and elevations of no more than 100 m take on the appearance of peaks, with deeply carved valleys and steep passes. This hilliness is explained by the violent action of the watercourses, which is far greater than atmospheric weathering. Aided by seasonal typhoons, with heavy and sometimes deluging rainfall, water erosion increases the number of V-shaped profiles (Berque, 1980).

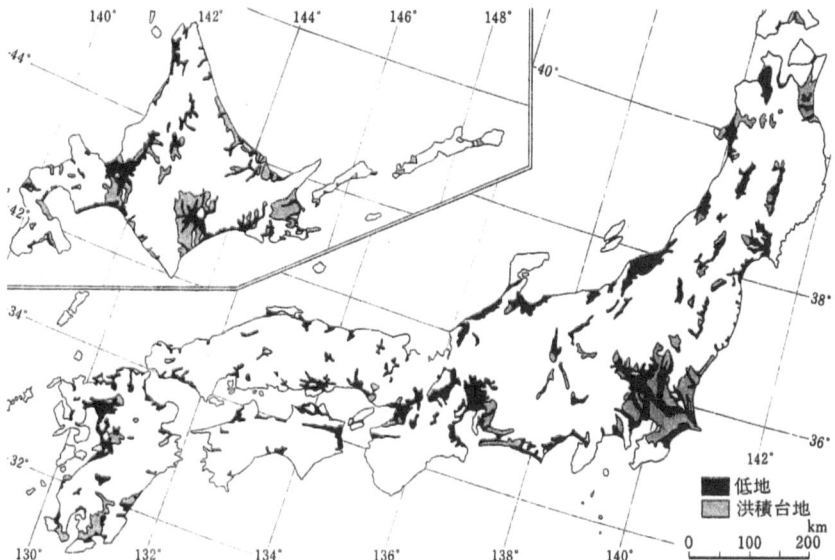

Map I.5 Topography of the mountains, intramountainous basins and coastal plains
Notes: Reconstruction of Neolithic topography before the alluvial advance onto the littoral and modern polderization. White: wooded mountains; black: lowland plains; grey: intramountainous basins and alluvial terraces.
Sources: Sasaki Kōmei, *Hatasaku bunka no tanjō*, Nihon hōsō shuppankai (1988: 256); after Fujioka Kenjirō, *Nihon chishi*, Daimeidō, 1982.

With such a terrain, most of the Japanese landscape is thus an unconducive environment for irrigated rice cultivation, as the gentlest slope requires terracing. In 1980, rice was grown wherever possible on the coastal plains and lowlands. Only 6% of rice fields were on slopes of more than 5°. In the early medieval period, however, some mountain populations were probably forced to farm unsuitable lands to fulfil their tax obligations to the imperial court. The mountainous zones, far from being unsuitable for dry crops, were on the contrary well suited to swidden-field crops, as these thrive on slopes of 30°.

Moreover, the wooded mountain areas had an abundance of plants available for gathering. In autumn, tree fruits and nuts were plentiful in the various environments. In the Fagus zone, there were acorns of *nara* oaks, beechnuts, walnuts, horse chestnuts and (sweet) chestnuts. The laurel-forest zone provided acorns of *kashi* and *shii* oaks, *kaya* (torreya) and other fruits such as the persimmon *kaki* and the akebia. Spring offered gatherers plants with edible parts, such as brackens and angelica. The mountain inhabitants extracted nutritious starches from the roots of lilies, yams, pueraria and bracken. Nuts and seeds have a calorific value similar to that of cultivated cereals (see Table A.5 in the Appendix). To sum up, while the lowlands were suitable for irrigated crops, the mountains also provided ample opportunities for obtaining plants.

1 Irrigated rice and dry crops

In early medieval Japan (8th–12th centuries), agricultural produce came, we think, from two types of cultivated areas: permanent fields and temporary (shifting) fields. The permanent areas consisted of irrigated rice fields, non-irrigated cereal fields (what we will call dry fields) and gardens with vegetables and fruit trees. Farming tasks followed one after the other throughout the year, at the times suited to each operation, and in parallel in the different areas. For example, the sowing of wheat took place at the same time as the irrigated rice harvest, in the 8th month (September) of the early medieval calendar. According to the calendar in use in premodern Japan, spring went from the 1st to 3rd month (February–April), summer from the 4th to 6th month (May–July), autumn from the 7th to 9th month (August–October) and winter from the 10th to 12th month (November–January). Farming tasks were divided equally from spring until autumn, at least in the central region (Kinai) around the two capitals of Nara and Heian (Kyōto), the area best documented by the archival texts.[1]

The public administration was tasked with overseeing the collection of taxes and, in order to achieve this, it encouraged farming activity. But contrary to Chinese practices, the Japanese government did not provide technical advice and did not distribute agricultural calendars or agronomy manuals before the 17th century. To stimulate agriculture, it repeatedly issued decrees to remind the local officials of all the provinces of their supervisory duty. Government encouragement consisted of rewarding zealous officials and having them undertake tours of inspection and write reports. The only technical recommendations that we know of, issued between the 8th and 12th centuries, deal with two points: the water-wheel and the rack for drying rice sheaves.[2]

The governmental decrees also urged the peasants "not to let the propitious time go by" for farming operations, without, however, making reference to an established calendar. In its admonitions, the administration seldom spelled out what tasks needed to be done, but once or twice it mentioned the sowing of wheat and buckwheat. Other than that, the almanacs distributed each year by the imperial court to all the provinces recorded the auspicious days for certain operations, such as tillage, the cutting of vegetation on waste ground and the repair of the irrigation channels in the rice fields.[3]

It may seem strange that the Japanese did not consult the Chinese agronomy manuals for technical questions. There was an abundant Chinese literature on the subject. It produced seventy-eight agricultural treatises and calendars up to the late 10th century, some having been ordered by the Chinese court and distributed on its orders to all parts of the empire.[4] Japan imported from China many technical works in various fields, and the most detailed agricultural treatise *Qiming yaoshu* (6th century) was in its libraries in the 11th century. Even today there remains a copy made in 1166 in the Kudaradera temple. The Japanese administrative and legal texts dealing with agriculture do not mention the Chinese technical manuals and refer only to the early medieval encyclopedias and classical Chinese literature, as well as the chapters on public economy in the Chinese histories.[5]

Can this be seen as ignorance or a lack of interest on the part of the Japanese authorities in the matter of farming techniques? The absence of any reference to the Chinese treatises is all the more surprising as the Japanese learned from China in other scientific fields such as medicine, botany and calendrical science. In these fields, they edited their own compilations from the 9th century, whereas they began to write agricultural treatises only in the 17th century. Some scholars explain this phenomenon by the geographical and climatic differences between the two countries. Up to the 7th century, China was mainly concerned with non-irrigated cereal crops in the north of the country, whereas the Japanese court focused on irrigated rice cultivation. Moreover, northern China experiences long winters and less well-defined seasons than Japan. It follows that Chinese technical directives could not be adapted to fit the Japanese environment. In this country, the seasonal changes of the flora and fauna would serve as the guide that called farmers to their task.[6] In Japan, the phenological calendar was in fact quite distinct, as can be seen in the early poetry. This evokes, for instance, the sowing of the seedbeds in spring, when the wild geese return to the Nordic regions, the transplanting of the rice in summer beneath the cuckoo's song and the harvesting of the rice in autumn when the lespedeza comes into bloom. Work in the fields ended, according to the poets, when the wild geese returned to pass the winter in Japan. Thus, in what follows, our informants will be the Japanese poets, as well as the administrators and officials in charge of rural affairs, since we have no agricultural texts as such.

However, more than the geographical differences between China and Japan, the Japanese way of seeing the world seems to us to be the main reason for the court's lack of interest in technical questions. In our view, the Japanese mentality put zeal before technical ability. According to the authorities of the early medieval period, the farmer did not have to reason, but to work. A rational approach coupled with an economy of human energy and the notion of productivity was only to appear, in agriculture, in the 17th century.

The ideal of the zealous farmer

The *Shinsarugakuki*, written in the mid-11th century, conveys to us the image of the zealous farmer, as idealized by the Japanese authorities of the time. Here, by way of introduction, is the portrait of this "ideal farmer". At the same time, the text surveys the agricultural vocabulary:

> Tanaka no Tomoyasu, supernumerary assistant in the government of the province of Dewa (present-day prefectures of Yamagata and Akita), who is the husband of the third daughter of the officer of the palace gate guard, residing in the right section of the capital of Heian, devotes himself solely to agriculture, with no other occupation. He finds himself at the head of a household that owns several *chō* (1.3 ha) and is a manager *tato* of some repute. Year after year, in times of drought and in times of rain, he prepares the spades *suki* and hoes *kuwa*, inspects the soil quality and repairs the harrowing-combs (harrows) *maguwa* and ploughs *karasuki*. He skilfully shows the farmers how to maintain the dams, dykes, canals and ditches, and, in the fifth month, he employs men and women to transplant *hashoku* the rice, after sowing *tanemaki* in the seedbeds *nawashiro* and tillage *kōsaku*. He cultivates early rice *wase*, late rice *okute*, non-glutinous rice *urushine* and glutinous rice *mochi*. For the grain harvest *karikai*, he outdoes everyone else and the quantities of pounded rice increase each year. Moreover, the wheat or barley *mugi*, soybeans *mame*, cowpeas *sasage*, red beans *azuki*, foxtail millet *awa*, common millet *kimi*, barnyard millet *hie*, buckwheat *sobamugi*, perilla *e* and sesame *goma* that he has planted ripen in abundance in the gardens *sono* and dry fields *hatake*. Each seed he sows *chirasu* in the soil in spring multiplies by the thousands in autumn, and is then stored in the granaries *kura*. He does not make the slightest mistake from tillage *tsukuri* in spring until the harvest *osame* in autumn. He delights in seeing the five grains *gokoku* ripen and in reaping an abundant harvest of them and never to have known losses due to drought, flood, insect infestations or the failure of a crop to ripen. He has never shirked his duty to welcome with a feast the inspectors of the land register and the taxes levied on the holdings. Naturally, he has never forgotten a single sheaf or a single measure in the payment of taxes including: the farm rent, regular tribute, the land-tax in hulled and unhulled rice, the tax in kind, the tax that replaces corvée labour, the products that replace the rice taxes, the exceptional tax, the equipment and provisions for the government envoys, the local specialities, the price of saké; nor forgotten anything at all in the management of the lands under his control, namely the handing out of seed, the management costs, the purchase of products for exchange, the management of the labour corvées, credit and gifts for the provincial administrators. Though he has the misfortune to belong to a household that is subject to taxes, he is nothing like a member of these pathetic families who are always there weeping, flattering and begging.[7]

This extract lists the work in the fields that we now discuss in detail, more or less following the annual cycle from the 2nd month (March).

Tillage

Every yearly agricultural cycle begins with the preparation of the soil, which involves breaking up the earth to enable it to bear the new crop. There are usually several tillings from the 1st month in the seedbeds, rice fields and dry fields. The number and depth of the dressings depend on the type of soil, type of crop, climatic conditions, etc.

Tillage is mentioned in a number of 8th-century poems, one of which says: "They have raked *aragaki* the rice field in front of my door and, under the bright sun, I wait for rain; I also wait for you." Another poet expresses himself as follows: "In the dug rice field *utsu ta*, there remains plenty of barnyard millet, but not removed *erabu* [it is abandoned; I too am abandoned] and tonight will sleep alone." From prehistoric times, the legendary emperor Nintoku (r. 313–399) and the empress reputedly composed poems about tilling *utsu* with the wooden hoe *koguwa*, carried out in a field of white radishes. Another legendary emperor mentions the iron hoe: Kensō (r. 485–487) addresses this poem to his brother on his arrival in the village of Shijimi in the province of Harima (Hyōgo): "Since you strike your fields *ta-utsu*, with hoes of Kibi iron, strike in your hands, all together, and I shall dance."[8]

These poems mention the hoe and the spade, the two basic tillage implements used for both irrigated rice cultivation and dry crops. They are made of carved wood, and hundreds of remains have been found in archaeological sites. The hoe consisted of a blade and a handle into which it was inserted and with which it formed an acute angle. The spade was made either of a single piece of wood, or with a handle inserted into an eye of the blade, but forming an obtuse angle. From the 5th century, iron implements became more widespread: the wooden blade was embedded in an iron cutting edge shaped like a crescent, called a "hoe-iron" or "spade-iron".

From the 7th or 8th century, tillage was part of the rituals of the Ise Shrine. In the 2nd month, a ritual hoe with an iron cutting edge was made and a priest tilled the sacred seedbed *mitoshirota o tagaesu*, accompanied by songs and dances. This ritual symbolized the start of agricultural work in the country. It was later taken up by the imperial court. The repositories of the Shōsōin in Nara have preserved two ritual spades, one dating from 758, used for the ceremonial tillage at the Nara court. This ceremony later took the form of an annual celebration at the palace. For example, in 864 and 866, all the members of the Heian court attended a ritual simulating the agricultural tasks *ta o tagaesu no rei*. This ritual is the origin of the agrarian festivals *dengaku* celebrated in the Japanese countryside from the medieval period to the present day.[9]

The examples mentioned refer only to the hoe *kuwa* and the spade *suki*, but on the occasion of the transplanting *taue* ceremonies and the rustic festivals at the court in 1127, two oxen were brought along, which suggests draught tillage or harrowing in the irrigated rice field (Figure 1.9). Some large aristocratic and monastic holdings or estates did in fact have harrows and ploughs, but in our view draught agricultural implements were still not widespread at this time among the rural population.[10] Nonetheless, the plough appears in a poem by Emperor Tsuchimikado (r. 1198–1210) about quadrupeds: "Nowadays, one wonders whether the ox has the strength to pull the plough *karasuki* with which the peasants till *tagaesu* their rice field." In poetry, the term *sukikaesu* probably meant "draught tillage", and the terms *shirokaki* and *aragaki* "raking";[11] the words *ta-utsu* and *uchikaesu* refer to digging with a hoe or a spade, and *tagaesu* to hand or draught tillage.[12]

It is not known whether the early medieval Japanese varied the depth of tillage, but we do know that they classified soils according to the expected yields into good, average, mediocre and poor quality. From the time of the minister Fujiwara no Muchimaro (680–737), people appreciated "the very black soil that characterizes a rice field of very good quality, where one can count on stable yields, even if there is drought or flooding".[13]

When was tillage done? The celebrations at the Ise Shrine and the imperial court took place in the 2nd month (March), but the fields were tilled several times during the first three months of the year. The fragments of three almanacs for the years 746, 749 and 756 in fact show the auspicious days for tillage *tsuchi okoshi* in the 1st, 2nd and 3rd months, but no day in the 4th month.

Tillage was also depicted in the image of the 2nd month in the calendars painted on folding screens that decorated the aristocratic residences. Though none of these paintings survive, the descriptions remain. The composition of *byōbu uta* poems about the screen images *byōbu e* was in vogue from the 10th century. For example, the chief of police Fujiwara no Tadakimi (?–968) describes one of these images as follows: "Scene of tillage in the middle of spring (2nd month); several people are present, gathered under a willow tree; they are celebrating the god of the fields." Other poems describe tillage in the 2nd or 3rd month.[14]

Apart from rice growing, our information on the other crops is restricted to the imperial vegetable gardens where wheat, barley and twenty-three vegetables were grown. Here tillage was done using a plough *kōchi*, hand tillage with hoes and spades *ryōri heiwa*, and raking *tsuchikai* possibly with a notched levelling board or a rake. Draught tillage was done once only for barley, wheat and beans, twice for fourteen vegetables, three times for Chinese chives, spring onions and radishes, five times for turnips and ginger *kurenohajikami* and seven times for wild chives. The first draught tillings were followed by a fairly intensive dressing with a hoe or spade, since one, two or three days' work would be required for the same areas. After that, ridges were formed *une-age*, *wake-une* for two-thirds of the crops. Raking was done after the sowing of melons, egg-plant and taro, in order to cover the seed or to clean (i.e. to weed or break up the topsoil). There is no mention of any draught harrowing, whereas two ox-drawn harrows *maguwa* formed part of the equipment for these gardens. According to Kōno Michiaki, harrowing was essential, particularly before transplanting water-celery and water-leeks.[15] In view of the interest of the regulation relating to the imperial vegetable gardens, the only document with technical information that we possess for the early medieval period, we will give a translation of it below.

The imperial vegetable gardens

The "Regulations of the Engi Era" (*Engishiki*) issued in 927 deal with the vegetable gardens managed by the Bureau of the Imperial Table. They provide the only detailed account of the methods used for cereal and vegetable crops in the early medieval period.[16] We therefore present a complete translation of the chapter.

> For the vegetable gardens, eleven oxen are to be provided by the Left and Right Offices of Horses. If an ox has died or is unfit for work, a request for its replacement is to be made to the Department [of Military Affairs]. The Department will give effect to the request after verification by the Ministry [of State Affairs]. Also to be provided are: 74 hoe and spade [blades], 40 hoe handles and 34 spade handles, to be replaced every two years (the old [blades] to be returned); 2 harrows (to be returned when they are damaged); 2 plough mould-boards and 4 ploughshares (to be asked for again if lost) [Figure 1.2] and 2 carts (replaced each year).
>
> The gardens are cultivated by 14 workers (a head gardener and 13 subordinates). There is a boat for river transport (3 *jō* [9 m] long), moored at the port of Yodo-tsu [in Yamashiro]. This boat is used to transport the vegetables from the gardens of Nara and Naiki [to the Heian palace].
>
> The gardens cover in total 39 *chō* 5 *tan* 200 *bu* [40 ha]. These are the vegetable gardens of Keihoku, 18 *chō* 3 *tan* [located north of Heian?], of Nara, 6 *chō* 8 *tan* 320 *bu* [at Kuse], of Yamashina, 9 *tan* [at Uji], of Naiki, 5 *chō* 5 *tan* 240 *bu* [at Kuse], of Hatsukashi, 4 *chō* 9 *tan* [at Otokuni], of Izumi, 1 *chō* [at Sōraku] and of Nara, 2 *chō* [at Nara?].[17]
>
> There are 460 fruit trees: 100 pear, 100 peach, 40 mandarin *kan[shi]*, 40 dwarf mandarin, 100 persimmon, 20 orange *tachibana*, 30 large jujube *ōnatsume* and 30 stauntonia *mube*; and a garden of blackberries *ichigo* of 2 *tan*. All this is to be cultivated in accordance with the regulation. If the quantities are insufficient, this will be dealt with by the Discharge Investigation Bureau, when the officials responsible hand over to their successors.
>
> There are irrigated fields of 6 *tan* 234 *bu* (to grow water-celery and water-leeks, located in the district of Otokuni).
>
> Work in the vegetable gardens:

- To grow 1 *tan* [0.3 acres] of barley *futomugi* requires 1 *to* 5 *shō* [12.75 litres] of seed and 14.5 days' work, namely:[18] 1 day for tillage with 1 ploughman, 1 ox driver and 1 ox; 1 day for hand tillage *ryōri* [with a hoe or spade]; 2 days to form the ridges; 1 half-day for sowing *tane-oroshi*, 2 days for harvesting, 5 days for selecting/threshing *erabu*,[19] 2 days for pounding *katsu*; (the same for wheat).

- To grow 1 *tan* [0.3 acres] of soybeans *mame* requires 8 *shō* [6.8 litres] of seed and 13 days' work, namely: 1 day for tillage with 1 plough-man, 1 ox driver and 1 ox; 1 day for hand tillage *ryōri heiwa*; 2 days to form the ridges; 2 days for sowing *uu* (in the 3rd month); 2 days for weeding *kusagiru*; 2 days for harvesting; 2 days for shelling.

- To grow 1 *tan* [0.3 acres] of red beans *azuki* requires 5 *shō* 5 *gō* [4.7 litres] of seed and 13.5 days' work, namely: 1 day for tillage with 1 ploughman, 1 ox driver and 1 ox; 1 day for hand tillage; 2 days to form the ridges (in the 5th month); 1 half-day for sowing *tane-oroshi*; 4 days for two weedings; 2 days for harvesting; 2 days for shelling.

- To grow 1 *tan* [0.3 acres] of cowpeas *sasage* requires 8 *shō* [6.8 litres] of seed and 13 days' work, namely: 1 day for tillage with 1 plough-man, 1 ox driver and 1 ox; 1 day for hand tillage; 2 days to form the ridges; 2 days for sowing *uu*; 3 days for weeding; 3 days for harvesting.

- To grow 1 *tan* [0.3 acres] of turnips *aona* requires 8 *gō* [0.7 litres] of seed and 32.5 days' work, namely: 2.5 days for five tillings with 1 ploughman, 1 ox driver and 1 ox; 1 day for hand tillage; 20 days to transport 120 loads of manure (6 *kin* [4 kg] per load), (that is, 6 return trips [error for 6 loads?] per day per person from the Left and Right Offices of Horses to the vegetable garden at [Kei]hoku; the same below for the other crops); 1 half-day for sowing (in the 7th or 8th month); 6 days for harvesting.

- To grow 1 *tan* [0.3 acres] of 'wild chives' *hiru* requires 3 *koku* [255 litres] of seed and 93 days' work, namely: 3.5 days for 7 tillings with 1 ploughman, 1 ox driver and 1 ox; 2 days for hand tillage; 3 days to form the ridges; 35 days to transport 210 loads of manure; 6 days for sowing (in the 8th month); 10, 8 and 7 days respectively for 3 weedings; 15 days for harvesting.

- To grow 1 *tan* [0.3 acres] of Chinese chives *mira* requires 5 *koku* [425 litres] of seed and 75 days' work, namely: 1 day and a half for three tillings with 1 ploughman, 1 ox driver and 1 ox; 2 days for hand tillage; 2 days to form the ridges; 35 days to transport 210 loads of manure; 6 days to select *erabu* the seedlings; 6 days to transplant them (in the 9th month); 21 days for three weedings (7 days each time).

- To grow 1 *tan* [0.3 acres] of spring onions *ki* requires 4 *shō* [3.4 litres] of seed, 1,200 seedlings and 87.5 days' work, namely: 1 day and a half for three tillings with 1 ploughman, 1 ox driver and 1 ox; 1 day for hand tillage; 2 days to form the ridges; 35 days to transport 210 loads of manure; 1 half-day for sowing (in the 8th month); 20 days for transplanting (in the 2nd month); 10, 9 and 7 days for three weedings.

- To grow 1 *tan* [0.3 acres] of ginger *kurenohajikami* requires 4 *koku* [340 litres] of seed and 78 days' work, namely: 2.5 days for 5 tillings with 1 ploughman, 1 ox driver and 1 ox; 2 days for hand tillage; 35 days to transport 210 loads of manure; 4 days to form the ridges; 4 days for sowing (in the 4th month); 9, 7 and 6 days for three weedings; 6 days to pull up *torimushiru* the ginger.

- To grow 1 *tan* [0.3 acres] of butterbur *fufuki* requires 2 *koku* [170 litres] of seed and 34 days' work, namely: 1 day for 2 tillings with 1 ploughman, 1 ox driver and 1 ox; 2 days for hand tillage; 20 days to transport 120 loads of manure; 2 days for sowing (in the 9th month); 2 days for the first weeding (in the 3rd month), 2 days for the second weeding (in the 6th month); 4 days for harvesting *karu*; replanting is done every three years.

- To grow 1 *tan* [0.3 acres] of thistles *azami* requires 3 *koku* 5 *to* [298 litres] of seed and 44 days' work, namely: 1 day for 2 tillings with 1 ploughman, 1 ox driver and 1 ox; 2 days for hand tillage; 20 days to transport 120 loads of manure; 2 days for sowing; twice 3 days for two weedings (in the 2nd and 7th months), 4 days for harvesting *karu*; 8 days to select/remove *erabu* the heads; the heads are transplanted every three years.

- To grow 1 *tan* [0.3 acres] of early melons *wasauri* requires 4 *gō* 5 *shaku* [0.4 litres] of seed and 46 days' work, namely: 1 day for two tillings with 1 ploughman, 1 ox driver and 1 ox; 3 days for hand tillage; 3 days to dig the furrows [for the manure?] between the ridges; 12.5 days to transport 75 loads of manure; 1 day to [dig and] tread 360 seed-holes *kura-i*; 1 half-day for sowing (in the 2nd month); 12 days to drive away insects; 3 rakings *tsuchikai* and 3 weedings, done in 5 days (in the first half of the 3rd month), in 4 days (in the second half of the 3rd month) and in 3 days (in the 4th month).

- To grow 1 *tan* [0.3 acres] of late melons *okuteuri* requires 4 *gō* 5 *shaku* [0.4 litres] of seed and 35.5 days' work, namely: 1 day for 2 tillings with 1 ploughman, 1 ox driver and 1 ox; 3 days for hand tillage; 3 days to dig the furrows between the ridges, 1 day to [dig and] tread 360 seed-holes; 1 half-day for sowing; 1 day for 3 rakings; 3 weedings in 10 days (in the 3rd month), in 8 days (in the 4th month) and in 7 days (in the 5th month).

- To grow 1 *tan* [0.3 acres] of egg-plant *nasubi* requires 2 *shō* [1.7 litres] of seed and 41 days' work, namely: 1 day for two tillings with 1 ploughman, 1 ox driver and 1 ox; 3 days for hand tillage; 1 half-day for sowing (in the 3rd month); 1 day and a half to select *toru* the seedlings; 10 days to transplant them (in the 4th month); 3 days for 1 raking (in the 5th month) and 3 days for the second raking (in the 6th month); 18 days for 3 weedings (6 days for each).

- To grow 1 *tan* [0.3 acres] of radishes *ōne* requires 3 *to* [25.5 litres] of seed and 18.5 days' work, namely: 1 day and a half for 3 tillings with 1 ploughman, 1 ox driver ox and 1 ox; 1 day for hand tillage; 1 half-day for sowing (in the 6th month), 14 days for harvesting.

- To grow 1 *tan* [0.3 acres] of lettuce *chisa* requires 3 *shō* [2.6 litres] of seed, 1,500 seedlings and 39.5 days' work, namely: 1 day for 2 tillings with 1 ploughman, 1 ox driver and 1 ox; 2 days for hand tillage; 2 days to form the ridges; 22 days to transport 132 loads of manure; 1 half-day for sowing (in the 8th month); 2 days to select the seedlings; 6 days to transplant them (in the 9th month); 3 days for weeding.

- To grow 1 *tan* [0.3 acres] of mallow *aoi* requires 2 *shō* [1.7 litres] of seed and 31.5 days' work, namely: 1 day for 2 tillings with 1 ploughman, 1 ox driver and 1 ox; 2 days for hand tillage; 2 days to form the ridges; 22 days to transport 132 loads of manure; 1 half-day for sowing (in the 8th month); 3 days for weeding.

- To grow 1 *tan* [0.3 acres] of coriander *konishi* requires 2 *to* 5 *shō* [21.3 litres] of seed and 28 days' work, namely: 1 day for 2 tillings with 1 ploughman, 1 ox driver and 1 ox; 2 days for hand tillage; 2 days to form the ridges; 22 days to transport 132 loads of manure; 1 half-day for sowing (in the 3rd or 8th month).

- To grow 1 *tan* [0.3 acres] of colza *ochi* requires 1 *shō* [0.85 litres] of seed and 28 days' work, namely: 1 day for two tillings with 1 ploughman, 1 ox driver and 1 ox; 2 days for hand tillage; 2 men to form the ridges; 22 days to transport 132 loads of manure; 1 half-day for sowing (in the 3rd or 8th month).

- To grow 1 *tan* [0.3 acres] of nothosmyrnium *soraji* requires 3 *koku* 5 *to* [298 litres] of seed and 35 days' work, namely: 1 day for 2 tillings with 1 ploughman, 1 ox driver and 1 ox; 2 days for hand tillage; 2 days to form the ridges; 22 days to transport 132 loads of manure; 3 days for sowing (in the 9th month); 2 days for weeding; 2 days for harvesting.

- To grow 1 *tan* [0.3 acres] of ginger *mega* requires 3 *koku* [255 litres] of seed and 35 days' work, namely: 1 day for 2 tillings with 1 ploughman, 1 ox driver and 1 ox; 2 days for hand tillage; 2 days to form the ridges (in the 9th month); 22 days to transport 132 loads of manure; 3 days for sowing; 2 days for weeding; 2 days for harvesting.

- To grow 1 *tan* [0.3 acres] of taro *imo* requires 2 *koku* [170 litres] of seed and 35 days' work, namely: 1 day for 2 tillings with 1 ploughman, 1 ox driver and 1 ox; 4 days to form the ridges and till with a hoe or spade; 3 days for sowing (in the 3rd month); 6 days for raking; 6 days for 3 weedings (2 days respectively in the 5th, 6th and 7th months); 4 days to dig up the taro; 10 days to select it? *erabu*.

- To grow 1 *tan* [0.3 acres] of water-leeks *nagi* requires 20 bundles of seedlings and 53 days' work, namely: 1 day for 2 tillings with 1 ploughman, 1 ox driver and 1 ox; 1 day for hand tillage; 20 days to transport 120 loads of manure; 15 days to plant the seedlings (in the 5th month); 15 days for 3 transplantings *hashoku* (5 days each time); 15 days to collect and harvest them (each of the three times).[20]

- To grow 1 *tan* [0.3 acres] of water-celery *seri* requires 5 *koku* [error for 5 bundles?] of seedlings and 44 days' work, namely: 1 day for 2 tillings with 1 ploughman, 1 ox driver and 1 ox; 1 day for hand tillage; 20 days to transport 120 loads of manure; 6 days for transplanting (in the 2nd month); 10 days to select the seedlings; 5 days to drive [error for harvest?].[21]

It may be noted that the terms used in this regulation differ from the vocabulary of the poets. This is a terminology specific to the compilers of the *Engishiki*, none of whom were agronomists. This regulation is found in the article on the Bureau of the Imperial Table that determined the deliveries of foods and the management costs. The description of the imperial vegetable gardens was therefore not intended to serve as a directive for the farmers, but it sets out the agricultural operations with the aim of quantifying the work. In order to designate these operations, the compilers, it would seem, have coined their own terms and have not consulted either the Chinese works or the Japanese dictionaries. Yet their description of the procedures seems reliable. Aoba Takashi points out that certain indications relating to the quantity of seed and to the planting calendar correspond exactly to traditional Japanese practices that were current up to the 1930s.[22] On the other hand, one may be surprised at the lack of manuring and weeding for wheat and barley. Similarly, harvesting is omitted for several crops. The text may contain a few errors by the compilers and some entries are less complete than others. However, apart from these few errors, this regulation does indeed appear to show the methods used for vegetable crops in the 10th century.

Fertilization

The texts concerning the methods of soil improvement are brief and their interpretations differ markedly. Let us first consider a legend about the village of Kōchi in Harima (Hyōgo):

In this village, rice is grown by sowing seeds, without spreading grass in the seedbeds. The reason is that the great deity of Sumiyoshi stopped in this village on her way [to the province of Settsu?], in order to take refreshments. Her attendants mixed up the grass that had been cut and spread *kariokeru kusa* by a peasant in a seedbed and fashioned a seat for the deity. The peasant, indignant, arrived and complained to the great deity. She replied after due consideration: 'The seedlings in your seedbed will grow as if the grass had been spread on it *kusa shiku*.' Since that time they have grown rice in this village without needing to spread grass in the seedbeds.[23]

Green manure

The above legend suggests that green manure was widely used, at least in the region of Hyōgo, in the first half of the 8th century. At Ise, the spreading of grass also seems to have been the norm, since the first deity prohibited it in the rice field that produced the morning and evening offering for certain rituals; and "since then, they do not spread grass *naegusa shikazu* and the rice field is no longer invaded by leeches".[24]

Spreading is also mentioned by Fujiwara no Tamenari (11th century): "After having spread the cut grass *nawakusa karishikite* in their small mountain rice field, they sowed the seeds kept in a reservoir in their small field of ten *shiro*." The writer uses the word *karishiku*, meaning "to cut and spread", which was to remain the technical term for green manure up to the modern period.[25]

A legend written down in the early 13th century relates that the peasants of a region south of the capital of Heian harvested *kariire* grass and leaves to improve their fields in such quantities that the hills around the village of Ide lost all their vegetation. It was therefore wilderness areas and probably also waste ground that provided green manure. It may be noted that the cutting of grass *kusakari* was done not only to obtain manure, but also to collect fodder.[26]

Ashes

Another method of fertilization was to collect vegetation from virgin lands, burn them in a heap and spread the ashes in the fields. In a manuscript of the journal of Fujiwara no Tametaka (1070–1130), an anonymous writer relates that "people gather brushwood *shiba o toru* to produce ashes that they scatter in the rice fields to fertilize them *koyashimu*".[27] We know that medieval people were well aware of the fertilizing value of ashes, as they would also set fire to waste lands with a view to swidden-field crops or the regrowth of useful plants. But in this case the ashes were produced on-site, whereas the method mentioned in the above manuscript suggests the gathering of fuel, followed by burning, then taking the ashes onto the cultivated land. It is, however, the only piece of evidence that we have for the early medieval period. This method has no equivalent in Europe, but is known by the name *citimene* in South Africa.[28]

One way of producing ashes on-site is to burn the stubble after the harvest, in the case of high cutting under the head, which leaves the stubble standing. This practice seems to have been widespread in medieval China, where people spoke of "ploughing with fire and fertilizing with water" *huogeng shuinou.*[29] It is possible that the Japanese also cleaned with running fire lands that remained uncultivated for several months before sowing or transplanting. Or they may simply have buried the stubble or vegetation during tillage, without setting fire to it. Could there be an allusion to the burying of vegetation in this poem by Minamoto no Moroyori (1068–1139)?: "After turning over the soil with the young cyperus shoots in my small mountain field, I leave the seedbed to destiny, after having brought water to it."[30] Perhaps the two methods, those of running fire and burying, are not mentioned in the early medieval sources because they were too commonplace and did not interest the public administration.

Animal manure

There are also a few scattered pieces of evidence relating to organic fertilizer. Animal manure was used in the imperial vegetable gardens, as can be seen in the regulations translated above. *Koe* or *kuso*, written like 'excrement' *fun* and provided by the court stables was used for sixteen vegetable crops. It is not mentioned for barley, wheat, beans, late melons, egg-plant, radishes or taro. But it is possible that the choice of crops to be fertilized was governed by the distance that separated the stables from the vegetable gardens. Perhaps the sixteen plants were grown in those closest to the court stables, namely the one at Keihoku north of the capital, which was the largest with an area of 18.3 *chō* (20.7 ha) and the irrigated fields of water-celery and water-leeks in the district of Otokuni. These vegetable gardens were in the province of Yamashiro in the vicinity of Heian-kyō, whereas the others were further south in Yamato. The quantity of animal manure seems quite large. Depending on the crop, 75, 120, 132 or 210 loads were estimated per plot of 1 *tan* (0.3 acres). With 6 *kin* (1 *kin* = 670 g) or 4 kg per load, this represented respectively 300, 480, 528 or 840 kg of manure per *tan*. For transport on the back of porters, the estimate was 6 loads (24 kg) per person per day's walk. If these figures are correct, the transporting of manure accounted for more than half of the work for the entire year in the imperial vegetable gardens.[31]

Animal manure is mentioned later in the *Shasekishū*, a 13th-century collection of legends. In one episode, horse manure *koe* is associated with the voice *koe* to recite the sūtra *Ninnōe kyō*, which attests to the reading *koe*, but up to the 12th century the dictionaries gave the reading *kuso* or *akuta.*[32]

In view of all these different references, historians give priority to green manure, as this method remained the most widespread up to the modern period. It is also attested for the medieval period by the disputes relating to the right of the peasantry to collect vegetation on waste lands, and it occurs in the agricultural treatises of the 17th–19th centuries. The archaeologist Kinoshita Tadashi thinks that large plank shoes *ōashi* were used from the medieval period to tread green manure into the fields, but that the *ōashi* discovered in sites dating from the 3rd century AD were used only for levelling the ground. However, the archaeologist Terasawa Kaoru is of the opinion that fertilization with grass was done using plank shoes from the 3rd century.[33]

For Furushima Toshio, the use of animal manure did not become widespread in Japan until livestock rearing was done more often in stables, that is, from the 16th–17th centuries, whereas in the early medieval period livestock rearing was probably limited to aristocratic landholdings and took place in the open. As for human manure, it is attested only from the 15th century. One can, however, point to toponyms such as Kusoda, "Field with excrement" which, in 735, designates a parcel of land belonging to the Gufukuji temple, this land being in Sanuki (Kōchi) on the island of Shikoku; and Kusooki, "Piled-up manure", part of a Tōdaiji estate in Echizen (Fukui).[34]

Generally speaking, the use of fertilizer remains difficult to assess for the early medieval period. Green manure requires waste lands or forests with their undergrowth and clearings close to the holdings, as well as transport techniques combined with sufficient labour. The collecting and spreading of vegetation was in fact an important task. The *Seiryōki*, a treatise written in the early 17th century, puts at fifty-two days a year the work of one man to cut the vegetation needed to fertilize an area of one hectare. This quite considerable amount of labour is confirmed by other examples. In 1686, the inhabitants of a village near Izu used 200 pack-horse loads of green manure per hectare; and in 1704, the farmers of a village in Nagano estimated 160–170 pack-horse loads for the same area.[35] However, were the rural households of the early medieval period, which had about ten members and few available adult males, capable of taking on this cutting work and such a volume of transport? The burning of the stubble in a field left uncultivated during the winter may seem much easier to us; but perhaps the waste ground adjacent to the fields provided additional green manure. Farmers also often gathered vegetation on waste lands, if we are to believe the many references to the cutting of grass *kusa kiru* in the 8th-century almanacs. The collecting of grass from waste lands remained a common practice in the premodern period, at least in the region of Aizu (Fukushima) in the 17th century.[36]

Ethnology knows of a similar case. At Kathmandu in Nepal, people have few livestock (the zebu) and have used green manure, at least until recently. In Europe, however, the practice of spreading vegetation without fermentation has no adherents, except in biodynamic agriculture. The fertilizing contribution of grass is its mineral, not its organic content, but vegetation releases minerals only when decaying during the fermentation process. Though the early medieval texts clearly attest to the spreading of grass, it may be asked whether there was not also fermentation caused by moisture in the atmosphere or by rain. Fermentation could occur, under certain conditions, in irrigated rice fields. In any event, there should perhaps be some acknowledgement of the existence of practices that have not survived, except in biodynamic farming.[37]

It may be noted that fertilizer is necessary for most dry crops. All growth depends on the three essential elements: phosphorous, potassium and nitrogen. Without fertilization, harvests can decrease by half or more from the second year, then gradually disappear. On the other hand, fertilizer is less important for irrigated rice growing, since the water contributes a certain amount of fertilizing material, such as potassium, calcium and magnesium. Though admittedly with lower yields, it was still possible, even without fertilizer, to count on yearly harvests of irrigated rice. As an 18th-century saying put it: "One harvests rice thanks to the vigour of the land, one harvests wheat and barley thanks to fertilizer." However, modern agronomists have paid the same attention to the fertilization of rice as to that of dry crops. By contrast, the early medieval sources refer mainly to the improvement of the rice fields.[38] One has the impression that, where fertilizer is concerned, the early medieval Japanese did not distinguish between dry and irrigated crops.

The sowing of dry crops

In the early medieval period, sowing and transplanting were done in parallel. The Japanese of this time distinguished, like the poet Ōtomo no Yakamochi (718?–785), "the transplanted rice fields and the sown dry fields" *ueshi ta mo makishi hatake mo.*[39]

The government issued decrees in the 8th–9th centuries aimed at encouraging dry crops, intended mainly as additional food reserves to guard against poor rice harvests. These decrees use the verb "to till and plant" *kōshu/tagaeshi-uu* for cultivation in general, the terms *hashu* and *hashoku* for sowing, and the word *uu* (mod. *ueru*) with two different characters in the sense of 'to sow and plant'. According to the decrees, wheat, barley, millets, beans and sesame are sown *uu*, while mulberry and varnish trees are planted (or sown?) *uu.*[40]

From 693, Empress Jitō (r. 686–697) encouraged the cultivation *uu* of the mulberry tree *kuwa*, ramie or hemp *karamushi*, the pear tree *nashi*, the chestnut tree *kuri* and turnips *aona*. From 715, the imperial court attempted to impose on the entire country the growing of wheat, barley and especially foxtail millet *awa*, by pointing out that the peasants went hungry when the rice harvest was insufficient. In 820, the court gave orders to proceed with the sowing of wheat and barley from the 8th month, citing the calendar in a classical Chinese work. According to two documents of a slightly later date, the sowing of wheat took place in the 9th and sometimes even in the 10th month. Moreover, in 839, the government ordered buckwheat *soba* to be sown in the 7th month (August), for harvesting in the 9th month (October), explaining that this cereal grows rapidly even in poor soil.[41]

According to these texts, sowing was done in August for buckwheat, and between September and November for wheat and barley. Buckwheat was therefore an autumn crop and the two other cereals were winter crops. It may be noted for comparative purposes that sowing was done at almost the same time in the early 17th century in Shikoku, that is, in July–August for buck-wheat, from September to December for wheat and from October to December for barley. Foxtail millet, in Shikoku, was a summer crop like rice.[42] The regulations of the 10th century also prescribed fodder crops intended for the stables of the palace guards, fixing sowing in the 2nd month (March) and harvesting no later than the 7th month, but without specifying the type cultivated.[43]

Cotton was introduced to Japan in 799 by a man from South-East Asia who explained the cultivation of this plant to the imperial court as follows: "The method involves first of all selecting fertile and sunny soil. Next make holes one inch (3 cm) deep and four feet (1.2 m) apart. Then wash the seed and let it soak overnight. Next morning, sow *uu* four seeds in each hole. Cover the holes by compacting the earth with the hands. The field must be watered every morning, as it must be kept wet. Weed the field while waiting for the shoots to come up." Later, the court had seed distributed to several provinces with orders to proceed with sowing in the 4th month (May). However, this directive of 800 seems not to have been followed, and cotton was not widely grown in Japan until it was reintroduced via Korea in the 15th century.[44]

Vegetable crops have left their mark in poetry. Sone no Yoshitada (?–ca. 1003) writes at the beginning of the 5th month: "In the gardens of the Komano plain where the melon has been sown *uu*, the vegetation is lush; summer is here!"[45] Many anthologies classify the poems by season, but Yoshitada groups his according to the months and even notes the beginning, middle and end of each month. Yoshitada, some of whose other poems we quote later, spent many years working in the province of Tango (the region between Kyōto and the Sea of Japan) and took part in four poetry contests *utaawase* in Heian between 977 and 1003. It may therefore be assumed that his notions of the calendar concern the central region. In another poem written in the middle of the 7th month, Yoshitada speaks of a field in which seeds have been sown: possibly an autumn crop.[46]

The province of Bungo (Ōita) in Kyūshū managed gardens of gromwell *murasaki*, but the time of sowing *maku* is not known. Other vegetables were cultivated from prehistoric times. The legendary emperor Jinmu writes about the sowing *uu* of Japanese pepper *hajikami* and the mythical emperor Nintoku mentions the sowing *maku* of turnips.[47]

The 10th-century regulations concerning the imperial vegetable gardens use the terms *tane-orosu* and *uu* for sowing. They note the time of sowing for a number of plants: sowing was done on ridges for soybeans and taro in the 3rd month, for ginger *kurenohajikami* in the 4th month, for red beans after the 5th month (when the ridges were formed), for mallow, cowpeas and wild chives in the 8th month, for coriander *konishi* and colza in the 3rd or 8th month, and for nothosmyrnium in the 9th month. Not sown on ridges were: thistles before the 2nd month, water-celery in the 2nd month, water-leeks in the 5th month, radishes in the 6th month, turnips in the 7th or 8th month, and butterbur in the 9th month. Water-celery and water-leeks were transplanted aquatic plants. Egg-plant was sown on ridges in the 3rd month and transplanted the following month; Chinese chives were sown on ridges at an unspecified time and transplanted in the 9th month; lettuce was sown in the 8th month and transplanted the following month; spring onions were sown in the 8th month and transplanted in the 2nd month. Thistles and butterbur were replanted every three years. For early and late melons, furrows were dug, probably for fertilizer, between the ridges and the seeds were planted, in the 2nd month for early melons, in 360 seed-holes that were covered with earth and compacted with the feet. The cultivated plots of melons, taro and egg-plant were raked after sowing; these plants were therefore "undersown".[48] In the almanacs distributed to the provinces by the imperial court, one finds mention of the auspicious days for sowing *tanemaki* in the 1st, 2nd and 4th months (the parts of the almanacs from the 5th to 12th month being lost).[49]

All these directives relate only to the imperial vegetable gardens. The rural population may simply have sown soybeans by the wayside, on the rice-field ridges and in the cereal fields, as was done in the 17th–18th centuries. Apart from permanent crops, the early medieval Japanese also practised shifting cultivation on swidden fields. In this case, they sowed *maku* the seeds by broadcasting them over ashes, on waste ground, and not in a tilled field. The following poem from the *Man'yōshū* seems to allude to direct sowing, without tillage, on waste ground: "The millet that I sow on Sanatsura hill, even if my beloved's horse grazes on it, I will not say to him: go away."[50]

One may compare the quantities of seed. For wheat and barley, the imperial vegetable gardens estimated 1.5 *to* (12.75 litres) for 1 *tan* (0.3 acres); for rice, the province of Yamato reckoned on 2 sheaves or 2 *to* (17 litres) for the same area. However, the yields differed markedly. Transplanted rice gave a harvest between 7.5 and 25 times greater than the quantity of seed, whereas the wheat harvest possibly came to only twice the initial quantity of seed. From these figures it can be estimated that on 1 *tan*, 25.5 litres of wheat and 127.5–425 litres of rice were harvested. The amount of millet would be between those of rice and wheat, since millet, like rice, produces a far higher number of grains per plant than wheat and barley.[51] Barnyard millet *hie* no doubt also gave a good ratio between sowing and harvesting, perhaps even higher than foxtail millet *awa*, owing to its very full spikelets.

The sowing and transplanting of rice

Our sources make frequent mention of sown and transplanted rice. The seedbed was a favourite topic of the poetry contests *utaawase* of the 10th–12th centuries. It forms a sub-heading in several poetic anthologies. We quote below, among others, the *Horikawa-in hyakushu*, a collection of poems composed during the reign of Emperor Horikawa (r. 1086–1107), completed around 1105, and the *Tango no kami ke hyakushu*, a collection of poems by six members of the Fujiwara family and two Minamoto, compiled by Fujiwara no Tametada (?–1136).[52]

Sone no Yoshitada, for his part, writes at the end of the 2nd month (March): "Cry of the departing wild geese; would their tears fill the rice fields laid out in rows? Water fills up the seedbeds *nawashiro*: spring is fast approaching." It was therefore time to proceed with the sowing of the seedbeds. The seed was normally set aside from the harvest of the previous autumn and kept in a reservoir *muro*, probably underground, dug in a corner of the field. The reservoir is mentioned by Minamoto no Toshiyori (1055–1129) in these words: "One thinks of the late rice cut in autumn, of spring when the seeds are soaked in the seed-well *tana-i*, [kept] in the reservoir." According to this poem, the seed was soaked *tanekasu* in water before sowing. Fujiwara no Chikataka (1099–1165) expresses it thus: "The lowly peasant who marks out with a rope the seedbed of his small field, soaks the seeds of early rice kept in a reservoir."[53]

Soaking consisted, at least in the Edo period (17th–19th centuries) of placing the seed in a straw sack and letting it soak for ten to twenty days in a pond or a rice-field channel (Figure 1.7). Minamoto no Kunizane (1069?–1111) and others called this a seed-well *tana-i*: "The lowly peasant who has raked his seedbed *nawashiro kaki* and prepared his ridge has now moistened the seeds in the seed-well."[54]

Soaking, which took place in the 2nd month, was intended to promote the germination of the rice. The seed was then left to dry for two to ten days.[55] In principle it was set aside from the autumn harvest, though in practice farmers were often obliged to borrow rice. In certain exceptional cases, the government, itself the main creditor, also gave assistance. In 723, in the 2nd month, the State gave 2 *koku* (170 litres) of seed to each family across the country as a way of encouraging rice cultivation, but we do not know the outcome of this directive.[56] The quantity of seed is known for the rice fields belonging to a shrine and to the palace guards that were located in the province of Yamato, in the district of Sōnokami (between the present-day cities of Nara and Tenri): at that time it was 20 sheaves of rice, equivalent to 2 *koku* (170 litres) of unhulled rice for an area of 1 *chō* (1.13 ha). This quantity corresponds to the practices of the 18th–19th centuries (Figure 1.8).[57]

Once the seed and soil of the seedbed or nursery had been prepared, sowing began. This operation took place in the 3rd month (April), which is why the granting of State rice fields to tenants ceased, according to the regulation, at the end of the 2nd month.[58] From the outset, it was necessary to pay careful attention to water management: "Even if I suffer for it, I will not go and store water, but will entrust the seedbed of my small field to the water from the mountains."[59]

After sowing they waited, all the while regulating the water level, for one and a half months until the seedlings reached a height of around 20 cm. At the beginning of the 4th month (May), Sone no Yoshitada was still waiting: "With [the vegetation] of the moors [as thick as] the weeds to be removed *kusa hiku*, will my seedbed grow [as quickly]?" Then Yoshitada says: "Sitting in my hut in the fields watching over [the seedbed]; today the fifth month has arrived; I must make haste to [transplant] the seedlings *sanae*; they will grow old."[60]

Transplanting

It was time to select the seedlings and transport them to the rice fields (Figure 1.10). A poet evokes this operation in these words: "Under the rain of the fifth month, it appears the sun has set; the road is long and I have not finished collecting the seedlings for my mountain rice field." The seedlings then had to be transplanted without delay: "The early rice in the reservoir, more visible than the seedlings, it seems that they hasten to plant it *uu* without having raked *shirokaki* the rice field."[61] Collecting and transplanting had to be completed in a very short time, ideally within a day. For this they formed groups called *yui*, by an exchange of work with neighbouring farmers. The monk Ryūgen (d. before 1120) writes: "I have but one rice field of ten *shiro* left; tomorrow I will only have to collect the seedlings without even having to ask for others' help *yui*."[62]

Transplanting requires a major investment in labour, but offers several advantages: a saving of seed, delaying the growth of weeds compared to the rice, and weeding facilitated by the alignment of the seedlings. This gives better yields than for rice sown directly in the rice field. A poem in the *Man'yōshū* refers to this work: "The rice fields we are transplanting, you will not transplant them [with me]; now that you have left the province, what will become of me!"[63] These words spoken by the wife of Nakatomi no Yakamori refer to his exile to the province of Echizen. The poem in fact mentions "the rice fields that people transplant together" *hito no uuru ta*, which suggests a system of mutual aid, although the 8th-century texts do not as yet speak of organized groups.

As with the seedbed, the rice field had to be irrigated before transplanting. An "auspicious poem" from the 12th century alludes to this in these words: "The transplanted rice field in the village of Yoshida, full of good water held back by a sluice-gate *seku*; one already sees the reflection of a year of promise."[64]

Transplanting was considered such a critical operation in rice growing that, in the eyes of the public administration, it gave the right to the harvest. For example, when changeovers of provincial posts occurred, the outgoing official received the harvest from his rice fields, even after his departure, if he had had rice transplanted into them.[65] Similarly, the officials of the capital were entitled to a holiday for the transplanting of their rice fields in the 5th month and another for the harvest in the 8th month. This does not mean that the entire population worked at the same time. The government acknowledged that there were geographical differences. In the province of Yamato, for instance, transplanting took place in the 4th month and harvesting in the 7th month, in the districts of Sōnoshimo (south of Nara) and Heguri (Ikoma-gun), but in the 5th–6th months and 8th–9th months respectively in the districts Katsuragi-no-kami (Gose) and Katsuragi-no-shimo (north of Heguri-gun). According to the early commentators on the regulation, these differences were due to the growing period and the soil quality in these regions.[66]

At a certain point in time, transplanting came to be thought of as women's work, but this was not yet the case in the 9th century. At a demonstration of work in the fields at the Unrintei residence, in the 4th month of 832, there were gifts for "the men and women" who did the transplanting. Previously, the court had repeatedly prohibited the drinking bouts of the men who transplanted the rice fields. Even in the mid-11th century, men can still be seen transplanting the rice fields managed by a large landowner in Dewa.[67]

The female role possibly goes back to a religious origin, judging by this legend from Harima: in the district of Sayo, the goddess Imotamatsuhime caught a deer, cut open its belly and sprinkled the seedbed with its blood. Thanks to this, the seedlings sprouted in a single night. Afterwards the goddess gave this order: "Young women, you will transplant these seedlings on the night of the 5th month *sayo*", hence the toponym of the district of Sayo.[68] This type of legend may be the forerunner of the simulated demonstrations of transplanting, organized each year from the 11th century at the imperial court as part of the agrarian festivals *dengaku*. Fujiwara no Michinaga (966–1027) attended a demonstration in 1023, in which fifty or sixty female transplanters took part. In 1127 and 1129, there were some twenty transplanters. At this date, the texts mention for the first time the term *taue*, this notion having been expressed previously by the verb *tauu*.[69]

Sei Shōnagon had the opportunity of seeing transplanting not as a spectacle offered at a gathering of nobles, but as work in the fields: "On my way to the Kamo Shrine, there were women transplanting rice *tauu*. They were wearing straw hats in the shape of a brand new tray *oshiki* and there were many of them, singing, bent over, then backing away to do something else. While I was curiously observing this, I was surprised to hear them making fun of the cuckoo thus: cuckoo, it is your song that makes us transplant the rice field."[70] This account allows us to imagine the transplanters moving back row by row. The cuckoo *hototogisu*, a summer bird, heralded the 5th month and so the time for transplanting. At that period, the literati considered certain birds and plants as markers of the agricultural calendar and of the seasons in general. Thus the bush-warbler *uguisu* was the bird of spring and the wild goose *kari* that of autumn. We read earlier in a poem by Sone no Yoshitada that the departure of the wild geese in the 2nd month heralded the time for sowing in the seedbeds.

Direct sowing

While transplanting is a very familiar image associated with rice growing, the texts also mention sowing in rice fields. The agrarian rituals celebrated at the Ise Shrine in the 2nd month included, in the rice field that produced the morning and evening offerings, "the first sowing of the rice field" *mita no tanemaki oroshi* (not *taue*). The ritual of the first day of the rat in the 2nd month next shows tillage, accompanied by dances and songs, then the harvest, and makes no mention of transplanting.[71] Furushima Toshio thinks that this ritual, found in an early 9th-century report from Ise, reflects the ancient practice of the prehistoric period, which was still unfamiliar with transplanting.[72] But Kinoshita Tadashi draws on ethnography to show that transplanting precedes direct sowing in history. According to him, transplanting was accidentally omitted in the Ise ritual.

Kinoshita quotes another text from Ise, dating from 1193, which also refers to sowing *hashoku*. In his view, the two Ise texts are corrupt and he supports this hypothesis by the fact that the order of sowing and harvesting is reversed in the second text.[73] We cannot decide in the case of Ise, though we saw earlier that green manure was spread there. The term *nawakusa* or *naegusa* denoting spread grass contains the semantic element of the seedbed *nawashiro*. Perhaps the ritual implies sowing, not of the rice field but of the seedbed?[74]

Be that as it may, the *Man'yōshū* poems also refer to direct sowing in a rice field, for example this one: "After cutting willow branches to fix in the soil, I sow the purified seeds; it is towards you, pure as you are, that my thoughts turn." According to another poem, sowing was done in a wet field: "Water is plentiful in my raised rice field where I have sowed the seeds; my work of removing *erabu* the abundant (barnyard) millet [is solitary]; so I will sleep alone." However, rice was also sown in dry ground: "I have gone out in search of a small field to clear and sow the purified seed, I have left home, my legs [protected by] a bandlet, [my feet] wetted by the ford."[75]

These poems cause problems for some Japanese historians. In their view, the direct sowing of rice is a marginal phenomenon.[76] Yet several notable examples come to mind. The best known is that of seedbeds. Fujiwara no Nakazane (1057–1118) composed this poem: "Near the sluice-gate that allowed me to store the mountain water, I drove in stakes, then I sowed my small field of five hundred *shiro*." The "small field" *oda* is sometimes associated with the seedbed *oda no nawashiro*.[77] The second possible example is dry rice, known today by the name *okabo*, sown in a fertilized dry field or in a swidden field. In fact, the "small field to be cleared" *araki no oda* mentioned above could be a patch of ground that has been cleared and burned, fertilized by the ashes. The following poem possibly refers to the same practice: "In Suminoe, they cleared *ta ni haru* the slopes, sowed the rice, and so, until the harvest, I did not see you." As for dry rice, Satō Yōichirō published his DNA analysis in the early 2000s. He found in more than ten archaeological sites of the Yayoi period, alongside temperate-type irrigated Oryza sativa japonica rice, rice of the same species but of a tropical type that is normally grown on swidden fields, and this in a proportion of 10–40% compared with irrigated rice. Satō also collected samples of tropical rice dating from the early and late medieval periods.[78]

The third possibility is rice sown directly (without transplanting) in wet ground, i.e. often flooded by surface or underground water. This type of terrain is near the groundwater table and is difficult to drain. Wet sown fields, called *tsumida* since the 17th century, have the disadvantage of needing a lot of seed and a great many weedings, but they allow lands where controlled irrigation is impossible to be brought under cultivation. Given that the surface of the soil a thousand years ago was much lower (50 cm–1 m) than the present-day level, one can easily imagine that much of the land was wet in early medieval Japan. For this reason, there were drainage campaigns in the 11th–12th centuries, according to Kuroda Hideo, at the same time as a wave of land clearing and a return to cultivation of lands that had been abandoned for a long period. At that time, direct sowing of rice with a red caryopsis *sekimai/akayone* was practised. Direct sowing was widespread up to the 20th century in some wet regions. It was only in the 1910s that the government undertook drainage works in many parts of Japan. The region of Shizuoka, particularly marshy, was not drained until after the Second World War. Old photos of this area show women up to their waists in the rice fields. Perhaps this is how we should picture the villagers of Sone in Hitachi, mentioned above, who cleared a reed plain to establish their rice fields.[79] This suggests that flooded fields requiring direct sowing were numerous in the early medieval period, but the absence of water control made their cultivation very difficult.

Irrigation

Water control is used to stabilize rice yields. The water level is gradually adjusted in the seedbeds, then in the rice (or paddy) fields as the rice ripens. A lack of water results in a partial loss of the crop. The water is renewed periodically, though care is taken not to use water that is too fresh and comes directly from mountain streams, as it retards growth. During this time, a lack of sunshine is just as bad as a lack of water. One waits until the summer heat has regulated the water temperature in the canals and rice fields. The paddies are also drained at some point after transplanting to aerate the seedlings and prevent rot, then they are refilled and the water level is lowered near the end of the rice-growing period. These details concern present-day methods. Whatever the precise method in the early medieval period, it required the provision of irrigation and drainage systems. These installations were one of the main preoccupations of early medieval governments.

The Japanese court undertook large-scale public works from the 7th century. The annals mention more than eight hundred storage dams and canals built up to 720. The government continued to support the construction and upkeep of dykes and reservoirs during the 8th century, requiring corvée labour of between ten thousand and three hundred thousand working days for each project. For their part, the large monasteries undertook irrigation works on their estates from the 8th century. The water-distribution systems have been extensively studied. Most interest has been in the political and social aspects and the collective nature of Japanese society has been linked to the history of irrigation management.[80]

We confine ourselves here to listing the types of hydraulic installations using examples from early medieval texts. The *Nihon shoki* relates an episode concerning the rice fields of the goddess Amaterasu and her brother Susanoo. When Susanoo saw that his sister had good rice fields, whereas he had been given inferior ones, he took his revenge: he destroyed the ridges, filled up the canals, pierced the water-pipes, sowed on top of the existing crop and drove stakes into his sister's rice fields. This legend names the main installations, i.e. ridges *a/aze*, canals *mizo* and water-pipes *hi*. Divine misdeeds are referred to as natural disasters in the prayer texts read at the imperial court during the annual purification ritual in the 6th month.[81]

Yet even when rice fields were equipped with these systems, water had to be brought from rivers and stored in artificial ponds, reservoirs and storage dams, all called *ike* ("pond"), as well as in dams or weirs, called *seki*. The struggle for water is attested by another legend, this one recorded in the geographical treatise of the province of Harima (Hyōgo). The god Iwatsuhiko and his sister Iwatsuhime wanted to bring water to the rice fields of two villages located on either side of a mountain. The god dug a trench on the summit so as to turn the course of the Minashigawa river northwards. Then the goddess built a dam *seki* using stakes driven into the river bed and had a canal dug leading south. Seeing this, the god again diverted the course of the river westwards. But the goddess had an underground water-pipe *shitahi* installed, thanks to which she finally managed to water the rice fields facing south.[82] This legend shows the problems posed by irrigation and tells us that dams were used to cut off water.

The poetry of the 8th century mentions the various methods. A number of poems can be quoted. "The water from the Sahogawa dam feeds my transplanted rice field; now that I am harvesting the early rice, will I be the only one eating it?" Water came either from rivers or from artificial ponds. Another poem extols the virtues of a ruler who supported the building of reservoirs that the writer compares to the sea: "The august lord, because he is a god, between the steep peaks, where the cedars rise, has made a sea appear." Reservoirs were reinforced by stakes: "I drive willow branches into the edges of the pond *ike no tsutsumi* near my small mountain rice field; whether I succeed or not, I will be with you." Holding or storage dams were also reinforced with stakes forming a palisade: "My heart is weak like the fragile palisade *shigarami* of a dam *ide* that collects fine seaweed." The stored water was then fed into canals: "The water of the pond where the wild geese gather, even if it overflows, will be directed towards a canal already dug." In some places, people used pipes made from tree trunks cut lengthwise, hollowed out and joined together. "My heart is heavy, like the invisible underground pipe that comes from the pond where the wild geese live; will I see you again today?"[83]

The various hydraulic installations are attested for the estates of the Tōdaiji in Nara. In the 8th century, this temple had canals dug on its estates in Echizen (Fukui), the longest of which ran for 7.5 km. The temple also installed pipes *hi* with an average length of 2–3.5 m to bring water from the canals to the rice fields.[84] Water-pipes no doubt existed from prehistoric times. Their introduction is attributed to the founder of the family Hida no omi, a name meaning 'Dignitary of the rice field with a water-pipe'.[85]

A number of maps of Tōdaiji estates are still in existence, and in them one finds notes and even drawings of irrigation systems. The plan of Minuma, in Ōmi (Shiga), gives us the oldest image of a sluice-gate, built on an artificial pond. This is probably a movable sluice-gate called *mito* in the early medieval dictionaries. The maps of Suka and Naruto show drawings of canals *mizo*, *ōmizo*, marshes *numa* and a spring *izumi* (possibly a well?). They also give, in several places, an illegible note which, placed away from the watercourses, could refer to the water-pipe *ihi* or *iri*.[86]

One cannot underestimate the extent of the irrigation systems at the level of the country as a whole. Even now, Japanese rice growing depends on the same installations. Today, the water for rice fields comes from rivers (74%), weirs and storage dams (17%), the subsoil (4%) and rain (5%). A region like Kagawa prefecture in Shikoku needs 2,500 storage dams to feed water to its rice fields. The largest, Mannō-ike, 20 km in circumference, provides water for 4,600 ha of rice fields. This artificial lake is said to date back to the 8th century. Its history is known thanks to a stele erected in 1020 and a 12th-century legend. Even today, the ceremonial opening of the sluice-gates of Mannō-ike is celebrated each year on 13th June to mark the start of irrigation of the rice fields.[87]

In the early medieval period, the hydraulic systems needed regular maintenance, both on large estates and on small individual holdings. Almanacs determined the auspicious days on which farmers could carry out these tasks. Towards the end of the 3rd month and the beginning of the 4th month, i.e. before transplanting, canals and inlet ditches, as well as dykes *teibō* had to be repaired. The almanacs also recommended just as often to "stop the gaps" *ana o fusagu*. Perhaps this involved stopping fissures – cracks and holes in canals and basins that were sometimes difficult to locate. This was done not only before transplanting, but also, repeatedly, from the start of the year.[88]

However, all hydraulic systems were useless if it did not rain. An early medieval poet wrote: "The days without rain follow one another; the transplanted rice fields and sown dry fields will fade and wither; I am heartbroken at the sight, and I await rain from heaven like a child waiting to suckle." Water was essential for the rice to grow, but too much caused the seedlings to rot. Fine weather was just as important. The annals provide evidence of many prayer sessions ordered by the imperial court, most of them in the 4th, 5th and 6th months. Between the mid-7th and late 8th century, the court prayed seventy times for rain and seventy-four times for it to stop.[89]

Later, prayers to ask for rain or to make it stop were regularized in various ways. In the time of Kanmu (r. 781–806), there were offerings at the Ise Shrine and celebrations in other large shrines. Then Buddhist rituals were added with sūtra readings at the palace, in the Daigokuden (Great Hall of State). In the 9th century, the divinities of the Ni(h)u and Kibune shrines made their appearance as tutelary gods of rain. In the 10th century, the Office of Divination began to celebrate the Five Dragons (*Goryū no matsuri*). Under Horikawa, prayers addressed to the Five Dragons were organized in the Shinsen-en garden. In their turn, the provincial governments presented offerings in the shrines under their control and organized Buddhist readings in the State temples. The official celebrations were the subject of two works: "Journal of prayers for rain" *Kiu nikki*, compiled by a monk around 1117, and "Treatise of prayers for rain" *Kiu hōki*, an account of nineteen sessions that took place between 875 and 1065. There is some doubt as to the customs of the rural population, but it is thought that there were even animal sacrifices.[90]

The quotations provided above give an idea of the hydraulic systems of early and late medieval Japan. The following installations can be identified:

collection and conservation of water:
- pond, storage dam or reservoir (*ike*)
- or weir, dam (*seki, ide*)
retention or releasing of water:
- dyke (*teibō, tsutsumi*)
- dyke, rice-field ridge (*a/aze*)
or sluice-gate (*suimon, mito*)
feeding/channelling of water:
- canal (*mizo*)
- or rill (*unate*)
- or water-pipe (*hi, shitahi*).

Weeding

While the care of hydraulic installations applied only to irrigated rice, weeding was a universal task, *a priori* necessary for all irrigated and dry crops. An 8th-century poet writes: "The stem of the pondweed that grows in the rice fields on the slopes of the Awa hills, I break it gently when I pull on it, but I beg you not to break off our conversations." This poem speaks of the stems *tawamizura* that are bound together when they are removed and that have been identified, among others, with the pondweed, Potamotegon distinctus A. Bennet, an aquatic weed. According to two other poems, barnyard millet was also removed from the rice fields. However, in the poetic vocabulary, weeds are "removed" *hiku*, while barnyard millet is "collected" *erabu*. This suggests that barnyard millet was gathered for food, even if it was only the wild variety (mod. *inubie*), Echinochloa crus-galli L. Nor can one rule out the possibility of the intercropping of rice and barnyard millet, as was practised in the Edo period in the region of Aizu (Fukushima) and elsewhere.[91] But another early medieval poet seems to regard this millet as a weed: "The ears have appeared; the millet *hiekusa* that mingled with the rice in the summer field has been removed and thrown away *hikisuterarete*; how quickly the time has passed!"[92]

Weeds appear in poetry under the names "grasses that mingle" with the crops *majiru kusa* and "field grasses" *tagusa*. Here is an example: "Lowly peasant, you wait for a break in the rains of the fifth month, to weed *tagusa hiku* the field where the rice has grown thickly."[93]

We do not know the methods, as the early medieval administrative texts are silent on this aspect of agricultural work. However, in the 17th–18th centuries, the first weeding of the rice was scheduled ten to twenty days after transplanting, then two or three other weedings during the summer. It was extremely time-consuming work done by hand or with a hoe (Figure 1.11). Horie Hisashi thinks that worn hoes were sometimes used in the medieval period to kill weeds.

In medieval China, weeding was done with small sickles.[94] The Japanese poems quoted above rather suggest weeding by hand *hiku*, but the dictionaries of the 10th–12th centuries record the verb "to cut grasses" *kusagiru*, which seems to refer to weeding with an implement, and the *Shinsen jikyō* (ca. 900) mentions a "hoe-spade" *kuwa-suki* used for weeding *josōki*. The nine "old hoes" *furuki kuwa* kept, in 761, in a granary with other tools, may also have been used for weeding.[95]

The "Regulations of the Engi Era" use the verb *kusagiru* for the weeding of vegetable crops. Those in charge of the imperial vegetable gardens allocated a relatively high number of work days to weeding, which sometimes followed raking done with a notched levelling board or a rake that could cover the seeds and do an initial weeding at the same time. The gardeners weeded everything except wheat, barley, radishes, coriander *konishi*, colza, water-leeks and water-celery (unless the text is incomplete). There was only one weeding for soybeans, cowpeas, lettuce, mallow, nothosmyrnium and ginger *mega*, two for red beans, butterbur and thistles, and three for Chinese chives, 'wild chives', spring onions, ginger *kurenohajikami*, the two kinds of melon, egg-plant and taro.[96] We do not know the customs of the rural population, but in principle cereal crops were impossible without weeding, except in the case of swidden farming.

The struggle against disasters

Weeds were not the only enemies of crops. The early medieval Japanese distinguished between natural disasters on the one hand, and damage that could be countered on the other. In the eyes of the government, the former included floods, droughts, typhoons, frosts and insects, and the latter the damage caused by birds and wild animals. The administrative regulations provided for the distribution of food and tax exemptions in regions hit by natural disasters.[97]

Insects

The most feared insect was the locust *inago*. The annals tell of cases of locust invasions that destroyed the harvest in one or even several provinces at once. This was a legendary plague. In mythical times, the god Ōkuninushi established precepts for "preventing the damage caused by birds, wild animals and insects *haumushi*" (literally, ramparts). These disasters found their way into the prayer texts for the purificatory rituals celebrated twice a year at the imperial court.[98] The method used to combat locusts is mentioned in the *Kogo shūi*: next to the rice-field inlet that leads into the ditch are placed a piece of ox flesh and a carved phallic symbol; grains of pigeon wheat *tsusudama* (Coix Ma-yuen Roman.), Japanese pepper *naruhajikami*, leaves of the walnut tree and salt are spread on the ridges. Besides this practice, steeped in exorcism, people called on heaven for assistance: in 874, for example, the court presented offerings to Ise, and in 1017 it ordered sūtra readings in all the provinces to avert the locust plague.[99] We also know of a case of human intervention: in the imperial vegetable gardens, a worker was given the task of "chasing away insects" *mushi o harau*, we do not know how, for twelve days, in the plot where early melons were being cultivated.[100]

Wild animals

As for wild animals, they came in autumn, especially at night, to ravage fields located away from dwellings or in the mountains. The poems of the 8th century, such as this one, call them "fields of deer and wild boar": "If our souls should meet, we would lie together, but my mother watches me, as one protects the small mountain rice fields from deer and wild boar *shishida*."[101]

A legend from the province of Bungo (Ōita) mentions a method for preventing the destruction caused by animals. It relates that a peasant was cultivating his rice field at the foot of Mount Kubi. Seeing that a deer would come and graze on the shoots, the peasant enclosed his field with a palisade *shigarami*. But the deer stretched out its neck, passed it through an opening and continued to eat the shoots. Then the peasant caught the deer and prepared to cut its throat. Whereupon the deer begged him to spare its life and promised that future generations would no longer graze on the rice in his field. The peasant spared it; from that time on, the rice ripened without suffering any damage and the place was called Mount Deer-neck Kubi. This legend mentions the protection of the field by a fence, though not a very effective one. In the 17th–18th centuries, the protective palisade was known in some mountainous regions. On Tsushima, a system of corvée labour assigned the young men to the construction and upkeep of field enclosures intended to prevent the ravages of wild boars, which regularly attacked the lands of this island. However, for the early medieval period, this method seems to us to be too onerous where small mountain fields were involved.[102]

Ethnology normally distinguishes three ways of countering the damage caused by animals: hunting, trapping or scaring them. In early medieval Japan, people scared away animals using light, smell and noise. Three poems from the *Man'yōshū* illustrate the method of a fire lit at night: "In the early morning mist, I hear the muffled croaking of the frogs under my deer-fire shelter *kahiya* and I long [to hear your voice]"; and: "In the early morning mist, under my deer-fire shelter *kahiya*, I hear the croaking of the frogs [as if they were secretly thinking of me]; if only I could tell her I am thinking of her"; and again: "The fire against the mosquitoes *kahi* set *oku* by the old man who watches over the rice field of the distant mountains burns slowly; so the fire in my heart [burns] for you."[103] These poems allude to the temporary shelters in which people stay, near fields located far from their home, probably to watch over the fires lit at night to scare away animals.

However, the ethnologist Nomoto Kanichi suggests another interpretation regarding this fire: he is thinking of the pungent deterrent. The ends of straw torches, tufts of animal hair or bundles of hempen cloth were singed and placed at the corners of the field, attached to stakes, in order to keep away animals by their smell. This device was known in the mountains of Gifu by the name *kabi* up to the 1950s. According to Nomoto, the etymology of *kahi* or *kabi*, written like "deer fire" and "incense fire", can be explained by the practice of pungent deterrents *kabi*.[104] But in our opinion, this deterrent does not require the field to be guarded at night, once the pungent pole has been placed there. In the third poem quoted above, the "mosquito fire" rather reminds us of "the incense used to repel mosquitoes" *kayaribi*. This incense would protect the old man in his hut and would therefore have nothing to do with the protection of the field. In this poem, the "fire" is "set" *oku*, not "lit". Mosquitoes annoyed everyone in summer, and so mosquito incense *kayaribi* is often mentioned in 10th to 12th-century poetry.[105]

Birds

Noise was used to chase away birds, which also caused enormous damage. A poet expresses it in these words: "In the small field tilled by Bodhi, see the crow, its eyelids swollen, after having [committed the sin of] pecking at the cultivated rice, perched on a pole."[106] To scare away birds, small wooden boards or pieces of bamboo were attached to a rope. One end was fixed to a pole stuck in the ground, and the other end was held by a person who shook the rope to bang the pieces of wood together. This device was known from the 8th century by the name of "small shaken boards" *hikita/hita*, and from the 12th century also by that of "crying child" *naruko* (Figure 1.8). The early medieval Japanese mostly associated this device with birds, but the shaken rope also scared away deer and wild boar. Here is what a 12th-century poem has to say: "On the transplanted mountain rice field, dirtied by muddy water, they have stretched [the rope] of small boards *hita*; and sleeves are wetted [by tears] for autumn has come."[107]

The writer of this poem uses the expression "to stretch *haeru* the rope". Other poems show that people "stretched a rope" *nawa* or *shime haeru* in order to guard *mamoru/moru* the field, without mentioning the noisy wooden boards. Perhaps they are alluding to a symbolic rope. According to some customs, a ritual place was marked by a rope *shimenawa* and a place was enclosed by a rope *nawa haeru* to indicate ownership. The early medieval poets may have been inspired both by the shaken wooden boards and the symbolic ropes.[108]

Whether it be to watch a fire or to shake the small wooden boards, some-one had to stay on the spot in a temporary shelter. Guarding a mountain field *yamada moru* and the solitude implied by this task are a favourite subject of early medieval poetry.[109]

The scarecrow in the shape of a dummy was also used to frighten birds. The scarecrow appears in poetry from the 10th century, for example in this poem by Sone no Yoshitada: "Here I am like a scarecrow *sohozu*, upright, watching over the rice field far from my home." It seems that the etymology of the term *sohozu* goes back to a legend from mythology relating to the god Sohodo who sacrificed himself in order to guard the mountain fields.[110] Yoshitada compares the irksome task of guarding the fields with the fate of the unchanging and solitary scarecrow.

Harvesting

After months of toil and anxious moments for the farmers, autumn brings the ripening of the rice. Many poems speak of the ears in the autumn fields and the auspicious time of the harvest. Here are three poems of the 8th century: "The ears *ho* in the autumn rice fields bend beneath the wind, to one side only; likewise, my thoughts are bent [solely] towards you, though you remain indifferent"; and again: "[On my way] to meet you, it seems to be the time of the early rice harvest; [in fact] the lespedeza is in bloom." A little later in the year, a poet writes: "After finishing the harvest of my share *kari-ba* of the autumn field, there are heard the cries of the wild geese *kari* heralding winter."[111] The last two poems evoke the harvest time. The lespedeza, an autumn flower, has inspired many poets. Its flowering heralded harvest time. By contrast, the wild geese arrived after the harvest from the Nordic regions to pass the winter in Japan.

The public administration placed the time of the rice harvest around the 8th month (September), while allowing for regional differences from the 7th to the 9th month. Officials were given leave at this time to attend the harvesting of their fields, after having been granted another holiday in the 5th month for the transplanting. As for the rural population, it was expected to be ready to deliver the rice taxes from the middle of the 9th month and at the latest at the end of the 11th month, for the early and late varieties. Generally speaking, the government considered the agricultural season *yōgetsu* to be the period from the beginning of the 4th month to the end of the 9th month, and the off-season *kangetsu* the rest of the year.[112]

We have textual evidence for the following cases. On the Kuroda estate in Iga belonging to the Tōdaiji temple, in a place corresponding to the present-day town of Nabari, in 1054, they began to harvest the early rice at the end of the 7th month. Similarly, in the 760s, on the 17th day of the 7th month, an individual residing in the district of Echi in Ōmi (Shiga prefecture) asked for twenty days' leave to harvest his rice fields. But harvesting was sometimes done much later. On another estate in Ōmi, this was carried out, in 873, from the 17th day of the 9th month to the 7th day of the 10th month, i.e. from 12th October to 19th November. At that time, they harvested 181, 245 and 251 sheaves of rice respectively over three days, but we do not know either the number of workers or the land areas.[113]

This disparity in the dates of the harvest can no doubt be explained by geographical differences and also by the custom of growing early and late rice simultaneously. According to Sone no Yoshitada, early rice *wase* and intermediate rice *nakate* ripened in the 7th month, and late rice *okute* in the 8th month. He writes in the 7th month: "While I am watching over the intermediate rice, with its hanging glumes, the ears appear in bunches." Then the poet writes a month later: "While I am watching over the late rice that grows pell-mell, will the lespedeza have already bloomed?" But late rice is also found much later in the year, in this poem by Fujiwara no Kiyosuke (1104–1177): "The lowly peasant, with his hat of woven sedge, harvests the late rice *oshine* in his pure white field, covered by the first snow that he has had no time to sweep." This poem was composed on the 10th day of the 10th month, i.e. 24th November 1175. In fact, it is very likely that the early medieval Japanese planted different varieties of rice together; this has become clear following discoveries of inscribed tablets with the names of the varieties of seed handed out by the local authorities to the farmers in some villages.[114]

High cutting or low cutting?

A question often asked and on which opinion remain divided concerns the height of the cut: were cereals harvested high on the stalk, i.e. 10–20 cm below the ear, or low down, i.e. in the lower third? In other words, were ears or stalks harvested? It is generally accepted that the evolution was from high to low, between the 3rd and 14th centuries, but there is no precise information for the intermediate period that interests us. Archaeologists offer as evidence the farming implements discovered in sites, while historians depend on the documents relating to storage. They rely on the following reasoning.

The harvesting knife, used since the Neolithic, allowed the ear to be taken off with a quick movement of one hand, holding only one stalk at a time. With the sickle, several stalks were cut at the same time, low down, the tool being held in one hand and a loose sheaf of stalks in the other (Figure 1.12). The harvesting knife is therefore associated with high cutting and the sickle with low cutting. Chronologically, the stone-bladed knife was replaced by a wrought-iron knife from the 3rd century AD in southern Japan (Kyūshū); then the iron knife spread towards the centre of the country and disappeared at the beginning of the 6th century. However, from the 8th–10th century there appeared another type of harvesting knife in a half-moon shape. The wrought-iron sickle existed from the first two centuries AD, but only spread throughout Kyūshū from the 3rd century, at the same time as the iron knife. It then moved up to the centre of the country around the 4th century and was still in use in the 20th century (Figure 1.12). In the periods from the 3rd to the early 6th century and from the 8th–10th century the knife and sickle therefore coexisted in the same regions. Archaeologists have discovered them in the same sites.[115]

Historians associate high cutting with the method of storage. The documents mention two types of storage: in ears counted in sheaves *soku* on the one hand; in grains measured by volume (*koku*, etc.) on the other. From this, scholars consider that sheaves of ears (still attached to short pieces of stalk) were stored as they were in the granaries, while long stalks were of necessity threshed and the rice stored in grains. In other words, in the granaries, high cutting produced ears and low cutting produced grains. It may be noted that high cutting allowed the ripe ears to be selected, stalk by stalk, while low cutting left open the option of recovering the straw for various artisanal uses. However, the administrative documents of the 8th century refer to the coexistence of the two storage methods: the yearly account registers (*shōzeichō*) for several provinces mention the number of granaries for grains and granaries for ears, with one or other predominating according to the region. The deeds of land sales note payments in rice, giving the amounts either in sheaves or in grains. Therefore both ears and grains were used. One can, however, observe that buying and selling move towards grains and that sheaves disappear from transactions in the first half of the 11th century. This time coincides with the disappearance of the harvesting knife, which is why for some scholars this date represents the definitive transition from high to low cutting. However, all these arguments do not take other possibilities into account. Archaeologists speak only of rice cultivation, but some tools could be used for both rice and non-irrigated cereal crops. Also known in ethnology is high cutting done with a sickle. Moreover, threshing the ears by pounding in a mortar is attested by the texts, meaning that storage in grains is therefore not determined by high or low cutting.[116]

Around the year 1000, Sei Shōnagon describes the following scene:

> For my retreat of the 8th month, I went [to the Kōryūji temple] in Uzumasa [west of the capital]. At that time I saw a great many people in the rice fields adorned with ears, moving about and harvesting the rice. We had already arrived [at harvest time] and I said to myself: 'Weren't they recently selecting [the young seedlings] at the time of my pilgrimage to Kamo?' Here, the men were holding the green stalks and harvesting *karu* the red rice at the base *ine no moto*. With some tool or other, they were cutting *kiru* the base of the stalks and appeared to be doing that very easily. How did they do it? It was interesting to see them lined up in rows after having spread out the ears *ho* on the ground, to see their field hut and many other things.[117]

While Sei Shōnagon was thus able to observe low cutting, the poets of her time, by contrast, write of the "cut ears" *kariho* that result from high cutting.[118]

The use of gleaning also attests to high cutting. The "Regulations of the Engi Era" prohibit the peasants working in the public rice fields "from bringing members of their family after the harvest to pick up the ears" *ho o hirou*. Gleaning was probably a privilege granted to the poor in rural communities. This is shown by the episode in the *Nihon ryōiki* of a poor devout woman who, in order to donate a Buddhist image to the Hatadera temple in Kawachi, was able to pay the artisan only with gleanings. Gleaning is also documented by poems such as this one: "In the mountain hamlet, the frost on the ridges of the field near the house has vanished, and the road where we pick up the fallen ears *ochibo* is unending." In the 18th century, it was usual on family farms in the region of Kaga (Ishikawa) for the elderly women and children to go gleaning after the harvest in the fields, in the drying areas and on the ridges. An illustration from the *Nōgyō zue* shows ears on short pieces of stalk, forgotten here and there, in and next to the fields. Furthermore, poetry relating to gleaning and "cut ears" continues until at least the 15th century, which seems to indicate the persistence of the high cutting of rice over the centuries.[119]

For ritual use, rice was not cut with a tool. At the time of the agrarian ritual in the 2nd month and the Harvest Festival in the 9th month, celebrated annually at the Ise Shrine, the ears were collected by hand *nukiho*, leaving the stalk standing. It was the same for the Great Thanksgiving when a new emperor was enthroned at the imperial court. For this occasion, the preparatory ritual included the collecting of ears in two rice fields of 6 *tan* (1.8 acres) located in the provinces of Ōmi and Tanba or Bitchū. Did the early medieval Japanese think that a cutting tool might offend the rice spirit? A legend from the province of Harima documents the stripping of rice by hand. It concerns the etymology of the name of the hill Tekari, meaning "Harvesting by hand": a deity, passing through this place, cut *karu* the grass by hand to make a seat; but some say that it was outsiders who had come here to live and who cut *karu* the rice by hand because they were not yet familiar with the use of the sickle.[120]

All this concerns rice. Harvesting of wheat and barley was done with high cutting, if one is to believe a poem by Sone no Yoshitada. According to him, the wheat harvest took place in the central region in the 5th month (June). The administrative documents place this time rather in the 4th month. Several actual cases may be mentioned. In Kii and Yamato, wheat or barley was cut respectively on the 18th day of the 4th month (2nd June) in 1181 and on the 24th day of the 4th month (12th May) in 1170. On another estate in Yamato, harvesting was done at the beginning of the 3rd month (1st–10th April) in 1101, but this seems very early to us. One finds similar dates for an estate in the province of Kii (Wakayama). Here the tax receipts for wheat or barley were issued, in 1048 and 1049, from the middle of the 5th month to the end of the 6th month. It may be assumed that harvesting was done one to two months before the receipts were issued.[121] In its turn, the central government prohibited several times, between the mid-8th and early 9th century, the premature cutting of barley and wheat in the 3rd month, in the region around the capital. This abuse involved selling unripe cereals to well-off families as fodder for their stables. One also sees standing wheat in the 3rd month, before the harvest, in a legend from the 9th century relating to the province of Izumi (Ōsaka). According to all this data, the 4th and 5th months seem to be the best time for harvesting wheat. The palace guards cultivated unidentified fodder plants that were sown in the 2nd month and harvested at the latest in the 7th month.[122] This may have been barnyard millet.

Harvesting of foxtail and barnyard millet was done at the latest in the 7th month, according to a 15th-century document of the Arata family from the province of Kōzuke (Gunma). For buckwheat, sowing was recommended, in the 9th century, at the beginning of the 7th month and harvesting in the 9th month. This data corresponds more or less to the modern calendar of the region of Kaga (Ishikawa). In the 18th century, barnyard millet was sown in the 2nd month and harvested, like foxtail millet, in the 7th or 8th month. Buckwheat was harvested in the 9th month. Wheat and barley were sown in the 8th or 9th month and harvested in the 5th month.[123]

The receipts mentioned above, dating from 1048 and 1049, also record the delivery of beans in the 11th month and, once, in the 1st month. For their part, the inhabitants of Owari (Aichi) harvested wheat and barley in summer (4th–6th months) and beans in winter (10th–12th months). At the court, soybeans were sown in the imperial vegetable gardens in the 3rd month. All this corresponds to the practice in the Kaga region. In the 18th century, beans were sown here in the 3rd month and harvested before the 9th or 10th month, the dates when shelling was done. But it may be asked whether there were any summer crops, as in present-day Japan. In fact, in the 10th century, the imperial kitchens served to the emperor "fresh" soybeans and red beans from the 6th to 9th month and cowpeas in the 6th and 7th months. Then, in the 12th century, fresh vegetables including soybeans and cowpeas were served in some aristocratic residences, on the occasion of the banquet of the 7th day of the 7th month. Taro is one of the "fresh vegetables" served to the emperor from the 9th to 1st month; in the imperial vegetable gardens, it was planted in the 3rd month and harvested after the 7th month (at the time of the third weeding).[124] In summary, it seems that rice, foxtail millet, barnyard millet, beans and taro were summer crops, buckwheat an autumn crop, wheat and barley winter crops.

The "Regulations of the Engi Era" note the number of work days per plot of 1 *tan* (0.3 acres) in the imperial gardens and give four different terms for the harvest: cut *karu* were: wheat, barley and nothosmyrnium in two days, butterbur and thistles in four days; picked *toru* were beans and ginger *mega* in two or three days, turnips in six days, radishes in fourteen days, wild chives and water-leeks in fifteen days; ginger *kurenohajikami* was pulled up *torimushiru* in six days; and taro was dug up *horu* in four days.[125] According to a register from the province of Bungo (Ōita) for 737, gromwell roots were also dug up; and poetry speaks of the peasants who dig up bracken *warabi*. The early medieval Japanese probably used digging-sticks.[126]

Poets often refer to the harvesting of wild rice *komo*, a graminae, the seeds of which were eaten and the stalks used for sparterie and making straw mats. These stalks were usually cut in autumn, but Sone no Yoshitada mentions the harvest in spring and summer at Yodo, south of Heian-kyō, a region noted for this plant. Another graminae, eulalia *kaya*/*susuki*, was harvested in autumn. Eulalia was favoured as a roofing material.[127] In these cases it is wild plants that are being harvested.

Drying

At first sight, it might be thought that the method of drying depended on whether stalks or ears were harvested. Yet this does not appear to have been so. We first give some examples from the 17th–18th centuries. Here is an extract from the *Kōka shunjū*, an agricultural treatise from the Ishikawa region:

> To dry *hosu* rice, [bound sheaves or loose sheaves?] are spread out in the field, if the ground is fairly dry. If the ground is wet and there is no [other] place to spread them out, [the sheaves] are hung on drying racks *haza*, to allow them to dry for seven days before bringing them in. For rice, but also for the other [cereals] harvested, it is advisable to carry out this operation quickly to prevent [losses] caused by natural disasters.[128]

Thus it seems that both ground-drying and suspension-drying were used. With the latter, the sheaves were hung from racks made of branches or stems of bamboo fitted together or one branch on two supports (Figures 1.1 and 1.5). (These racks were called 'parrots' in some parts of France.) The choice of drying method depended on the type of ground, dry or wet. In the above description, sheaves with their stalks are probably meant, but sheaves of ears that had only a short stalk were also suspended, for example in agricultural rituals. The *Hyakushō denki* tells us that "before the harvest, they hang up ears *ho-kake*; they stick two stems of bamboo in the ground, on the east and west sides, and stretch [a rope] between them. Then they cut five [loose sheaves of ears with short stalks] of rice that they bind into five small bundles. They hang them at five points along the rope in order to present them as offerings to the tutelary deity of the place."[129] So even ears could be suspended thanks to their short stalks: the expression 'small bundles' is used to refer to loose sheaves of cut and bound ears.

The *Nōgyō zue* shows, for the Ishikawa region in the 18th century, images of drying on the ground, on racks and also in stacks. After a period of drying in the field, the rice was transported to the dwelling or the place of work, then piled up into stacks that could be two or three times the height of a man. The ears were turned inwards and the whole stack covered with a straw mat, to await threshing. After threshing, new stacks were formed with the remaining straw.[130]

The drying methods of the early medieval period are documented by a government decree issued in 841:

> Ministerial decree for the purpose of introducing a device for drying rice ... From what one hears, the peasants of all the provinces, when they are working in the fields, rely too much on the sun and forget about the heavy rains. However, if the sky refuses to clear and the rain does not stop, the peasants will surely suffer from hunger, even with the rice spread

out *oku* in the courtyard right before their eyes. The people, who are ignorant, are very [likely] to get into this situation. But the inhabitants of the district of Uda (Ōuda-chō and Haibara-chō) in the province of Yamato (Nara prefecture) fit tree branches together in the middle of their rice fields and hang up the harvest *tanemomi o kakehosu* to dry it. This way of drying cereals is like exposing them to the heat of a fire. In the vernacular, they call this a rice-drying rack *inabata*. Drying racks are found here and there at present in several provinces. We order that this device be introduced into all the provinces. No effort must be spared to improve the well-being of the people. Jōwa 8th year, 9th intercalary month, 2nd day.[131]

This date, i.e. 19th October 841, suggests there was a harvest of late rice; that year it was possibly at the end of the autumn typhoons, hence the concern about rain. The decree tells us that ground-drying was done in the courtyards of dwellings and suspension-drying in the fields. The authors of the decree use for the former the term *ine*, rice with its stalks, and for the latter the word *tanemomi*, literally "unhulled grains", which here, however, means the harvest in general.

Another government directive, issued in the early 10th century, adds to our information. It condemns an abuse by inhabitants near the capital of Heian that involved forwarding harvests to their patrons in the capital, without paying the rice taxes levied by the province. The order is worded as follows: "The government officials of Yamashiro made their tour of inspection of agriculture near the capital at the start of the early rice harvest. The farmers were cutting the rice as fast as they could. Some were assembling racks *inabata* and hanging up the [sheaves] in rows *kaketsuranu*, others were piling up *tsumioku* the rice behind the wicket-gate [of their courtyard], still others were transporting it dragging along [a container?]. The officials knew this was forbidden, but they were unable to prevent it", then when they returned two days later, no one was left and the rice had vanished.[132]

The date of this document, the 3rd year of the Engi era, 9th month, 4th day, i.e. 27th September 903, seems to us too late for early rice, which is mentioned in the text. Rice was either hung up to dry, as above at Uda, or piled up *tsumioku* in the courtyard and not spread out on the ground *oku*, as at Uda. This expression reminds us of the offerings of rice piled up *tsumioku* "like a mountain" in honour of the deities and of a poem that mentions rice *ine* piled up *tsumioku* in a village at the foot of Mount Takakura.[133] The verb *tsumioku* is associated, according to our interpretation, with "stack-drying", known today as *tsumi-boshi*, i.e. in stacks.

Kōno Michiaki, who has analysed the decrees, does not mention stacks, but Hotate Michihisa gives an example of stack-drying for the year 1115. He quotes the *Denryaku*, according to which a wealthy landowner piled up rice *ine o tsumu* in a courtyard of his residence. The stacks formed in this way were called *inamura* in poetry. A legend relating to Mount Inadane compares the shape of this mountain to "piled-up rice" *inazumi*. The ethnologist Orikuchi Shinobu also made the connection between piled-up rice and stacks in the Great Thanksgiving rituals.[134]

What is stacked rice inazumi?

This notion appears in the early medieval toponyms for Inazumi meaning "Stacked rice". Cut and stacked rice occurs in an old song that accompanied rice pounding: "At Sakata on the way to Ōmi (Shiga), see the stalks of rice *ine* harvested and stacked *karitsumite*; we pound the rice at the start of a new and virtuous reign." Here the female pounders were preparing for the enthronement ceremony. A poet also uses this expression: "Tonight I will go into my temporary hut to watch over the late rice *okute no ine*, cut and stacked *karitsumite* in my mountain field." This suggests that poems evoking *karitsumu* are referring to stacks.[135]

Another poem speaks of the dwellings at the foot of Mount Ōkura in Ōmi in these words: "It is indeed the countless ears cut and heaped up *kariho o tsumite* in autumn that have given their name to the mountain of the Large Granary, Ōkura." Here the reference is to ears and not to rice on the stalk. The possibility of stacks must therefore be excluded. The poem alludes rather to the storing of the harvest in ears in the granary. This notion applies to another toponym, Hozumi, meaning "Heaped-up ears". The word *tsumu* in fact has two meanings, 'piling up in an orderly way' and 'stacking in bulk', depending on whether it involves stalks as in Inazumi or ears as in Hozumi. But some scholars indiscriminately associate piled-up stalks and stacked ears with the harvests stored in granaries, unaware that rice was stored in ears (or in grains), never on the stalk. Further, the texts quoted describe rice piled up in the field or in the courtyard of a residence, not in a granary. How then are we to interpret the piled-up stalks, if not in stacks?[136]

The confusion comes from the fact that there is not always a clear distinction between the terms *ine* and *ho*. However, the texts quoted distinguish them and the protocol manual *Gōke shidai* explains the meaning: "When cereal is harvested, one obtains rice *ine* (with its stalk); when the ear *ho* is cut, one obtains ears for storing in a granary *ei*."[137]

Let us return to the other methods of drying. The illustrated scroll *Ippen shōnin eden*, painted in 1299, shows images of sheaves lying in the field and sheaves arranged in shocks shaped like cones, as well as images of stacks which, however, can sometimes represent stacks of straw.[138] A female poet describes one of these methods as follows: "[In front of this] hut covered with rice straw, one sees in the very narrow courtyard the rice from the nearby field, cut and left to dry *karihosu*." She describes ground-drying in the courtyard, but the verb *karihosu* by itself does not appear to denote a specific method of drying. The monk Ryūgen, for his part, recommends suspension-drying in these lines: "In a cramped dwelling, rather than putting the rice *ine* out to dry *hosu* every morning, it is better to join racks *hade* together and hang up *kaku* [the sheaves]."[139] The term *hade* is similar to the names *haze* and *haza* in some modern dialects. It occurs in a poem by Sone no Yoshitada: "The ears of wheat *mugi no ho* cut and suspended *karihosu* on the drying racks *hade* by a mountain-dweller; at this moment, [seeing this] I am heartbroken."[140] These are not rice sheaves, but small bundles of wheat or barley ears, showing that suspension-drying was not restricted to sheaves with long stalks.

Another writer uses the official term for this device: "While the drying racks *inabata* were lined up as far as the eye could see, above the fields near the houses, covered by the morning mist, an autumn evening breeze had already arrived." *Inabata*, written here in *hiragana*, in the decree quoted above contains the characters for rice *ine* and loom *hata*, possibly referring to a system of wooden sticks joined together. The reading *inabata* in the decree is from a gloss in the earliest manuscript of the *Ruiju sandai kyaku* dating from 1266. However, the dictionary *Iroha jiruishō*, compiled in the second half of the 12th century, gives the reading *inaki* and explains that it refers to a tree *ki* from which rice is suspended. This reading is also found in the name of the district of Inaki in Owari (Aichi), designating both a shrine and a family that originally came from this place. The practice of hanging sheaves from trees is attested both by the poetry and the iconography of the late medieval period.[141] In summary, we can say that the drying of cereals was done using different methods: ground-drying, suspension-drying (in sheaves or in small bundles of ears) and stack-drying.

From ears to grains

The harvest goes through several processes before it is ready for consumption: threshing, hulling and, depending on the cereal, a second hulling (or milling) and husking (or whitening). We give an overview before examining the Japanese data. The first operation, threshing, involves getting the grain out of the ear. Ethnologists distinguish three basic processes:

- Hitting or beating: the sheaves are struck with the hand against a hard surface or in a container; this is called 'beating off' (or 'whipping out') and is done only with long stalks. Or else the stalks or ears, spread out on the ground in a suitable area or on a straw mat are threshed with an implement such as a long stick, mallet or flail.
- Rubbing: the stalks or ears are trampled by the hooves of animals, crushed by a roller or, for small quantities, crushed between the palms of the hands.
- Stripping: the stalks are pinched to separate the grains; cereals with loose inflorescence (rice, oats, barnyard millet) are drawn between two long chopsticks tied on one side and squeezed with the hand on the other. This implement is called *mesorias* in Asturia, Spain; or else cereals are drawn through a threshing-comb, but in the case of wheat and barley both methods only separate the ear from the stalk.

Horio Hisashi suggests that hitting and rubbing are suitable for cereals with compact inflorescence, like wheat, barley, millet (*setaria*) and rice of the indica variety (long rice) and that stripping is suitable for rice of the japonica variety (round rice). As indica is more usual in China and japonica more widespread in Japan, this explains the fact that medieval Chinese iconography shows the threshing of rice with a flail and Japanese iconography of the 17th–18th centuries shows threshing with a comb, followed by hulling with a flail.

Hulling and husking

The grains are then hulled, i.e. separated from their outer husk, normally by pounding. The winnowing that follows serves to separate the grains from the chaff. This is done using a winnowing-basket and, in the modern period, using other implements that aerate the grain.

Then follows a second hulling: for some cereals, the glumes or thin inner husks of the grain must be removed by pounding in a mortar or using a hulling-mill. The grain is then sieved and winnowed again to remove any refuse. In this way one obtains barley, millets and rice ready for consumption, as well as wheat, barnyard millet and buckwheat ready for grinding into flour. But the rice is still "whole"; it can be husked or polished by pounding.[142]

The operations of threshing, hulling, second hulling and husking can all be done, if need be, by pounding with a pestle in a mortar (Figure 1.5). These two implements, for Leroi-Gourhan, are the most widely used throughout the world for preparing wheat and millets and for shelling beans.[143]

Above, we have summarized the basic operations common to all societies. As regards Japan, historians in fact think that, until the late medieval period, the entire processing of rice was done at the one time by pounding in a mortar, which implies at least intermediate winnowings and sievings (Figures 1.3, 1.12 and 1.13). It may, however, be asked whether processing was the same for rice on the stalk and rice in the ear. For rice on the stalk, this no doubt required the help of someone who held the loose sheaves with the ears at the bottom of the mortar. For rice in the ear, the ears had to be removed from the mortar as soon as they had been threshed, before continuing to pound the grains. Historians think that pounding in a mortar was the most widespread method of threshing and hulling. They base this on the presence of mortars and pestles, both in documents and in archaeological contexts and note the almost total lack of evidence of other implements.[144] But their reasoning requires some explanation.

The Japanese archaeological heritage consisted, in 1990, of about a hundred vertical pestles *kine* and some forty wooden mortars *usu*, dating from the 3rd century BC to the 9th century AD. There were also a few winnowing-baskets *mi*, an object identified as a stick for threshing cereals and some fragments of pestles with levers *karausu* (Figure 1.4) that were operated with the feet.[145]

Pounding in a mortar was a very widespread method of processing cereals. It is described in poetry as an arduous and thankless task: "My hands chafed by pounding the rice *ine tsukeba*, this evening will my young master take them again and lament my fate?" One imagines the peasant woman who spends hours dropping into the mortar a large heavy pestle, 1.5–2 m long, which she holds in both hands (Figure 1.5).[146]

Threshed rice, unhulled, had to be delivered annually by village communities to the government of their province. An early commentary on the administrative code of 701 states: "The rice-field tax is rice converted to grains *momi*." The regional authorities recorded the rice they received in grains *momiyone*. They also had the rice received as ears *kachishine* threshed by pounding, using their own workers.[147] It may be assumed that this operation served both to thresh the ears and to remove the grains from their outer husk, and that it was followed by winnowing. This is shown by a legend relating to the village of Nukaoka, "Rice-bran hill", which is as follows: one of the armies of two warring deities began to pound rice *inetsuku* for a meal; the glumes (or husks from the grains) *nuka* piled up to the point where they formed a hill; after sieving *hiru* they also formed a mound.[148]

The next stage of rice processing is documented for the rice paid in taxes. The provincial governments stored the rice in the public granaries, either in ears or in unhulled grains. This rice was a source of capital for the public authorities. According to their needs, they then had it pounded, winnowed *hiru* and sieved *hifuruu*, in order to obtain hulled rice *yone*.[149] The provincial governments also delivered each year to the imperial court a small quantity of hulled rice, called "pounded rice" *shōmai/tsukiyone*.[150] Rice was therefore hulled by pounding, according to the following standard: one sheaf *soku* of rice gave 1 *to* (8.5 litres) of unhulled grains *momi* and 0.5 *to* (4.25 litres) of hulled rice, the pounding ratio being fixed at 50%.[151]

Hulled rice included "white" or husked rice *hakumai/shirayone* and "black rice" *kokumai/kuroyone*, the former being fully pounded at 50% and the latter, less so, corresponding to "whole" (brown) rice (mod. *genmai*).[152] At the court, rice of both qualities was eaten, but "white" rice was most especially used by the Grains Office, which supplied all the court offices and the imperial table for daily meals; "black rice", it seems, was distributed to low-ranking employees.[153]

In public accounting, there was mention of the cost of "cleaning or polishing" *shiragu* the grain, a word containing the semantic element "white" *shiro/shira*. A poem alludes to the whiteness: "While the flower of the month of the hare (the deutzia of the 4th month) dedicated to the deities opens all white, the pestles continue to whiten the grain." The *Utsuho monogatari* describes a scene in the kitchen of a wealthy landowner, where eight pounders took turns on four mortars to polish the rice *yone shiragu* to be dried, and in the metal workshop attached to this residence they pounded the rice to feed the fifty workers, using mortars equipped with pestles that had mechanical levers operated by the feet *karausu*. In the kitchens of a temple estate at Fujiwara-kyō, nine women pounders *kachime* each husked 5 *to* (42.5 litres) of rice per day. Generally speaking, the task of pounding fell to women.[154]

Besides the pounded rice *shōmai* delivered by the provinces, the court had other reserves with the harvests from the government-run rice fields *kanden*, stored in the Grains Office that supplied the tables of the imperial family and the officials in service at the court. Eight female pounders *tsukiyoneme* attached to this Office each pounded three sheaves of rice a day. They obtained 0.5 *to* of hulled rice from 1.2 sheaves, meaning that the husking ratio was higher than the standard imposed in the provinces. (The 3 sheaves thus gave 1.25 *to* or 10.6 litres of rice.) The equipment in the Grains Office included: three mortars 90 cm high and 78 cm in diameter, ten pestles, three large tubs 3.6 m long by 90 cm wide, and twenty winnowing-baskets. Since the straw was recovered, the pounders processed rice on the stalk. They did the threshing with pestles and mortars, or possibly in the three tubs. With the former, one person had to hold the loose sheaves of rice, enabling the other person to pound the ears; with the latter, the loose sheaves could be laid in the tubs to be threshed, then the grains hulled in the mortars. The large tubs in the Grains Office may have corresponded to the "horizontal mortars" *yokousu* mentioned in the *Nihon shoki*. This type of elongated pounding tub existed in Europe in the 19th century. Be that as it may, in the Grains Office, processing of the rice was done at the one time, from the sheaf to the husked grain, but in the provinces processing was almost certainly done at different times, since the rice was stored as ears or as unhulled grains, with hulling and husking done later according to need.[155]

While all the operations for the public administrations were carried out by pounding in a mortar, this certainly does not mean that the early medieval Japanese were unfamiliar with other methods. The court lady Sei Shōnagon observed these tasks, after having earlier witnessed transplanting and harvesting in the fields. She relates her unexpected visit to the residence of Takashina no Akinobu (?–1009) in these words: "He said to me: 'When one goes somewhere, one should see what people do', and he had brought in what they call rice on the stalk *ine*, invited some girls from the neighbourhood, quite pretty young peasant women, and had five of them thresh *koku* the rice. He also asked two other girls to pull [the ropes] to turn a rotating device such as I had never seen. They did this while singing a song in such an amusing way that I began to laugh."[156] Sei Shōnagon thus observed threshing and hulling.

Koku: which method of threshing?

Here threshing *koku* was not done by pounding, otherwise our author, Sei Shōnagon, would have used the term *tsuku*. She would certainly have recognized the pestle and mortar, implements far too widespread for her not to have known them, but she does not give details of the method of threshing that she was able to observe. The same term is found in a poem by Minamoto no Yukimune (1064–1143): "In a mountain hamlet, seated on a stool threshing *koku* the late rice, the rain catches me unawares." Another poet,

Sakanoue no Korenori (?–930?), composed this while admiring a screen decorated with a painting of a farming scene: "The rice of the harvested mountain field has been put out to dry *karite hosu*; the geese arrive crying, tears well up like picked out grains *kokitaru*, in an autumn melancholy." Both poets use the verb *koku* without explaining the process of threshing.[157]

The implements used for threshing *koku* are known to us only for the Edo period. In use then were two long threshing-sticks *kokibashi*, or else two short bamboo tubes *inekan* between which the stalks were drawn while being squeezed (Figures 1.3 and 1.12).[158]

A method without any special tool is beating off (or whipping out); it involves striking the sheaves against a solid object such as a large stone, wooden board, ladder or the edge of a tub or large container. Without a container, a threshing area is required, for example in tamped earth, or the use of a straw mat to collect the grains, as shown in Edo-period iconography. This is still practised today in Yunnan using devices such as a ladder or large tubs. However, as Kōno Michiaki has pointed out, it is for cereals with compact inflorescence, which is not the case for japonica rice.[159]

Another method of threshing without an implement is known thanks to a legend about the village of Chiho (Thousand ears) in the district of Usuki (Mortar and pestle) in Kyūshū (Miyazaki prefecture, Usuki-gun). It relates that the two tutelary deities of the place, called Ōsuki (Large spade) and Osuki (Small spade), saved the inhabitants from the darkness caused by the gods, by urging them to pull off a thousand ears of rice, collect the grains *momi to nasu* and scatter them in the four directions. The local people then rubbed the ears between their hands *temomu* and scattered the grains, after which the sky opened and the light of the sun and moon filled the village. The process mentioned here is threshing by crushing between the palms of the hands, probably used for small quantities. This method is also recorded as a verb *tamomu* in the dictionary *Shinsen jikyō* compiled around 900, a work that is less influenced by the Chinese dictionaries and therefore closer to Japanese customs than the other early medieval dictionaries.[160]

The 10th-century *Wamyō ruijushō* sometimes gives purely Japanese characters and pronunciations, quoting "what people say" *yo ni iwaku*. This is the case for another method of threshing, called "rice-roasting" *yakigome* in "the vernacular". The dictionary explains: "they roast *yaku* rice *ine* to turn it into hulled rice *yone*", and again: "hulled rice is obtained by roasting and pounding new non-glutinous rice *atara/shin urushine*". But the term *yakigome* occurs elsewhere as roasted rice for eating in the protocol for the banquet of the 5th day of the 5th month at the court and in some aristocratic residences.[161]

However, the verb used by Sei Shōnagon is *koku*, meaning 'to thresh by drawing, extracting, pulling off or stripping', which differs from rubbing or roasting. Kōno Michiaki thinks this refers to hand threshing, with each ear being drawn and squeezed between the fingers and/or the palm of the hand. The Chinese character *koku* was in fact read *mushiru*, meaning 'to pull off or remove by hand'. But rice has very sharp phytoliths that cut even hardened skin. This method is hardly practicable for a large number of loose sheaves of rice. And this technique is not attested by ethnography. We therefore think that, in the scene described by Sei Shōnagon, as in the poems, threshing *koku* was done using two short bamboo tubes, which could have escaped her notice, or using threshing-sticks, although neither is attested in the sources until the 16th–17th centuries.[162]

As well as threshing, Sei Shōnagon observed hulling that was done using a rotary device (equipped with ropes) pulled by two people *kurubeku mono futari shite hikasete*. Kōno Michiaki, like Furushima Toshio, identifies this device with the wooden hulling-mill *suriusu*, consisting of a fixed lower wheel and a revolving upper wheel operated by ropes (Figure 1.6). The *suriusu* appears in the early medieval dictionaries and Kōno has found drawings of it in two 12th-century documents, an illuminated sūtra and the painted scroll *Jigoku zōshi*. Just how widespread was the hulling-mill in early medieval Japan? For his part, Fujii Kazutsugu has pointed out that the almanacs recommend maintenance of hulling-mills *suriusu* (together with that of mechanical pestles *karausu*) from the 8th century. The *tengai*, introduced to Japan in 610 by a Korean monk and mentioned in the "Code" of 701, may correspond to the hulling-mill, since the character *gai* of *tengai* was read *suriusu* in the early dictionaries. A hulling-mill is found among the personal effects of a cattle breeder attached to the Ninnaji temple in 1261, and later, in the 15th century, these mills were part of the equipment of large holdings. A poet of the time gives a picture of these tasks reminiscent of that of Sei Shōnagon: "Among the simple homes *tami* in the middle of the fields, I have seen the lowly peasant girls threshing rice in the courtyard and heard the sound of the hulling-mill *inekoku niwa ni suriusu no oto*." So this poet also observed first threshing, then hulling. Since here again it is women's work, this supports our hypothesis that Sei Shōnagon witnessed threshing with short bamboo tubes or threshing-sticks,[163] a scene reproduced, with women, in Edo-period iconography (Figure 1.12). It is therefore clear that the early medieval Japanese knew implements other than the pestle and mortar for threshing and hulling.

All this concerns rice. What methods were used for dry crops? Even in the 18th century, they were still said to differ in each region and from province to province. Practised among others were threshing with an implement, beating off, pounding, roasting of ears (on Tsushima) and rubbing with the hands (for small quantities). It is interesting to note that the roasting of wheat was also recommended by the early agronomists of Rome and China.[164]

In the imperial vegetable gardens of early medieval Japan, barley and wheat were threshed *katsu*, probably with a long stick or other implement. The verb *katsu* is also read *tsuku* 'to pound', but in this case one expects the usual character for the verb 'to pound' *tsuku*. The pestle and mortar are absent from the equipment of these gardens. By contrast, in the imperial kitchens there was a "place for making wheat flour" *muginoko-dokoro*, where the wheat was pounded *tsuku* in a mortar to make noodles *muginawa*. In the imperial vegetable gardens, soybeans and red beans were shelled by beating *utsu* with an unspecified tool. We think it possible that the many fulling-blocks *kinuta* discovered in archaeological sites across Japan could have been used for thrashing beans, as well as for threshing straw.[165]

Storage

The population fed itself all year long with summer crops harvested in autumn (rice, millets) and winter (beans), autumn crops (buckwheat), winter crops harvested in early summer (wheat, barley), seasonal vegetables, plants collected in spring and tree fruits and nuts gathered in autumn. It was there-fore necessary to store what would keep and also put some of it aside as seed until the next season. The proportion of seed could be as much as one-seventh of the rice harvest and half the wheat harvest. We give examples from four archaeological sites to show the different storage methods.

The site of Yamada-Mizunomi, at Tōgane in Chiba prefecture, has yielded up many dwellings dating from the 8th–9th centuries, in the form of 140 pit-buildings and 52 pillared-buildings. Of the buildings revealed by postholes, some, constructed at ground level, were used as dwellings; others, with a raised floor, were used as granaries. The raised floors of the granaries are recognizable by small holes left in the soil by the posts that supported the floor. The same site also has twenty-three oval pits roughly one metre by two and one metre deep, grouped together away from the dwellings. Given the absence of grave goods, some scholars have identified these pits as underground silos. Silos are found in other sites of the same period. They were constructed in such a way as to prevent rain from entering.

The site of Hiraide is in the mountains near the town of Shiojiri in Nagano. There are 123 pit-buildings and just four pillared-buildings dating from the 4th–11th centuries. Unlike Yamada-Mizunomi, several have one or even two silos inside. Generally speaking, interior silos disappeared throughout the country around the 7th–8th century, but exterior silos survived for a time. On the other hand, building no. 22 of the Hiraide site, with an almost square layout of 6 by 5.40 m, has an interior space 2.40 by 1.50 m long, separated by a ditch and marked by small postholes in the soil. This type of space has been identified by archaeologists as a dividing wall with a raised floor bordered by a partition-wall, the boards of which were fixed in the enclosure ditch, and that could be used for storing foodstuffs.

The two sites mentioned are in the north-east of the country, far away from the early medieval capitals of Nara and Heian. Another site, that of Gūke-Imashiro, dating from the 8th–9th centuries, is located in the central region, at present-day Takatsuki between Kyōto and Ōsaka. Here archaeologists have discovered 140 pillared-buildings and no pit-buildings, which is hardly surprising as these disappeared from the central region in the 8th century. They did, however, survive elsewhere until the 13th or 14th century, notably in the eastern regions. At Gūke-Imashiro it has been possible to identify four residential groupings, each comprising a main house and annexes. Each grouping has a granary with a raised floor and a well. Such areas correspond to the residences of local notables, recorded in the deeds of sale that have come down to us from that time.

Yet another site, that of Kōnosu near Nabari in Mie, is located on the edge of the central region, east of Nara. About 40 pit-buildings and 61 pillared-buildings date from the 7th and the first half of the 8th century. Of the latter, eleven appear to have had raised floors and thus to have been used as granaries; but the number of granaries relative to the dwellings is far fewer than at Gūke-Imashiro. At Kōnosu, pits have also been discovered that have not, however, been identifiable with certainty as silos.[166]

The four sites reveal different methods of storage: underground silos, the corner set aside in a dwelling and the granary. Another method known from previous periods probably continued: storage in pots and large earthenware jars. A few examples from before the 7th century can be given, as no data is available for the following period. Round pots buried in underground silos have been found in various sites. In several cases they contained unhulled rice, sometimes charred: in one instance ears of rice and, in another, grains of rice, foxtail and barnyard millet. Other silos had no receptacle, but had an inside layer of burnt clay or a red coating (of cinnabar?). The site of Hashibara in Nagano has provided a total of 46.80 litres of charred rice together with beans and millets deposited in the corner of a building. These grains could have been kept in a very large earthenware jar 54.20 cm high, the fragments of which have been found here. This jar had not been buried, but had probably been placed in the dwelling. Experimental archaeology has shown that an almost constant temperature and humidity could be maintained in silos built for the purpose or provided with a receptacle, as long as they were covered with a wooden board.

Storage in granaries and underground silos was practised simultaneously in the same regions. The Japanese thus adapted in one way or another to climatic factors. But the quantitative report reveals a social factor: given that three to five silos were counted for one dwelling instead of one granary for several dwellings, the silo would seem to be for individual use and the granary for collective use. The length of storage may also have played a role in the choice of method.[167]

Among the methods of storing cereals, granaries are mentioned in the archives, whereas silos are not, meaning that we do not even know the term by which the early medieval Japanese may have referred to them. By contrast, the granary is evoked in poems like this one: "No granary on the river bank to store the harvested rice, earlier transplanted by my wife, dressed in a red garment, [its sleeves] wetted."[168]

The texts deal mainly with the practices of the public authorities. The provincial governments were concerned above all with filling their public granaries *ōkura/shōsō* with the grains received in taxes from the population. In 701, the administrative code stipulated that these taxes would amount to two sheaves *soku* and two bundles *ha* of rice for an area of 1 *tan* (0.3 acres), i.e. the State calculated rice in ears, but in 719 the government ordered that rice be supplied in grains *momi*, giving as the reason the rapid deterioration of ears. This reasoning seems surprising as it is known that ears keep better, at least in Europe and China. In Europe, grains were far more easily attacked by insects than ears.[169] In Japan, rice was stored after 719 in two ways: in grains and in ears, depending on the period, circumstances and place. Twenty-five yearly account registers for several provinces, dating from 730–738, are still in existence. These registers provide us with information about the public granaries.

A count of the public granaries noted in these twenty-five (albeit fragmentary) registers gives 1,500 buildings. It can be estimated that the total number of public granaries for the entire country was between six and ten thousand, taking into account that each of the sixty provinces managed one to two hundred granaries. The buildings mentioned in the registers include granaries with rice in ears *kaikura*, granaries with rice in grains *momikura*, granaries with dried rice *hoshiiikura* (drying being a very early preservation method known in other countries), granaries with foxtail millet in ears *awakaikura* and granaries with millet in grains *awamomikura*. Cereals in grains were stored with their husks (keeping them better aerated). The estimated reduction in the edible volume after winnowing and sieving was 6–10%, the proportion increasing as the grain aged.[170]

All the buildings no doubt had raised floors, but were constructed in a variety of ways: of boards *itakura*, of billets *marukikura*, with mud-brick walls *tsuchikura*, with interlocking horizontal beams *azekura*, in a pair of two granaries with a single roof *futagokura*. They were often roofed with thatch or shingles, sometimes with cypress bark *hiwadabuki*, sometimes with tiles *kawarakura*. Houses on posts *ya* roofed with thatch *kusaya*, with, in one case, plastered walls *nurikabeya* were sometimes converted into granaries. The large monasteries had the same types of granaries to store the income from their estates. Very long granaries *nagakura* and *hōsō* were rare.[171]

The register for the province of Izumi dating from 738 gives specific examples of the methods used. Granary no. 1 on the east side has an almost square layout of 5 by 4.80 m, and is 3.15 m high. There is an opening *soko*(?) in the floor(?) measuring 1.53 by 1.20 m. In this granary was stored a volume of 891 *koku* (75,735 litres) of rice in grains with their husks, from which 81 *koku* had to be deducted after winnowing. The grains were stacked on a bed formed by 130 small bundles of rice ears. The stored volume had a height *tsumidaka* of 3.10 m, meaning that the granary was filled almost up to the ceiling. The other granaries in Izumi followed the same pattern, though some were not as full and others were empty.[172]

The register of the public granaries of Etchū, drawn up in 910, thus two centuries later, provides us with other details. This document is fragmentary, like that of Izumi, but here one sees granaries that are far larger in size: they are 8–12 m long by 5–9 m wide and are 3–4.50 m high. They can hold 1,000–6,000 *koku* (85,000–510,000 litres) of grain. Like the granaries of Izumi, they contain either grains or ears, and two granaries are filled with dried rice. The grains are stacked on a bed of ears 15 cm thick. Reading these registers, one has the impression that ears were stored loose. But for dried rice they used cloths in which quantities of 5 *to* (42.5 litres) were wrapped. Beans were probably stored in straw sacks *tawara*.[173]

The storage capacity shows marked differences between Izumi and Etchū. On the one hand, it may be supposed that a gradual change could have occurred between 738 and 910; on the other hand, there were administrative irregularities. From 714, the government was aware of the difficulty of managing granaries of very disparate sizes when it came to calculating the quantities of the reserves. It therefore laid down three standard sizes to be adhered to for new buildings: large granaries of 4,000 *koku*, medium-sized granaries of 3,000 *koku* and small granaries of 2,000 *koku*. Yet in Izumi just two of the twenty-nine known granaries meet this standard. Perhaps it was only later that the provincial governments built larger granaries like those of Etchū.[174]

Moreover, some of the reserves had aged. In Izumi, the granaries contained, in 738, grain rice stored since 734 and the beds formed by ears went back to 722. At the same period, other provinces recorded reserves older than five years. In Etchū, some stocks went back several decades, others nearly a century. This is why the documents distinguish between recent reserves *atarashiki ine* and old stocks *mukashi no ine, furuki momi*.[175] The regulations stated that rice and foxtail millet could be kept for nine years, dried rice for twenty years and the other cereals for two years. The central government required the regional administrations to use the old reserves before the recent stocks, to use ears before grains and to thresh the stocks of old ears first. All this did not prevent the damage caused by insects and rot, especially of dried rice. Every year, it was necessary to replace rice that had been attacked by insects *mushihami*, dried rice that had spoiled *ashihoshiii/akubi* and rice that had been charred as a result of fires *yakinokori no momi*.[176] Similar storage problems almost certainly affected the large aristocratic and monastic estates; but for the general populace it was rather the opposite. Often the household reserves were exhausted and there was no rice left to eat or to sow the rice fields in the spring, which meant that the peasants were obliged to obtain a loan of rice every year.

The management of agrarian space

Fallowing and idling

A few general points deserve our attention, now that we have followed the cycle of agricultural work from the preparation of the soil through to the storing of the harvest in granaries. These concern the problems of fallow, idle and waste land, double-cropping and the conversion of irrigated fields to dry fields and vice versa. Agronomists distinguish between fallow land – land that is tilled but not sown in order to prepare it for the next season's sowing, and idle land – a field left untilled in order to recycle the minerals in the soil with a view to cultivation in the near future. They differ from waste land, which is considered to be unused or abandoned land.

With fallowing, it is first of all important to distinguish between dry and irrigated crops, as a fallow period is in fact detrimental in the case of irrigated rice fields. A rice field left idle for one or more years becomes porous: the roots of self-sown vegetation damage the bed of the rice field formed by the sole of the plough, which subsequently makes management of the water level impossible. Experimental archaeology has shown that a rice field left idle for a year loses roughly half its productivity.[177] By contrast, fallowing is beneficial for dry crops. Without a period of non-cultivation, yields decrease markedly from the second harvest because the minerals in the soil become exhausted. This point is not mentioned by the early medieval texts and even by Edo-period agricultural treatises. Japanese works show no interest in fallowing, whereas they give a great deal of attention to manuring, another method of soil improvement.

It is all the more remarkable that the *Aizu nōsho* (18th century) recommends fallowing every second year on non-irrigated land for poor soils. This treatise uses the expression "half-idle" *kataarashi*, which seems to imply a cultivated plot and uncultivated land side by side. This principle is documented by an archaeological discovery. The site of Kuroimine in Gunma was buried following a volcanic eruption in the second half of the 6th century. It has preserved a fossilized field with ridges, next to a plot with levelled ridges and thus resting. On the other hand, the encyclopedia *Wakun no shiori*, published from 1777 to 1887, defines *kataarashi* as "a rice field or dry field left uncultivated or idle every second year" and is therefore unaware of the difference between dry and irrigated crops.[178]

In the early medieval period, the administrative code of 701 instituted the distribution of rice fields by the State to the population of the country by way of public holdings, and it tolerated the granting of areas twice as large where soils were poor *ekiden*. The "Code" also provided for the confiscation of lands left uncultivated for three years and their redistribution to other applicant farmers. In other words, the State accepted idling for one year, though seldom for more than two years. The "Code" uses the term "alternating fields" *ekiden*. According to the 9th-century commentaries, this term refers to land sown every second year *toshi hedachite kōshusu*, but this Japanese explanation is taken from a Chinese classic, the "Rites of Zhou", written between the 8th and 5th century BC.[179]

The rule laid down by the Japanese "Code" was applied in certain cases, even though it deprived the State of tax revenues. In 821, the peasants of two districts in Kawachi (Ōsaka) note their difficulty in paying the rice-field taxes, arguing that these fields produced only 200–300 sheaves for an area of 1 *chō* (1.13 ha) and that this volume was decreasing year by year. "This is why we cannot sow the rice fields that were cultivated the previous year", they said, and asked to receive twice the area by way of alternating fields *ekiden*, basing their argument on the regulation in the "Code" and on the fact that this regulation had previously been applied in favour of another province, that of Harima (Hyōgo). Six years later, in 827, the neighbouring province of Izumi (Ōsaka) lodged a similar request on behalf of three districts, referring in its turn to the examples of Kawachi and Harima. Izumi requested a total of 500 *chō* (565 ha) of *ekiden*, explaining that as good rice fields were scarce, it was necessary to allow for losses in the event of drought.[180]

Apart from these two petitions, which obtained favourable responses from the government, alternating fields *ekiden* are not mentioned again in the texts. Also, reading the two decrees, idling seems to us to be misunderstood and seldom practised. In fact, the yields from a rice field left idle for a year do not increase, but are more likely to decrease, unless special care is taken. Furthermore, the figure of 200–300 sheaves of rice per *chō* is not outside the standard of the time; on the contrary, it is within the limits of yields estimated by the government for lands of mediocre quality. The second petition, in which the province of Izumi claims to want to guard against droughts, clearly shows a lack of understanding of idling, since its role is not to guard against bad weather but to renew the soil. In both cases, it seems to us, the regional governments cleverly managed to obtain additional arable areas. They quoted the regulation in the "Code". This regulation, copied from a Chinese work, was not intended for rice fields but for the dry fields of northern China.[181]

While the term *ekiden* disappears from the texts, *kataarashi*, "half-idle", occurs in poems composed around 1200: "What a sad sight, this rice field harvested last year, half-idle land, while they collect the seedlings from the seedbed beyond the heavenly river." "In the small half-idle rice field, behind my mountain dwelling, I do not even encircle with a rope the plot cultivated last year." In both cases, the "half-idle" plot was cultivated the previous year.[182]

"Half-idle" lands kataarashi

In the land surveys, land maps, inventories and other administrative documents of the 8th–12th centuries, large numbers of uncultivated plots are evident among the fields. On the monastic estates, cultivated dry fields *hatake* and cultivated rice fields *kenkaiden* were interspersed with other plots lying fallow or idle for a year *toshi-are*, temporarily fallow or idle *ima-are*, permanently idle or waste *tsune-are* and abandoned or waste lands *kōhai*. In fact, the uncultivated areas often accounted for at least one-third and sometimes exceeded one-half of holdings, but these areas decreased around the 12th century.[183] However, the administrative terms do not take account of circumstances and the technical aspect of fallow or idle lands.

To explain the large number of waste lands, Toda Yoshimi, in 1967, put forward an interpretation that may well raise certain questions. Toda distinguishes three types of waste or idle lands: uncleared virgin lands, lands abandoned after having been cultivated and temporarily uncultivated fields, i.e. lying idle for a short time, these last corresponding to the *toshi-are* of the texts.

Quoting the definition of the dictionary *Wakun no shiori*, mentioned earlier, Toda explains that *kataarashi* means "idle or uncultivated *arashi* or *are* every second year *kata-*, or *toshi no kawarugawaru*". He therefore seems to think that the "idle lands" of the texts refer to fallow fields being prepared for sowing. Toda also accepts the explanations given in the decrees of 821 and 827, but we noted earlier that these decrees, like the *Wakun no shiori*, ignore the technical reality of rice fields. Thus Toda maintains, incorrectly, that idling or fallowing is necessary in "unstable arable lands" (*fuantei kōchi*, in what sense?), rice fields and dry fields. He even thinks that fallowing can solve the porosity of rice fields.[184]

At this point, it is necessary to further clarify certain technical terms. Both fallow land and idle land are temporarily uncultivated and are not sown, but a fallow field is being prepared for sowing by one or more tillings, as François Sigaut has explained. In other words, while fallow fields are tilled, idle fields are not. The early medieval Japanese term *arashi/are* indicates the wild nature of uncultivated land which, in the Japanese climate, is rapidly invaded by self-sown vegetation. Duration is expressed in Japanese by the semantic elements *tsune* (permanent), *ima* (temporary) or *toshi* (for a year). The term *arashi* does not imply any technical notion, merely the physical aspect of a piece of land. Moreover, idling or fallowing certainly does not apply especially to poor or "unstable" soils, but it enables the yields of soils of any quality, even good, to be maintained or increased while reducing the investment in labour, and this only for dry crops, as the 18th-century treatise *Aizu nōsho* rightly points out.[185]

In view of all this, how can we account for the many uncultivated plots scattered among the holdings of early medieval Japan? The failure of the hydraulic systems may be a factor in the case of rice fields. For example, during a dispute with the public authorities of the province of Kii about the use of irrigation equipment, a monastery on Mount Kōya pointed out that the rice fields had been abandoned *kōhai*, because they could not be sown owing to a lack of water *mizu no tayori nashi*. This seems to have been a frequent occurrence, since a decree of 902 states, with regard to abandoned fields *kōhai*: "The areas brought under cultivation and the uncultivated areas *are* change every year, and the hydraulic situation *suiriku no tayori* changes every day." In these instances, idling or fallowing was not intentional but accidental.[186]

One can also point to management problems, such as the shortage of labour. There is, for example, the desertion of the rice farmers, in 988, in the province of Owari. The cause, in this case, was attributed to the administrative failings of the governor, while in other cases it could have been due to tax evasion or a shortage of labour, to negligence at the level of monastic or individual holdings, or sometimes possibly to a ruse for obtaining tax exemptions. Be that as it may, since idling or fallowing was detrimental to rice growing, it could not have come from a desire to increase yields.[187]

However, it is necessary to re-examine the duration of fallowings and the tillings of irrigated rice fields. Rice was sown in the 3rd month, transplanted in the 5th month and harvested in the 8th, 9th or 10th month. In the early medieval period, the rotation of rice followed by a winter crop was still uncommon, which meant that the rice field was left uncultivated for seven or eight months. During this time, there were one or more tillings, attested for seedbeds in the 2nd month and for rice fields between the 1st and 5th month. Poets mention the tilling of rice fields *ta* and that of "idle rice fields" *arata*.[188] The word *arata* contains the semantic element *ara* meaning 'uncultivated', 'wild', 'lying idle', with no differentiation as to the length of time involved. We therefore think it possible that "idle rice fields" refer to reclaimed abandoned fields or else simply to rice fields which, although appearing to be uncultivated in winter, await sowing in the spring or transplanting in the summer.

For dry crops, two hypotheses are possible: accidental or intentional idling, the latter intended to renew the soil. The main thing is to know whether idle land was tilled and thus prepared for sowing like fallow land. Dry fields, in effect, were left uncultivated for several months of the year. Unlike Europe, there was no knowledge of a three-course rotation system over two years, a practice born of hot dry summers and cold wet winters. The climate of Japan, hot and humid in summer, cold and dry in winter, favoured a single annual harvest, double-cropping being restricted to certain regions from the 9th century. Tillage is mentioned especially for the rice fields, but is also attested for the imperial vegetable gardens. Obviously dry fields were also tilled, both on aristocratic holdings and on small individual holdings. In this sense, one may consider the period of the year during which the land was unproductive, and that came in between the harvesting and sowing of dry crops, as a kind of catch-fallowing. The early medieval Japanese may also have used fallowings of one or two years in order to renew the soil. This is shown by the discovery of tropical rice made by Satō Yōichirō in archaeological sites dating from the Yayoi to the Heian period. This is non-irrigated rice grown in dry fields. An investigation of the site of Magari-kanekita in Shizuoka, dating from around 400 AD, has also revealed the existence of adjoining rice-field plots, some overrun with weeds and the others cleaned.[189] This suggests the presence of plots of cultivated rice fields and idle rice fields on the same land. This cluster corresponds to the description of temple lands mentioned above. There were probably also fields left idle for an indefinite period corresponding to lands abandoned due to management problems.

With regard to the examples of uncultivated soils recorded in the documents, these can be classified into three types: cultivated fallow or idle fields, temporarily idle land for reasons of poor management, and intentionally abandoned land. But let us not forget that the latter two could provide the vegetation for green manuring of the adjacent plots and, in this sense, could be very useful for crops. None of this, however, appears in the early administrative terminology, which takes no account of circumstances and seems to be ignorant of the technical aspects of tillage and manuring as they relate to idling. In the final analysis, we find ourselves, for early medieval Japan, faced with the rather unusual case of a historiography that does not distinguish clearly between idling for dry crops and that for irrigated crops.

Double-cropping

How far back does the practice of annual double-cropping go and how wide-spread was it? This question has interested a number of scholars.[190] We give here the textual references collected by Kimura Shigemitsu. A petition titled "Report of the district officials and local notables of the province of Owari", *Owari no kuni gunji hyakushōra gebumi*, addressed to the imperial court in 988 by low-ranking officials on behalf of the inhabitants of Owari, sets out thirty-one grievances concerning the abuses attributed to the governor of this province. Article fifteen accuses the governor of levying wheat or barley as rice-field taxes, in addition to the rice tax. The text mentions wheat or barley harvested in summer *natsu mugi*, i.e. from the 4th to the 6th month.[191] The inhabitants of Owari started from the premise that dry crops were not subject to taxation.[192]

In Owari, it was a question not of dry fields, but of a winter crop grown in the rice fields after the rice harvest. Double-cropping *nimōsaku* meant successive cultivation of a main crop (rice) and a secondary crop in the same field. It was probably not yet widespread, but is nonetheless attested in several cases. In 1118, a dispute brought into conflict the intendant and the cultivator of land connected with the Ise Shrine, because the cultivator had on his own initiative sown wheat or barley in the rice fields in the 9th month, after the rice harvest. Following the manager's complaint, Ise recognized the cultivator's right to sow a winter crop. The subject of the dispute was therefore the privilege of a second crop after rice.[193]

A similar event occurred on the Zentsūji and Mandaradera temple estates in Shikoku. In 1123, the intendant requested an exemption from the tax on the "spring rice fields" *haruta* that were sown with wheat or barley and safflower. This suggests a combination of rice with dry winter crops. Yet, twelve years later, these "spring rice fields" are clearly distinguished from the rice fields allocated for early rice *waseda*. Thus the "spring rice fields" no longer carried rice, but only winter cereals. Then in 1145, these rice fields are referred to, in the relevant document, as dry fields with temporary idling *toshi-are* and permanent idling *tsune-are*. In other words, they had winter crops alternating with idle fields in summer. Kuroda Hideo concludes from this that the "spring rice fields" originally carried rice in summer, wheat, etc. in winter, that they remained uncultivated in summer (for lack of water or some other reason) and were therefore recorded as permanent fallow lands, which nonetheless continued to be sown in winter. Permanent fallow lands thus referred to the absence of cultivation in summer. In short, the "spring rice fields" were in fact turned into dry fields. This is an example of double-cropping that resulted in the abandonment of the main crop.[194]

The attraction of the second harvest lay in the fact that it belonged in its entirety to the farmer, for both rice fields and dry fields. However, things changed very quickly for double-cropped dry fields. Let us take the example of two Tōdaiji estates. In 1160, there were two inspections of the dry fields, in spring and in autumn, on the Yuge and Ichii estates of this temple, inspections intended to fix the two taxes to be paid during the year. The other estate, that of Kataoka in Harima, paid a tax in spring in wheat or barley, and in autumn in oil. Perilla was probably grown in summer.[195]

Three other holdings in Yamashiro also paid two taxes in 1169, 1181 and 1214. In one case, it was specified that the first tax was to be used to make wheat noodles *muginawa*. The Kamakura shogun later established the legality of the double tax on dry fields in the regulations *Tsuika hō*.[196] This involved two non-irrigated harvests. The shogun nonetheless banned the tax levied on the second harvest after irrigated rice. When certain complaints reached Kamakura from the provinces of Bizen and Bingo in 1264, a decree stated: "the peasants of all the provinces sow wheat or barley after harvesting the rice, and they call this rice-field wheat/barley". In reply, the shogun granted full enjoyment of the fruits of the winter crops to the farmers on the estates under shogunal administration.[197] Still, it is possible that outside his jurisdiction estate owners imposed double taxation. While other evidence of rice growing associated with a winter crop is later than the 12th century,[198] it does not exclude the possibility that this practice existed much earlier. The double-cropping of rice fields could easily have escaped mention in the administrative documents, since it did not involve either legal deeds or taxation.

We find a reference to this practice in the 9th century. The *Ryō no shūge*, a compilation of commentaries by various authors relating to the administrative code of 718, mentions double- and even triple-cropping in the clause concerning the leasing of rice fields. As this text is the subject of considerable debate among historians, we first give a translation:

> The lease of rice fields is limited to one year. The *Kekki* says: how is the one-year limitation to be understood? If the lease period begins in autumn this year, is it permissible to lease the rice field [from autumn] in order to grow [rice in it in the spring of] next year or not? Reply: it cannot be prohibited; this can be understood from the meaning of the following part of this regulation. The *Shuki* says: how is the lease limited to one year to be understood? If the rice field is sown two or three times, while respecting the term of one year, can the lessee's wishes be agreed to? Reply: there is no rule; it depends on the lessee's wishes … Question: for the lease, is there a precise time of the year or not? And again: for lands cleared by individuals, is it admissible to give them an indefinite lease? Reply: it is not prohibited to enter into the lease in autumn in order to execute it the following year.[199]

This clause can only be interpreted in the sense of the leasing of rice fields where there is the possibility of a winter crop before rice, whereas the payment concerns only the rice harvest. When the lease began in autumn, the contract year included the sowing of wheat or barley in the 9th month, harvesting in the 4th month, the transplanting of rice in the 5th month and the rice harvest in the 8th month. A third crop of buckwheat, sown in the 7th month and harvested in the 9th month, seems unlikely.

In his article on double-cropping, Kawane Yoshiyasu suggests that this practice dates from the 12th century, and he thinks the 9th-century text is unrealistic and the product of the author's imagination. Yet, to "imagine" double-cropping, it was necessary to have some knowledge of it, at least in theory. What is more, Kawane wrote these lines in 1965, at a time when dry crops were still little studied. In his turn, Kimura Shigemitsu, a specialist of dry crops, gives examples of double-cropping, but he thinks that the few sources relating to it do not allow us to consider double-cropping as being widespread from the 12th century.[200]

Japanese scholars seem to regard annual double-cropping as a rare phenomenon. Yet it was practised in Egypt and in north-eastern India from the 2nd millennium BC. In China, it is attested from the Tang (618–907) and, under the Northern Song (960–1127), double-cropping of rice and wheat became widespread in the region of Jiangnan. We share the opinion of Isogai Fujio, according to which the 9th-century Japanese text shows that this practice was known at the time in Japan. The second yearly grain harvest was very attractive to farmers, since it was exempt from taxation up to at least the 13th century. Moreover, at that time, rice cultivation was insufficient to feed the population of the country. If double-cropping is hardly attested by the texts, this is because it fell outside the control of the public administration; and if it were subject to taxation from the 13th century, this meant that by then it had already spread to some extent, at least in certain contexts. It must not be forgotten, however, that double-cropping needs intensive tillings and higher quantities of fertilizer than normal, since wheat and barley are hungry crops.[201] Also, it was not practicable in very wet soils. Cereal double-cropping was thus subject to stringent requirements as regards labour, manuring and geographical conditions.

The conversion of fields

The "spring rice fields" in Sanuki were gradually turned into dry fields in the 12th century, as noted above. Field conversions are attested from the 8th century. A map dating from 735 of the Gufukuji temple estate, also in Sanuki, shows "dry fields that had become rice fields" *hatake narita* and "cleared rice fields now bearing dry crops" *imahatake konden*. One also finds in other documents "dry-irrigated fields" *hatakeda*. This type of land had previously been regarded as a field with alternating crops, i.e. it would have had irrigated or dry crops depending on the water supply during that year.[202]

We give some ethnographical data before interpreting the textual references. In the middle reaches of the Shirokawa river at Kumamoto in Kyūshū, during the Meiji era (1868–1912), colza was grown in winter, foxtail millet and rice in summer. Ridges were prepared in the fields after the colza harvest, with between one-third and one-half of the land given over to irrigated rice and the rest to millet, depending on the water supply at the time. But it was first necessary to go over the ground dozens of times with an ox-drawn harrow.

This practice, called *hata narita* or *tanari hata*, had been known in this region since the 17th/18th century. Even if the comparison with the Gufukuji lands is tempting, the investment in labour required by this method seems to us impracticable in early medieval Japan.[203]

The agricultural treatise *Nōgyō zensho*, written in 1697, states: "The dry and wet fields must be moved every year to allow the soil to rest. If the areas are insufficient, crops must be alternated on the same lands. One may, in certain places, turn a rice field into a dry field for one or two years to renew the soil." The writer recommends this method for the profitable commercial crops that were becoming popular in the 17th century, such as cotton. He in fact thinks that the land will again be suitable for rice growing after one or two years of dry farming, but he also explains that this procedure requires a very good knowledge of the soil and great zeal.[204] This example also seems to us to be hardly applicable to the early medieval period.

Outside of Japan, Sasaki Kōmei mentions having observed the alternation of dry and irrigated crops in Nepal in the 1950s and 1960s. Around Pokhara, a region 2,000 m above sea level, they grew rice and sowed *shikokibie*, Eleusine indica Gaertn. var. coracana Makino, in the rice fields in years of drought.[205]

The above three instances of irrigated and dry crop rotation are very special cases, brought about by exceptional circumstances and requiring a great deal of labour. Normally, it is not advisable to drain a rice field, as the soil becomes porous and no longer keeps the water level even. The rice field thus loses the regenerative effect of the water. Dry farming is possible in a drained rice field, but not the reverse. In other words, the soil structure necessitates a long-term decision. Irrigated and dry crops are not interchangeable.[206] Of course, all this concerns the rotation of crops over one or more years, and not irrigated and dry double-cropping in a single year, since the rice field is drained every winter.

What then is the direction of field conversions in early medieval Japan? After the example of the Gufukuji estate, mentioned above, we can list the following cases. In 846, there is a "dry-irrigated field" *hatakeda* on a small individual holding in Yamato. This is probably a dry field turned into a rice field. The Japanese term could also be read "a dry field and a rice field", but enumeration is normally expressed in the texts in inverse order: *ta-batake*. For the years 873 and 887, two surveys of the same land in Yamashiro, belonging to the Kōryūji temple, show that several plots, originally irrigated, were later drained and vice versa. In 905, the Kanzeonji temple in Chikuzen, Kyūshū, records 25 *chō* as rice fields (of mediocre quality) in a batch of 49 *chō* regarded as dry fields two centuries earlier, in 703. The 25 *chō* had therefore been turned into dry fields at an unspecified time between 703 and 905.[207]

We have more precise information thanks to the land survey of the Eizanji. It records, in 990, four small plots of 1 *tan* (0.3 acres) of dry fields scattered among the rice fields of its estate in Yamato. Later, the dry fields increase and decrease periodically, reaching a total of 10 *chō* (11.3 ha) in 1025, an increase due to the acquisition of new lands, but also to the conversion of several rice fields. For the Tōji, the land survey records in 1062 the condition of the cultivated plots on its estate at Hirose in Yamato. There in addition to rice fields and dry fields are marked several plots called "irrigated-dry field" *tashiro* also read *tabatake*, plots where some rice fields have probably been turned into dry fields, i.e. the opposite of *hatakeda*. Other *tabatake* are found on the Hida estate of the Tōdaiji in Yamato. In 1169, the inspection register for this estate mentions dry fields, including on the one hand converted rice fields *tabatake*, and on the other hand rice fields *ta* that may be interpreted as irrigated dry fields sown with rice.[208]

The cases mentioned confirm the transformation of rice fields into dry fields and of dry fields into rice fields. For the latter, Kimura Shigemitsu suggests a process of land clearing.[209] In fact, cleared lands in principle first carried dry crops and could then grow irrigated rice if hydraulic systems had been put in place in the meantime. For at least four years, cleared virgin land normally produces only dry crops, and then only through annual rotation, for instance of wheat, millets and beans. The land can be turned into a rice field after the first dry crops. This conversion therefore takes place in the context of new land acquisitions. One can mention the example of three small plots belonging to the Tōji: in 915, the temple demanded their return after they had been incorrectly recorded by the district office as public fields. The temple explained that it had cleared this land and that "the hydraulic conditions, *mizu no tayori ari*, have recently made it possible to turn it into rice fields".[210]

It may be noted that, generally speaking, expressions such as "depending on the hydraulic conditions" *mizu no tayori*, or "depending on the wetness or dryness" *suiriku no ben* of land do not refer to the alternation of irrigated and dry crops, but have to do with the choice at the time, which is governed by the water supply. This fact ties in with the case, noted by Kuroda Hideo, of the "spring rice fields" that carried rice in summer and dry cereals in winter, but remained uncultivated in summer owing to a lack of water and ended up becoming dry fields.[211]

The interpretation of Kimura Shigemitsu concerns the conversion of dry fields to rice fields and that of Kuroda Hideo the transformation of rice fields into dry fields. They are quite unusual cases. Next to these examples, there are more mundane cases of too little or too much water, such as the following. In 928, a man sold his land of 1 *tan* (0.3 acres), consisting of a rice field of 200 *bu* and a dry field of 160 *bu* (1 *tan* = 360 *bu*). Since this plot was previously recorded as a dry field of 1 *tan*, the owner explained that "as the hydraulic conditions were favourable, we could not keep all of it as a dry field and turned part of it into a rice field". The opposite case is documented in 830 for rice fields granted as a public holding to the peasants of the province of Awa in Shikoku: in a plot of 10.2 *chō* (12 ha) of rice fields were scattered several dry fields, because "there is no way of irrigating these plots, as they are located too high up on ridges".[212] These two examples, it seems to us, show that the conversion of fields in one or other direction had nothing to do with the farmer's preference, but was very often dictated by the hydraulic conditions and the geomorphology of the terrain. This appears to be the case as much for individual holdings as for temple estates. Moreover, some converted fields of the Gufukuji temple, mentioned above, were located in an old river bed and had thus been subject to the topographical changes of the area.

Be that as it may, the government was wary of the conversion of rice fields to dry fields, for the simple reason that the latter were not taxed. It formulated the prohibition of this practice by decree in 840. Nevertheless, the conversion of fields continued. It produced the toponyms "Dry-irrigated field" *Hatakeda* and "Irrigated-dry field" *Tabatake*, which are found in some land surveys.[213]

In summary, it may be said that in early medieval Japan there was probably no voluntary and systematic alternation of dry and irrigated crops, but only occasional conversions from one to the other. The conversion of fields could have been the result of specific practices such as land clearing and annual double-cropping, but in the other cases it must have been dictated by the water supply.[214]

Agrarian developments in the late medieval period (13th–16th centuries)

In this chapter, we have followed the agricultural operations month by month and in parallel for rice cultivation and dry (non-irrigated) cereal crops. The main work in the fields has been mentioned, from tillage to the storage of the harvest, taking in fertilization, sowing, irrigation, weeding, struggles against disasters, harvesting, drying and hulling. The historical sources give more information for the 8th–12th centuries than for the following period, thanks to well-preserved legislative and administrative documentation accessible in the collections of historical texts. Moreover, the poetry evoking peasant life has enabled us to retrace the agricultural operations from the cultivators' point of view. For the period from the 13th century, the documentary situation changes along with the social environment of the country, owing to the decentralization of political power. With multiple hierarchies divided between the aristocratic, military and religious élites and encompassing diverse regional and local interests, the historical sources available to us are equally scattered.

The late medieval sources were mined by the first generation of agronomist-historians and their researches resulted in "The Medieval History of Irrigation" (1943) by Hōgetsu Keigo and "The History of Agricultural Techniques" (1947, 1949) by Furushima Toshio. The latter is still the standard reference in Japanese. However, characterized by an evolutionistic approach, Furushima's work places significant technical progress and generalized increases in yields in the late medieval period. His data was used and completed by Kuroda Hideo (1984) and Kimura Shigemitsu (1992), but the image of increased yields was modified and toned down by these two scholars. Other interpretations have since been added, but there have been no major discoveries since the works mentioned. In this section, we take account of the Japanese researchers who have studied the archival texts, in order to give an idea of developments during the late medieval period.

For the period from Kamakura (1185–1333) to Muromachi (1392–1573), general works on Japanese history still often refer to the rapid expansion of agriculture, as set out by Hōgetsu and Furushima, and even speak of intensive agriculture. They note in particular the following technical advances: the spread of iron tools, the spread of draught ploughing, the increase in dry crops, and improvements in hydraulic systems and fertilization.[215] Kuroda and Kimura have noted two other important aspects: double-cropping and improvements in crop species. We discuss these later in the chapter.

Concerning the spread of iron tools, the archaeological sites of all periods have yielded relatively few hoe and spade cutting edges, and it was long thought that iron tools were not widely used. It is now known that the iron was recycled, because of the few forged objects but a lot of scoria or scrap from the forge discovered in sites from the 6th century. Before that, iron objects were imported from Korea, in the first centuries before and after our era. The techniques for extracting and forging iron were transmitted from the continent to Kyūshū from the 5th century. The Japanese first forged harvesting knives, then sickles and the U-shaped cutting edges of hoes and spades that were fitted onto wooden heads. All these wooden implements tipped with iron spread from the west of the archipelago towards the centre and then to the east of the country. They were widespread in the central region and as far as the area of Tōkyō and Chiba from the 7th to 8th centuries. Thus the progress thanks to iron in manual tillage and harvesting dates from the early, not the late medieval period. Spades and hoes with iron-tipped wooden heads remained in use up to the 20th century (Figure 1.6), albeit with temporal and geographical variations in the shape and weight of the tools.[216]

The variants of the iron-tipped hoe and spade are shown, with drawings, in the "Handbook of Useful Farm Tools" (*Nōgu benriron*), written in 1817 by Ōkura Nagatsune (1768–?). This treatise states at the beginning of the first chapter: "The hoe is the most important agricultural tool in Japan as in China; it is our daily companion both for the cultivation of the one hundred grains and for that of the various vegetables; without the hoe we cannot till the fields."[217] It may be noted that hoes and spades tipped with iron, as well as sickles with an iron blade formed the basis of the tillage and harvesting tools over the centuries.

Farming equipment

The spade, hoe, harrowing-comb (a harrow in the shape of a large wooden comb) and plough or ard (scratch-plough) (Figures 1.2 and 1.9) were the tillage tools of the "zealous farmer" of the 11th century, as he is presented to us in the *Shinsarugakuki* (see above, p. 14). The 12th-century *Konjaku monogatari* relates the story of "A brother and sister who went to live on an unknown island in the province of Tosa". These two inhabitants of a village located on a bay in the district Hata-gun sowed rice in a seedbed; then, at the time of trans-planting, they loaded the seedlings into a boat, engaged "transplanters" *uebito* and took away "all their goods, starting with their food supplies", that is: "the harrow, plough or ard, sickle(s), hoe(s), axe(s) and the spade used to prepare the edges of the rice fields *tatsuki*". Another version of the same story in the *Uji shūi monogatari*, ca. 1221, adds to these goods cooking-pots and large iron pots.[218] It is clear that tools and utensils formed part of the family inheritance. It is precisely because of this status that they are found in the administrative documents relating to disputes or confiscations.

We can now "visit" a few landholdings of the 12th–15th centuries, thanks to several administrative documents (see Table 1.1). These texts show lists of confiscated goods, including farming implements, weapons, kitchen utensils, textiles and food reserves. This data makes it possible to place the owners of these goods in their social milieu. They are local proprietors who are fairly well-off, and one is a poor woman farmer.

The farm of the "master of Sakura" *Sakura no shujin*, in 1170, seems as well stocked as the one in the story mentioned above. He keeps a horse as a draught animal for his plough, but at the same time works with a hoe. In fact, tillage was done using both a harnessed plough and hand tools, according to the layout of the field or the quality of tillage required. The plough did not exclude the hoe. It should be noted that the Japanese language uses a single term *karasuki* for the plough that turns over the soil with its mould-board and the ard or scratch-plough that slices through the soil with its ploughshare and has no mould-board. The master of Sakura may be considered a proprietor, that is, owning a parcel of land and/or having fiscal rights over the plots belonging to other landowners or cultivators, because he is able to keep a draught animal and therefore till quite large areas, and also because he seems to be under the protection of a temple, the Kōfukuji in Nara, to which he supplies rice. When his goods are confiscated, he has in his granaries a capital of 9.5 *koku* of rice, equal to between 800 and 1,600 litres or 700 kilogrammes to 1.4 tonnes of rice. Of this reserve, 3.5 *koku* are available for his household use and 6 *koku* are payable to the temple as an "offering" that probably constituted a kind of rent.[219]

Another landowner also had to pay an "offering", this time to the Hiyoshi Shrine. In 1291, this man, called Tametoki, had to supply the shrine with 21.5 *koku* of rice, i.e. over three times the offering from the master of Sakura. Tametoki's household capital of 10.3 *koku* was also triple that of the man from Sakura. Tametoki was a wealthy landowner, for he had in addition a significant monetary capital of 35 strings (each of one thousand coins) reserved for the shrine and 17.5 strings of private capital. This man may have been a moneylender and, considering the 5.8 *koku* of unhulled rice, he may also have loaned out seed. His inheritance was quite significant. By way of military equipment, he had a set of armour and three horses. He was a samurai who was trained for possible warfare. His wardrobe reveals his social obligations. He wore the samurai outdoor costume *hitatare*, a kimono with a straight-falling

Table 1.1 Farming implements of the 12th–15th centuries

The master of Sakura, landowner, 1170 in Settsu (Ōsaka)

Farm tools	1 plough or ard, 1 hoe
Livestock	1 horse
Food reserves	9.5 *koku* of rice, 6 *koku* being payable as an offering to the temple; 4.5 *koku* of soybeans; 30 skewers of dried persimmons
Utensils	1 cooking-pot tripod
Clothing, textiles	3 pieces of hempen cloth, 4 pieces of ramie cloth, 3 light jackets (with loose sleeves), 3 formal pieces of headgear *eboshi*
Weapons	2 bows, 2 quivers
Source	*Ichijōin monjo*, Kaō 2.11.28 (1170), "Kōfukuji-ryō Settsu no kuni Kanan no shō Sakura no shujin" (coll. Kyōto Daigaku)

The large landowner Tametoki, 1291 in Kii (Wakayama and Mie)

Farm tools	not mentioned
Livestock	not mentioned
Food reserves	31.8 *koku* of rice (21.5 *koku* being payable as an offering to the Hiyoshi Shrine), 5.8 *koku* of unhulled rice (seed?), 7 *koku* of barley, 3 *koku* of wheat, 3 *koku* of soybeans, 2.2 *koku* of red beans, 1.5 *koku* of buckwheat; capital in coin: 52.5 strings of coins (35 strings being intended for the Hiyoshi Shrine)
Utensils	27 large and small cooking-pots, 13 large and small iron pots, 23 cooking-pot tripods, 27 large and small earthenware pots
Clothing, textiles	2 padded night-watchman garments, 5 everyday outfits with small sleeves, 3 padded everyday outfits with small sleeves, 18 samurai outdoor outfits, 24 summer jackets, 25 pieces of hempen cloth, 2 black shirts, 120 ounces of ramie wadding, 60 squares of hempen wadding
Weapons	1 set of armour, 1 set of armour undergarments, 11 bows, 9 quivers, 3 saddles, 3 bits
Source	*Kōyasan monjo mata zoku hōkanshū* no. 1566 (Kii no kuni Arakawa-shō)

A poor peasant woman, Kuro no Kamiko, 1347

Farm tools	1 axe, 2 hoes, 1 adze
Livestock	none
Food reserves	0.5 *koku* of rice, 1 *koku* of millet
Utensils	3 large and small cooking-pots, 2 cooking-pot tripods
Clothing, textiles	3 light jackets, 2 padded jackets, 2 summer jackets, 3 pieces of hempen cloth
Weapons	none
Source	*Tōji hyakugō monjo*, Shi 10–12, Jōwa 3 (1347)

Hyōe no Jirō, landowner, 1425 in Yamashiro (Kyōto), Kōzuke-shō

Farm tools	1 mortar, 3 pestles, 1 hulling-mill, 1 plough or ard, 1 harrow, 1 axe, 1 hoe
Livestock	1 ox
Food reserves	30 bundles of rice grains, 1 sack of soybeans, 1 sack of millet, 30 bunches of dried turnips and taro leaves, 1 tub of fermented soybean paste
Utensils	2 large iron pots, 3 large and small cooking-pots, 4 large and small tubs, 1 cooking-pot tripod
Clothing, textiles	not mentioned
Weapons	2 lances, 1 bow, targets, arrows
Source	*Kyōō gokokuji monjo*, book 4, no. 1109, Ōei 32 (1425)

The master of Izumi, landowner, 1450 in Wakasa (Fukui)

Farm tools	2 mortars, 1 hulling-mill, ploughs
Livestock	3 oxen
Food reserves	1 *koku* of soybeans, 1 *koku* of rice, 6 units(?) of rice in the husk, 1 sack of unhulled rice, 40 bundles of rice grains, 1 tub of fermented soybean paste
Utensils	1 tea mill, 2 tea bowls, 1 tray *oshiki*, 1 tea-water kettle (*tedorigama*), 5 *tatami*, 17 sliding doors, 1 cooking-pot tripod, 1 large earthenware oil jar, 3 large and small cooking-pots, 2 pots, 6 large and small tubs, 1 basin
Clothing, textiles	1 pair of trousers, 1 pillow
Weapons	1 lance
Source	*Tōji hyakugō monjo*, Shirakawabon, 140, Hōtoku 2 (1450)

Note: The *koku* is a measure of capacity that varies according to the region and period; it corresponds to 85 litres in the 8th century and to 180 litres after 1624.
Source: Kuroda (1983: 49–53); Amino (1993: 329–336).

neck ("*hitatare*"), which was normally worn with "samurai headgear" *samurai eboshi*, and he owned no fewer than eighteen of these formal costumes. The kitchens of his residence were equipped for official banquets, with twenty-seven cooking-pots *nabe* and thirteen large iron pots *kama*.[220] At his receptions, Tametoki served not only rice but also noodles, as shown by the 7 *koku* of barley and 3 *koku* of wheat stored in his granaries. Despite his high standing, Tametoki, called a "rogue" *akutō* (bad fellow) in 1291, was stripped of his possessions. Yet among these possessions there were no farming implements or draught animal. Whether or not he owned either of these, they were not counted as part of the family inheritance from an administrative point of view. It may be assumed that this man was a large landholder, from a far higher social class than the master of Sakura, who engaged in financial affairs and administered the Arakawa estate in the province of Kii, not overseeing the farming work himself, but calling upon a manager who assembled, in his turn, a servile labour force with its own tools.

Medievalists distinguish between absentee domanial lords *ryōshu* from Kamakura or Kyōto or religious establishments in the capital or the provinces, large resident landowners *zaichi ryōshu* who live on the estate over which they have fiscal and/or administrative rights, and resident medium and small landowners who manage parcels of land and were called *tato* up to the 12th century, later *myōshu hyakusei*. However, the very notion of ownership was ill-defined and there were multiple ownership structures. Possession involved either the land, or a life-interest only, or again fiscal or administrative rights. Landowners formed hierarchies in the provinces throughout the country based on a population of farmers who worked the owners' lands all their lives.

The "master of Izumi", *Izumi no taifu*, belonged to the resident landowner class. In 1450 he lived in a house where the main pavilion had a raised floor with one *tatami* room and seventeen sliding doors *to* opening onto the external gallery that encircled the building. This man organized tea receptions. He ground the tea-leaves with a mill *chausu* to prepare the powdered tea for formal receptions and probably served the bowls of tea with side dishes on trays *oshiki* placed on the floor. Yet, with three cooking-pots and two other pots, his kitchens were far less well equipped than those of the large landowner Tametoki, mentioned above, and his reserves too were far less plentiful. He belonged to a lower social class than that of the large landowners. His granaries contained 1 *koku* each of rice and soybeans, other quantities of different rice, and he had wooden tubs *oke* for preserved foods such as vegetables dried and macerated in fermented soybean paste. When his possessions were confiscated, only a lance and a few clothes were found. The master of Izumi kept three oxen and owned one or more ploughs. His rice was pounded with pestles and two wooden mortars and he also owned a hulling-mill, which suggests considerable quantities of rice; but like Tametoki he had no simple agricultural tools such as hoes, spades, axes or sickles. One may well ask why the sickle, a valuable tool with its iron edge, did not appear among the household goods for the landholdings mentioned here. Be that as it may, the master of Izumi no doubt directly managed the lands placed under his control, while organizing corvée labourers or paid farmers.

A quarter of a century earlier, in 1425, the landowner Hyōe no Jirō managed a similar landholding, but slightly lower in status than that of the master of Izumi. Jirō had an ox, a plough and a harrow, as well as an axe and a hoe. These tools were used either by himself or his servants. However, the presence of the mortar with three pestles and a hulling-mill suggests the processing of cereals in quantities requiring additional outside labour. The food stocks, military equipment and kitchen utensils were similar to those of the master of Izumi, with one important difference: Jirō did not belong to the samurai class whose members could host formal tea receptions.

These few examples illustrate the diversity in the types of late medieval holdings. They concern the landowning and managerial classes. By contrast, the vast majority of the population, made up of small farmers, appears only as a labour force in the historical documentation on the agrarian economy. The common people, who were illiterate, had no voice or at least no written voice. These individuals probably lived like the peasant woman Kuro no Kamiko in 1347. She worked every day with an axe, an adze or one of her two hoes. She ate more millet than rice and cooked the grains in one of her three cooking-pots fixed on a tripod and set over the fire in the hearth of her home.

The peasant condition is the subject of a legend in the *Jizō bosatsu reigenki* (11th century, with 15th-century additions). An old farmer *denpu*, of relatively low status but an honest man, lived near the Great Shrine of Izumo. As happened at the beginning of each year, the estate manager assembled workers to till his lands and his agent came to recruit the old man to sow the fields. It was the rice transplanting season, at the beginning of the 5th month, but the old man, who was ill in bed and unable to move, refused the corvée. The agent, in a rage, seized the old man's tools and ordered him to get to work. Four months later, in the 9th month, the old man died. It was the harvest season. On harvest day, miraculously, seventeen or eighteen Jizōs turned up in place of the old man to do his work.[221] This tale, with its moralizing content, illustrates the practice of seasonal recruitments and the obligatory services owed by each person, equipped with his tools. Judging by the example of the peasant woman Kuro no Kamiko, these tools included at most one or two hoes and one or two axes. The spade seems to have been far less common than the hoe, judging by the number of occurrences in the source texts mentioned in this book.

The plough, harrow and draught animals

On the landholdings mentioned above, three owners keep livestock and have one or more ploughs. Japanese historians are divided in their opinions regarding the spread of harnessed ploughs and the extent to which their use developed during the late medieval period. No statistical data is available, but according to Kuroda Hideo ownership of draught animals was restricted to landowners who managed quite extensive lands, namely those who contracted out the rights of cultivation, taxation, etc. to subordinates. For his part, Kōno Michiaki, a historian and ethnographer of harnessed ploughs, thinks that ploughs and harrows were widespread in the countryside from the Heian period and that they played an essential role in land clearing.[222]

Several factors must be taken into account when considering the development of draught tillage, beginning with the division of rice fields into small plots. Small areas do not allow ploughs or harrows to be used. Because they require a 100% horizontal water level, irrigated rice fields were terraced and small in size; they remain so in present-day Japan. The topography does not provide farmers with perfectly level surfaces that would allow them to cultivate large rice fields of around one hectare. With an uneven relief, each field was small and the plots of a single landowner or farmer were, moreover, scattered over the countryside. The land surveys of the 8th century and the detailed survey maps of the 8th–9th centuries show small areas of around 0.26 acres. As some of them record the farmers' names, their fields can be identified on the detailed surveys: their plots were located several hundred metres minimum and seven kilometres maximum from each other.[223] It may be assumed that this situation did not fundamentally change during the late medieval period, since the topography remained unchanged. Large surfaces became part of Japanese agriculture only at a much later date, from the time of the colonization of Hokkaidō in the Meiji period, this island being covered by extensive plains. As for Japanese history before Meiji, this means that a farmer who used a draught animal for tillage moved from place to place for several hours during the day, with his plough on his shoulder and his animal led by its halter to reach each of his plots to be tilled. Leaving aside the rice fields, the non-irrigated fields that grew dry cereals or leguminous plants were more suited to ploughing, once their production became part of the tax system, namely from the 13th century. The plough was also useful for land clearing and would therefore be of interest to investors or estate managers.

Moreover, the plough, the shape of which is known from archaeological excavations and late medieval iconography, had a long and fairly large sole and thus did not allow the depth or quality of tillage to be varied. It is perhaps for this reason that in the imperial vegetable gardens the soil was turned with a hoe or a spade, even after tillage with a plough, and that later on, before the 19th century, the plough was abandoned. Its shape did not evolve from the 8th to the 17th century, which shows that this implement was not systematically used.[224]

Livestock rearing presents another problem. Compared with European countries, this practice has had a very limited impact in Japan. Until the Meiji period, Japanese people did not eat the meat of livestock, did not drink milk, did not slaughter or even castrate animals and did not use ox leather. Nor did they use mills or other devices operated by animal traction, and animal manure was not widely used. Under these conditions, an animal was useful only for tillage, during the first three months of the year; but it had to be fed for twelve months. The Japanese countryside did not (and still does not) provide green pasture and the straw from rice fields did not provide sufficient food, so that people had to cut vegetation in order to feed an ox or draught horse throughout the year. They already spent several months over the summer cutting green manure to spread on the rice fields. With this kind of human investment, one may ask whether livestock rearing did not cost more than it was worth; but pack and draught animals were also used for transport.[225] In this sense, livestock rearing was worthwhile for a certain type of landholding that needed to move large quantities of harvests and taxes in kind without having to call upon transport services, which were however quite widespread in the late medieval period. In short, it would seem that livestock rearing and implements such as the plough and harrow did not really evolve during the late medieval period and that their role was restricted to large estates. Draught animals probably had no place at all in the lives of the poorest Japanese peasants over the centuries.

Fertilization, hydraulic systems and double-cropping

As regards fertilization, there was no significant progress. In fact, green manure, namely the spreading of plant matter in the fields, remained the principal method up to the introduction of commercial fertilizer consisting of vegetable oil or dried fish meal from the 17th century (see above, pp. 21–25 and below, p. 84). Ashes, which were the typical fertilizer used for shifting cultivation on swidden fields, occasionally provided a supplement for some permanent crops.[226]

The hydraulic systems, considered earlier in our discussion on irrigation, consist of reservoirs, ponds, dams, sluice-gates, canals, dykes, ditches and underground water-pipes. These installations were in use over the centuries and still are today. By contrast, the water-wheel or noria, known in Japan since at least the 13th century, hardly made any inroads (see below, pp. 83–84). Thus one cannot speak of any significant advances in water management for the late medieval period. However, both fertilization and water management are crucial for the practice of double-cropping.

The Japanese call "double-cropping" *nimōsaku* two successive crops in the same field, namely irrigated rice in summer as the principal crop and wheat or another dry crop in winter as a secondary crop. Double-cropping has been known since the 9th century and specific cases appear in the administrative documents of the 12th–15th centuries.[227] It was central to the studies of the first generation of agronomist-historians who presented the late medieval period as a time of major technical advances. Later, Kuroda and Kimura did more detailed analyses of the medieval documents and highlighted the difference between double-cropping, crop rotation and the practices of short- and long-term idling (see above, pp. 56–60). As a result, they modified the view of double-cropping and also brought in factors concerning geographical, climatic and technical constraints. Since their studies, double-cropping has ceased to be considered responsible for a so-called explosion in yields.

Isogai Fujio has devoted a book, published in 2002, to the questions of yields in relation to double-cropping. He is the first Japanese historian to draw attention to the chemical composition of fertilizers. In Japan, as in Europe, manure has three basic components: phosphorous, potassium and nitrogen. These three elements are recycled in the soil through tillage, the addition of manure or the addition of water. Yields are mathematically proportional to the presence of these elements, so that an irrigated rice field can reach 60–70% of its normal yield, even without fertilization, thanks to water. But a dry field will produce only 20–30% of its normal yield without fertilizer. In other words, the success of a secondary crop depends on the addition of a very large quantity of manure and intensive tillage, and the next rice crop, too, will require excellent water management and the addition of large quantities of manure, since the field will have been subject to the weight of another crop. We have seen that neither the methods of fertilization nor those of tillage or water management saw any significant advances during the late medieval period. In fact, Isogai has found in the administrative documents cases of reduced yields in fields under double-cropping.[228] We therefore think that double-cropping was subject to constraints and could only develop under specific, indeed exceptional conditions. In short, as Isogai states, double-cropping was a marginal phenomenon and did not bring about a generalized increase in agricultural production.

As regards the bringing under cultivation of arable lands, Kimura draws attention to a gradual shrinkage of the uncultivated areas within landholdings. Whereas in the 8th century even the great monastic estates included large tracts of waste ground that could account for a third or even more of the registered lands, Kimura notes a net increase in the rate at which lands were brought under cultivation on the late medieval holdings. He quotes the figures in administrative documents, but does not mention any technical aspects. Kimura also quotes the travel account of a Korean delegation to Japan, dating from 1420, which expresses its admiration for the practice of triple-cropping rice, wheat and buckwheat,[229] a practice not attested in the Japanese sources. Yet even with a gradual increase in the area brought under cultivation on some estates in the late medieval period, one cannot speak of intensive agriculture in relation to Japan before the 17th century. We should remember, however, that the calories produced in relation to the area cultivated are far higher for irrigated rice growing than for dry cereal crops. In this sense, land utilization in Japan may be considered intensive in the same way as was the case in other rice-growing societies.

The data given above rather suggests a lack of technical progress and an unchanging late medieval period. Certainly there were fluctuations in the agricultural practices during that time, but no real innovation can be observed or any net increase in overall rice production. This is no doubt a result of factors pertaining to the social environment. Late medieval Japanese society was made up of landowning élites who sub-contracted management rights and at the lowest level a life-interest in small plots in exchange for labour. The rural population for its part was dependent on the élites and was obligated to do paid or unpaid seasonal work. It may be assumed that, as in medieval Europe, the local élites had a vested interest in keeping this cheap labour in a dependent state, rather than developing processes that reduced the investment in human effort. Productivity (in the sense of increased returns for the energy invested) was probably not a major concern of the élites. A social environment characterized by cheap labour is not conducive to inventiveness or rationalization. It was not until the 17th century that the central government encouraged collective efforts to develop new technologies and achieve production levels far higher than those of the late medieval period.

However, this situation did not prevent the owners of small and medium-sized holdings from being interested in an increase in yields that could augment their immediate revenues. In this sense, it is necessary to point out an innovation for the late medieval period: the introduction of the rice variety *taitōmai* from South-East Asia.

Improved varieties

An increase in rice yields was in fact achieved over the centuries by the selection of varieties. Selection was practised in Japan from the 8th century, as shown by the names of the many varieties inscribed on the tablets used for loans or grants of seed to farmers.[230] Peasant households received from the State "household fields", a type of tenure subject to various rice taxes. These early varieties persisted and, in the Heian period, there appear in the administrative documents names such as "monks' grains" *hōshiko*, a naked (unbearded) rice "like monks", or "a thousand grains for one" *chimotoko* ("a thousand ears for one stalk"?). The Kamakura period saw the development of varieties intended for saké brewing. According to Kuroda Hideo, the new varieties were able to circulate and spread from region to region by means of the tax networks or commercial credit.[231]

A new variety appeared around the 13th century: "rice of the great Tang" *taitōmai*. Unlike the traditional white rices that were Oryza sativa japonica, it was a "red rice" *akamai*, that is, with a red husk, of the indica variety. This rice of Vietnamese origin came to Japan via China, hence the name "rice of the great Tang". The cultivation of *taitōmai* appears in the Japanese sources from 1308. This rice spread into the centre and west of the country and became popular with farmers at all levels right up to the modern period. Less resistant to wind and not as fragrant, it nonetheless offered undeniable advantages. Judging by the present-day *taitōmai* variety, it was more resistant than the traditional varieties to the ravages of climate and insects, needed little fertilizer, was able to adapt to less fertile soils and had a rapid growth rate. This rice provided stable harvests. Above all, it gave far higher yields than the old varieties.

We have no figures for the yields, but Kuroda Hideo has found a very evocative modern song about this rice. Called "Transplanting rice" (*taue zōshi*), in the 20th century this song accompanied rustic dances across Japan. It sang of this new rice comparing it with the traditional variety "a thousand grains for one" in these words: "Arrived yesterday from the capital, the rice (*taitōmai*) with black seeds: three handfuls of seedlings give eight *koku* of rice; these grains of happiness, for three *gō* sown they give three *koku*."[232] Since a thousand *gō* was equal to one *koku*, the yield was as high as 3:1; and with 0.01 *koku* per handful, the yield for three handfuls (0.03 *koku*) gave 8 *koku*, that is, 267:1. These are clearly poetical and abstract figures. Moreover, yields were increased by the selection and manipulation of this variety over several centuries, but the important point is the comparison. In the 8th century, the yield was 7.5:1 to a maximum of 25:1 depending on the quality of the soil on the landholdings managed by the State.[233] The yields of "grains of happiness" of the *taitōmai* variety, sung about in the 20th century, gave a multiple of the traditional varieties of early medieval Japan. The new variety certainly brought about an increase in yields from the early medieval period, at least in the centre and west of the country and on holdings that had access to this seed and had suitable lands.

The evolution of acreages and the use of farmland in general

The 8th century and the 17th–18th centuries are usually considered to be times of significant agrarian expansion in Japan. These periods were in effect the theatre of land clearance campaigns sponsored by the government. The following figures may be quoted regarding the arable acreage of the country:[234]

- 862,000 *chō* (974,000 ha) in the 10th century (*Wamyō ruijushō*, ca. 930) or 909,000 *chō* (920,800 ha, *Iroha jiruishō*, 12th century) or again 946,000 *chō* (1.1 million ha, *Shūgaishō*, 1341, which largely reproduces the preceding figures) in the 12th century;
- 1,635,000 *chō* around 1600 (1,622 M ha, *Keichō sannen daimyō chō*); – 2,960,000 *chō* in 1700 (2.95 M ha, *Chōbu geso chō*).

However, these figures only partially reflect reality. Kimura Shigemitsu has pointed out that the dictionaries of the 10th–12th centuries do not record all the arable lands, just those of taxable rice fields. It was only after the agrarian reforms of Toyotomi Hideyoshi (1537–1598) in his last years that non-irrigated fields, i.e. permanent dry fields, came into public accounting. The figures prior to 1598 therefore do not take into account the areas of dry fields. In other words, the figures available to us fluctuate: around 900,000 ha of irrigated rice fields alone in the 12th century, then 1.6 M ha for all arable lands (irrigated and non-irrigated) and an expansion of these same areas to 2.9 M ha in 1720.

Historians have also identified another wave of agrarian expansion that took place from the second half of the 11th century to the end of the 12th century. A count has been done, from the 1040s, of cases of estates established on waste lands that investor land clearers brought under cultivation. One hundred and fifty cases of newly formed estates have been counted, although no actual acreage figures are available. It is now known that these were in fact lands that had been left idle for a long time and were brought back under cultivation. Thus there was no real expansion of agricultural lands.[235] The development of arable lands needs to be re-examined based on this data. According to Kimura, there was little expansion of agrarian lands during the late medieval period, even taking account of the colonization of the eastern provinces by the shogunal government in Kamakura from the end of the 12th century.[236] But, while awaiting more detailed analyses, it can be surmised that the total area, irrigated and non-irrigated combined, did not expand to any significant extent until the 17th century and that it then almost doubled up to 1720. This course of events can be explained on the one hand by a lack of technical progress during the late medieval period and on the other hand by the agrarian expansion and rapid technological advances after 1600.

The demographic figures also depend on estimates. Present-day diction- aries estimate the population of the Japanese archipelago as follows: 161,000 inhabitants at the end of the Neolithic; 600,000 around the 3rd century BC; the population would have increased rapidly from this time, thanks to irrigated rice cultivation and the waves of migration from the continent (China, Korea); it would have reached 5.6 million inhabitants in the 8th century. For the year 1600 the population is estimated at 12.3 million. The first reliable national census undertaken in Japan in 1721 gives a figure of 31 million people. All of the assessments prior to 1721 rely on a food base produced by a supposedly rice-growing monoculture that ignores the other cereal crops. In our view, the current demographic estimates are unreliable and in need of fundamental revision.[237]

That dry crops have played an important role is clear from the figures for the Edo period. In 1720, the land survey records 1,317,000 *chō* (1.3 M ha, or 45%) of dry fields and 1,643,000 *chō* (1.63 M ha, or 55%) of rice fields. A national average of around 40% of dry fields and 60% of rice fields following the land clearing of the 17th–18th centuries may be supposed. For example, in 1650, the province of Musashi (Saitama prefecture) had 55% of dry fields and 45% of rice fields. In 1720, the proportion of irrigated and non-irrigated areas differed markedly depending on the region. In some regions it was 30–40% of rice fields compared with 60–70% of dry fields.[238] This means that dry cereal crops were widespread and even dominant in certain regions of modern Japan. In our view, they were even more so in the early medieval period, where we are dealing not with a rice-growing monoculture, but with a cereal polyculture. In Japan, monoculture dates from the mid-20th century and came about following the changes in land use instituted by the prime minister Tanaka Kakuei (1918–1993).[239]

In conclusion, it is important to note that in all periods the agrarian land- scape was multifaceted. Generally speaking, this landscape was shared, until the 20th century, between irrigated fields, dry fields and shifting swidden fields (see Chapter 2). While the medieval period saw the emergence of double- cropped rice-growing plots, the exact opposite also existed in parallel: rice fields that were flooded and never drained. These were permanently flooded fields, impossible to drain, which did not allow transplanting and were sown directly with seed. As the seedlings were never thinned out during the growing period, yields were poor. And in order to prevent seedlings from being suffo- cated because of the sowing density, the unwanted seedlings were "removed" *tsumu* before ripening. The name "Cleaned field" *Tsumida* appears for the first time in the 17th century, but according to Harada Nobuo the presence of these flooded fields is certain for all periods because the Japanese archipelago was covered by extensive humid and marshy zones until the large-scale agricultural land drainage campaigns undertaken by the Meiji government. In these permanently flooded rice fields, the farmer found himself, for sowing and harvesting, up to his knees, waist or neck in mud, as shown by the photos of rice farming on the broad Kantō plain, dating from the Meiji period. For Harada Nobuo, much of the rice growing up to the modern period consisted of these flooded fields, with poor yields, and the well-equipped medieval landholdings described in the administrative documents belonged to a very small élite.[240] Generally speaking, no major increase in agricultural production can be seen during the late medieval period. It was only from the 17th century that there was real growth, thanks to the technical innovations that characterized the Edo period.

The technical innovations of the Edo period (17th–19th centuries)

We have already referred a number of times to the agricultural treatises of the Edo period (1603–1867), in order to fill in the gaps in the medieval texts and to better understand certain processes. Since we refer to them later in connection with modern technical innovations, we give here an overview of the most useful texts for our purpose. Of the one hundred or so agricultural treatises written from the 17th to 19th centuries, we have found these to be of most interest.[241]

The *Seiryōki*, "Notes of Seiryō/Kiyoyoshi", brings together episodes in the life of a warrior named Doi Kiyoyoshi (1546–1629), head of the Uwashima domain in Iyo (Ehime), Shikoku. Of interest to us is book 7, titled *Shinmin kangetsushū*, which talks about the customs of the tenant-farmers on the estate and is considered the first Japanese agricultural treatise. It is attributed to Doi Suiya (Saneyoshi) and is thought to date from 1628. Under the management of Suiya, from 1558 to 1575, Kiyoyoshi brought prosperity to his domain, thanks to a benevolent policy towards the tenant-farmers, who built a shrine to his memory after his death. The *Seiryōki* (of which we in fact quote only book 7) takes the form of a conversation between Kiyoyoshi and his intendant and agricultural adviser Matsuura Sōan. Book 7 is divided into chapters dealing with the following subjects: the various crops, the calendar from sowing to harvesting, soils, fertilizer, cultivation techniques, the investment in labour for each operation and the way of managing the tenant-farmers. The writer emphasizes the improvement of crop species, manuring and soil renewal as methods of increasing yields, but the work is imbued with a moralistic spirit that places the farmers' dedication above technical competence.[242]

Half a century later, the *Hyakushō denki*, "Peasants' Tales", written between 1681 and 1683, shows a more scientific spirit, emphasizing returns and productivity. This anonymous text deals with the practices of the regions of Mikawa (Aichi) and Tōtōmi (Shizuoka). With fifteen books, it is one of the most detailed agricultural treatises. It covers nearly all the cropping operations and devotes one book to farming implements. Compared with the *Seiryōki*, this work reveals a fundamentally different attitude towards techniques, as can be seen from the example of tillage. The *Seiryōki* recommends one tilling with the hoe in winter, followed by two tillings with an ox-drawn plough in spring; it also calls for a single harrowing using an ox before transplanting in the irrigated rice field. The effectiveness of these tillings depends, according to the *Seiryōki*, on the farmer's skill in guiding the ox. This text distinguishes skilful, moderately skilful and unskilful farmers based on the criteria of zeal and dedication. By contrast, the *Hyakushō denki* has the field tilled with a new and heavy hoe that penetrates deeply into the soil, and then hoed a second time with an old, light, worn hoe, the thinned cutting edge of which allows the roots of weeds to be easily cut and buried. This work distinguishes the weight and size of hoe-heads suitable for the different soil qualities (light, heavy, sandy, clayey, pebbly). Thus the *Seiryōki* reflects the point of view of the proprietor and the *Hyakushō denki* that of the farmer.[243]

The *Aizu nōsho*, "Agricultural Treatise of the Aizu Region", also considers things from the standpoint of the farmer. Written in 1684 by Sase Yojiemon (1630–1711), the work concerns the domain of Aizu (Fukushima) and in particular the village of Makunouchi (Aizu Wakamatsu) where the writer, a local notable, took on the function of intendant *kimoiri* for the estate authorities. It is a mountainous region, located in the north-east of the country and with a cold-temperate climate. The three books deal with rice growing, some other crops and general questions relating to the agricultural calendar, the natural vegetation and housing.

The author of the *Aizu nōsho* later wrote simplified versions intended for wider circulation at the regional level. The *Aizu uta nōsho* of 1706 presents the work in the fields in 575 *waka* poems and includes an appendix with an annotated and illustrated list of eighty-three agricultural implements, the drawings now being lost. However, the monograph *Inawashiro Kawahigashi gumi yorozu no fūzoku kaichō* (1685) dealing with the area around the village of Makunouchi has drawings that correspond to the descriptions given in the *Aizu uta nōsho*. Only two copies of the *Aizu nōsho* remain, dating from 1724 and between 1789 and 1800 and published in the collection "Nihon nōsho zenshū". This text has a paragraph on implements, but another version of the *Aizu nōsho* dating from 1748 includes a detailed chapter on approximately one hundred of them, based on notes by Sase Yojiemon or was added to his text by an anonymous author. This chapter on implements was published by Sasaki Takeo.[244]

The works named above are based on the experience acquired by the writers in their native region. In 1697, Miyazaki Yasusada (1623–1697) completed the first theoretical treatise, *Nōgyō zensho*, "Manual of Agriculture", inspired by the Chinese agronomy manual *Nongzheng quanshu* (Jap. *Nōsei zensho*) published in 1639 by Xu Guangqi. While quoting entire passages from the Chinese text, Yasusada describes the situation in the Ōsaka-Kyōto region, traditionally the most fertile and most developed in the country, and he also puts down his own observations made in the course of his travels through several provinces. The *Nōgyō zensho*, in ten books, gives practical advice for each crop, listing separately 150 cultivated plants. The writer wished to provide a general manual intended for wide circulation. His is the first printed agricultural treatise and Yasusada was still alive when it was published.

The *Nōgyō zensho* became a reference work that influenced later agronomists, including Tsuchiya Matasaburō (1642–1719). Intendant *tomura* of the Maeda domain in Kaga (Ishikawa), Matasaburō wrote *Kōka shunjū*, "Spring and Autumn Crops", in 1707, ten years after the *Nōgyō zensho* appeared. For thirty years, Matasaburō had the task of encouraging agriculture in his region. The work, in seven books, deals with the various farming operations and devotes the seventh book to implements, all drawn, giving the measurements, price and function of each one. After finishing it, Matasaburō took up his brush to illustrate the agricultural calendar. In 1717, he completed the 150 coloured drawings of the *Nōgyō zue*, an educational work showing the complete annual cycle of farming tasks.

The *Seikei zusetsu* (1806), "Illustrated Encyclopedia of Management", is another very valuable document for its illustrations. This encyclopedia of natural history in one hundred books was an ambitious project, begun by Shimazu Shigehide (1745–1833), daimyō of the Satsuma domain (Kagoshima) in Kyūshū. He entrusted it to Sō Han (1758–1834), a physician of Chinese descent. A team worked on it for thirty years, drawing on Chinese, Japanese and even European sources. However, part of it was lost in a fire. In 1805 and 1806, Sō Han published the first thirty books dealing with farming techniques. Many drawings show technical details, notably those of book 13 devoted to farming implements (Figures 1.2–1.4).

The *Nōgu benriron* (1822) is by Ōkura Nagatsune, one of the most celebrated agronomists of the Edo period. Born into a farming family in Bungo (Ōita), Kyūshū, Nagatsune witnessed two great famines in 1783 and 1787. From the age of twenty-nine, he travelled to Kyūshū and Shikoku and settled in Ōsaka. He was interested in the growing of sugar-cane, introduced by Satsuma, and in the cultivation of the "wax tree" *hazenoki*, Rhus succedanea, this being the subject of his first work, *Nōka eki*, an enlarged edition of which was printed in Edo in 1810. Using the observations he had made during his travels between Ōsaka and Edo, Nagatsune wrote the *Nōgu benriron* in 1817 and published it in Edo in 1822. In 1825, he settled in Edo and wrote several monographs about specific crops. In 1833, he introduced sugar growing to the province of Suruga (Shizuoka); he was then sixty-five years old. In the following years, he received invitations from various domains in the Kantō region. In 1859, he completed his last work, *Kōeki kokusankō*, dealing with commercial crops and manufactures. The *Nōgu benriron* is a catalogue of forty-two implements with their regional variants, all represented by drawings showing measurements and weights, accompanied by a commentary. Thanks to this collection of data, the writer explains how to obtain better results by the choice of the most suitable tool, while saving on human effort. He displays rationalism and shows an interest in productivity that we have already encountered in the *Hyakushō denki*. But this attitude was still rare in a society imbued with Confucian ethics where zeal and loyalty were valued above all else.

Table 1.2 Agricultural treatises mentioned in Chapter 1

Title	Date	Author	Region of Japan
Seiryōki	ca. 1628	Doi Suiya	Iyo
Hyakushō denki	1681–1683	anonymous	Mikawa[1]
Aizu nōsho	1684	Sase Yojiemon	Aizu
Aizu uta nōsho	1706	Sase Yojiemon	Aizu[3]
Nōgyō zensho	1697	Miyazaki Yasusada	a general work about Japan
Kōka shunjū	1707	Tsuchiya Matasaburō	Kaga[3]
Nōgyō zue	1717	Tsuchiya Matasaburō	Kaga[2]
Seikei zusetsu	1806	Sō Han	Satsuma[3]
Nōgu benriron	1822	Ōkura Nagatsune	a general work about Japan[3]

Notes:
1   Includes a chapter on tools.
2   Includes illustrations on tools.
3   Includes a chapter on tools with illustrations.

Technical innovations

The Edo period is characterized by technical innovations and agrarian reclamation campaigns supported by the shogunal government and daimyō, including land clearing, hydraulic installations and drainage works. These measures brought about a significant increase in the arable acreage and yields.

Here are a few figures. In the Keichō era (1596–1615), the country produced 18 million *koku* of grains on lands totalling 1.6 million *chō* (hectares). Production rose to 25.78 million *koku* in the Genroku era (1688–1704); and the arable areas amounted to 2.95 million *chō* in the Kyōhō era (1716–1736). The cultivated areas had therefore doubled in a little over a century.[245]

While agricultural operations have remained fundamentally the same since the early medieval period, they have been able to give better yields thanks to the introduction of new implements and new methods at all levels: tillage, irrigation, manuring, weeding and the processing of grains after the harvest. We now briefly retrace the major technical innovations of the Edo period.

Tillage

From the early medieval period, the aristocratic estates (holdings) used, together with the hoe and spade, the square plough *karasuki*, equipped with a mould-board and a very long and fairly broad sole, which archaeologists call a long-soled plough *chōshōri* (Figure 1.9). This quite heavy implement, introduced from China no later than the 7th century, would have been preceded, in the 6th century, by a lighter unsoled scratch-plough or ard, but no trace of it remains, either written or material. The long-soled plough was in principle better suited to dry lands than to rice fields. Yet it persisted in Japanese rice cultivation until the 17th century, possibly owing to its stability, which made it relatively easy to handle. But according to some, this heavy plough, represented in the iconography of the late medieval period, tilled the soil only superficially: it penetrated to a depth of just 5–7 cm. As long as little fertilizer was used, this shallow tillage made it possible to consolidate the bed of the rice field and also help maintain its imperviousness, but once manuring was increased, surface tillage became totally insufficient. This is how specialists often explain the return to the hoe in the 17th century, a return documented by the *Hyakushō denki* and *Nōgu benriron*. But Kōno Michiaki draws attention to geographical differences and maintains that the plough remained in use in the centre and west of the country. He notes that it was used in the region of Ōsaka, a region with high yields, up to the mid-20th century. Be that as it may, it was perhaps due to the stimulus of European sciences that there was developed in the 19th century, in Kyūshū, the very short-soled plough *karasuki* and triangular unsoled scratch-plough *kakae-mottatesuki*, which made it possible to penetrate 15–20 cm into the soil (Figure 1.2). The idea for this scratch-plough is attributed to Hayashi Onri (1831–1906), an agronomist from Fukuoka, who made a major contribution to various agricultural improvements. Through his initiative, these implements were distributed systematically in the eastern part of the country: in 1875 and 1878, delegations of technicians were invited respectively from Fukuoka to Iwate, from Ōita to Miyagi and from Kumamoto to Aomori.[246] After a return to deep tillage with the hoe, it was therefore only at the end of the 19th century that the short-soled plough and scratch-plough became widespread in Japan.

Irrigation

When it comes to rice-field irrigation, various hydraulic installations have been used over the centuries, such as canals, dams, dykes and weirs, but people have generally not had access to a mechanical device for conveying water. The water-wheel, a device with paddles used to draw water to be poured onto land above the reservoir, did not become widespread in Japan until quite late, even though the court had tried to introduce it in the 9th century. In 829, a government edict proclaimed: "Following the custom in China, water-wheels *suisha/mizuguruma* are built in this country in areas that lack dams and canals. This makes it possible to obtain a return from those lands that have no water … The population [of our country] must be instructed to build these devices in order to support agriculture. Depending on the possibilities of the various regions, they will be operated with the hands, with the legs or by oxen." According to Kōno Michiaki, the Japanese legislators drew on the Chinese annals of the Tang dynasty to formulate the decree, but their directive was not followed up.

The "water-wheel" *mizuguruma* is listed in the 12th-century dictionary *Iroha jiruishō*. Another man-powered device, the square-pallet chain pump, is listed under the Chinese name of "dragon-bone wheel" *ryūkotsusha* in the 15th-century

Japanese dictionary *Satsujōshū*, but it was reintroduced from China in the 17th century and did not become widespread in Japan until around the second half of the 18th century. The water-wheel, however, does not appear to have replaced the traditional system of bucket irrigation known since the early medieval period. The swinging tub *tsurube* or *mizukae-oke* attached to four ropes made it possible for two men to very easily draw and siphon water (Figure 1.11). Besides the man-powered systems, medieval Japan also knew a water-powered mechanical wheel: the noria, a bucket wheel or pot-chain pump. Kimura Shigemitsu draws attention to a poem by Minamoto no Toshiyori (1055?–1129) who speaks of the noria *mizuguruma* installed on the Setagawa river in the hamlet of Takami south of Lake Biwa, and to an image of this noria in the *Ishiyamadera engi emaki* (14th century). Other poems from the 12th and 14th centuries are known, but the noria almost certainly remained an uncommon device that did not come into widespread use before modern times.[247]

Weeding

Hayashi Onri is credited with the idea of weeding claws *ganzume*, finger-protectors in the shape of small tubes cut in bamboo. Weeding was traditionally done by hand or with non-specialized tools, such as a worn hoe, like that recommended by the *Hyakushō denki*. The Edo period saw the appearance of specialized equipment, such as claws, a stirrup-hoe *kusa-kezuri* and various types of weeding hoes with small heads of different shapes, documented, among others, by the *Seikei zusetsu* and *Nōgu benriron*. These implements had, however, been known in China for several centuries and are shown in the 14th-century *Nongshu*.[248]

Fertilization

Fertilizer is the subject of particularly long discussions in the Japanese agricultural treatises. The Edo period saw added to the different kinds of plant and organic manure, dried sardine or herring meal *hoshika* and the dregs of vegetable oil *abura-kasu*. Unlike traditional green manure, where the quantity was always limited by what farmers could collect, these new fertilizers could be bought from merchants. Cultivators thus had the possibility of obtaining large quantities, especially useful for commercial crops like cotton and for crops with two harvests that were increasingly grown in the west of the country.[249]

Threshing

To thresh rice, *kokibashi*, two long wooden chopsticks or *inekan*, two short tubes or sections of bamboo tied respectively on one side, were used. The top of the rice stalks was squeezed between the sticks or between the short bamboo stems and the grains were extracted from the ears by pulling the rice stalks (Figures 1.3 and 1.12). We considered earlier (see pp. 47–53) whether this method was already in use at the beginning of the 11th century, in the time of Sei Shōnagon. During the Genroku era, threshing-sticks were replaced by the threshing-comb (Figure 1.3). Developed in the Ōsaka region, this implement gained favour in all the provinces in the space of fifty years. With the threshing-comb, one person could thresh 23–30 sheaves of rice in half a day, compared with 9–12 sheaves with chopsticks, according to a work written in the Tenpō era (1830–1834). Ihara Saikaku (1642–1693) says the comb had the popular name "replacement for old people" *gokedaoshi*, because old women had the task of threshing rice with chopsticks before the introduction of the threshing-comb. Owing to its good returns, this implement was officially called "thousand-bunch thresher" *senbakoki*, or simply "rice thresher" *inakoki*. It may be noted that all the names of implements had regional variants. The threshing-comb for wheat and barley *mugikoki* resembled that for rice, but, with more widely spaced teeth, it pulled off only whole ears. The actual threshing was done, according to the *Hyakushō denki*, by beating with a flail *karazao*, with a wooden mallet or by treading with the feet[250] (Figures 1.3 and 1.12).

Hulling

In the early medieval period, the most widely used method seems to have been pounding with a pestle in a wooden mortar, intended to separate the grains from their husks (Figure 1.5). Other methods were also known, but these are seldom if ever mentioned in the early medieval texts: threshing grains of rice with a flail and the use of hulling-mills *suriusu*, methods that are shown in modern iconography (Figures 1.6 and 1.12). In the second case, two wooden grinding wheels were superposed horizontally, with the movable upper one flush-mounted above the fixed lower one. The upper wheel made half-turns in both directions, worked by four ropes pulled by two people seated opposite each other. A scene of this kind occurs, around the year 1000, in Sei Shōnagon's "Pillow Book". Here is what the *Hyakushō denki* has to say about it:

> As hulling-mills did not exist in ancient times, each grain of rice was skinned with the nails, or else the grains were placed on a stone and rubbed with another stone to hull the rice. Later, a more practical method was developed. Wooden mills *suriusu* were used to hull rice. This at once improved [the yield]. Then, some Chinese arrived in Nagasaki around the first year of the Kan'ei era (1624), and they showed the Japanese how to make clay mills. From then on, this device was made in Japan and wooden mills were abandoned.

These were clay or "earth" hulling-mills *tousu*, made of wicker or wooden staves, filled with dried clay. Unlike wooden mills, the upper wheel of the *tousu* was worked in a full rotation by means of an articulated horizontal lever (Figures 1.3 and 1.13). According to the information given in the *Hyakushō denki*, clay mills hulled triple quantities, but they also resulted in losses, because 1.5 *shō* (2.7 litres) out of 1 *koku* (180 litres) of grain was reduced to flour and other quantities were spoiled and could no longer be husked. Thus the spread of clay mills probably remained limited to large estate holdings, since the losses were too great for individual holdings. These mills do not appear in the *Kōka shunjū*. Wheat and barley, even at this period, were always hulled by pounding in a mortar.[251]

Winnowing and sieving

After threshing and hulling, the grains are mixed in with their husks and chaff. Winnowing serves to remove this refuse. From the early medieval period, winnowing involved tossing the grains into the air using a wicker winnowing-basket *mi* (Figures 1.5 and 1.13). This operation was greatly facilitated and speeded up after the introduction of the winnowing-machine *tōmi* from China, around the Genroku era (1688–1704), a device consisting of a box with a fan inside worked by hand with a crank-handle (Figure 1.4). After hulling, there remain unhusked grains and refuse that are removed by several successive sievings, using sieves *furui* in the shape of flat baskets with different sized holes (Figure 1.3). From the Jōkyō era (1684–1688), this operation was also simplified thanks to the invention, this time Japanese, of the device called "thousand-*koku* sieve" *sengokudōshi* (Figure 1.4). The *Saizōki*, written by Ōhata Saizō (1642–?), gives a list of the grain-processing implements of a small-scale farmer: threshing-sticks, flail, clay hulling-mills, sieving device for wheat, mortar with levered pestle and stone mill, the last two implements being used for husking rice and grinding wheat.[252]

Polishing and milling

We return to the wooden mortar *usu* with its vertical pestle *kine*, used in Japan since prehistoric times. To the simple pestle was added the hammer-pestle *yokogine* with a perpendicular handle, between the 14th and 16th centuries (Figure 1.4). These implements performed all the grain-processing operations up to that time. In the Edo period, they were used to husk rice or to pound wheat eaten as whole (uncrushed) or crushed grains. Also known, since the 8th century, was a variant, the "levered pestle", i.e. a mechanical tipping pestle with a horizontal lever worked by the feet, called the "exotic mortar" *karausu* (Figure 1.4). However, the *Hyakushō denki* says that the mechanical pestle only became widespread during the Keichō and Genna eras (1596–1624). In fact, this device seems to have been forgotten and then re-introduced to Japan. A more mechanized version is the pestle operated by the water-wheel, a device of Chinese origin. Éric Trombert discovered the earliest image of it, dating from the 12th century, on a mural at Mount Wutai in China. In Japan, there is written evidence of it relating to the inauguration ceremony of the Great Buddha in Nara in 1195. It is also known that some large aristocratic hold-ings in the 17th–18th centuries had a battery of pestles operated by a water-wheel, which husked large quantities of rice, but apart from this, the hydraulic pestle does not seem to have come into general use. Another Chinese device was never intro- duced to Japan: mills powered by draught animals. They aroused the curiosity of the Japanese monk Jōjin, who went to China in 1072. He notes in his travel diary: "Two donkeys, eyes blindfolded and muzzles bound, were harnessed to a stone mill *ishiusu* for making wheat flour; the mill turned all by itself, without help from anyone."

In Japan, the pestle and mortar were the milling implements used to reduce cereals to flour. Foods made with flour, like wheat noodles *muginawa*, were eaten at the imperial court from the 8th century. However, the mill *ishiusu* or *hikiusu* seems not to have become widely used in Japan until the 17th century, although it was known earlier, as shown by the remains of grooved stone mills discovered in archaeological sites of the early medieval period. These were small mills, the upper wheel being turned with a hand-lever (Figure 1.3). Mill wheels differ from hulling-mills, mentioned earlier, by their material, stone, and by their function: milling reduces the grains to flour, but hulling does not crush or damage the grains. It may be noted that the earlier and modern Japanese language does not distinguish these different implements; it uses the same term *usu* for the mortar, hulling-mill and grinding-mill, so that the meaning can only be known when there is a prefix speci- fying the type. As for the water-mill, so common in medieval Europe, it did not find its way into Japanese practices. In Japan, the small hand-mill and mortar seem to have been able to satisfy the need for cereal flour.[253]

Overall, some new methods brought an increase in yields in the sense of a greater volume of production, and others a rise in productivity linked to a saving in human effort. For Horio Hisashi, deep tillage and commercial fertilizer are the main keys to progress in the Edo period.[254] These practices resulted in the spread of two annual crop harvests and of commercialized crops, responsible for the economic development of the agricultural sector.

To conclude, we present, by way of documentation, a work that shows the farming processes of the Edo period: the *Ehon tsūhōshi*, published in Ōsaka in 1729, by Tachibana Morikuni (1679–1748). A pupil of Kanō Tanyū (1602– 1674), Morikuni wished to establish an iconographical model for Japanese illustrators, with his work in ten books depicting the customs of everyday life. This wish is reflected in the title, *tsūbō/tsūhō* meaning 'ready money'. The *Ehon tsūhōshi* did in fact have a wide circulation and certain scenes were often repro- duced. The first book shows the agricultural calendar for rice cultivation in fifteen images, all except the first reproduced below. Morikuni set out his ideas about agricultural iconography in another work called "Images of Til- lage in Our Country", *Wakoku kōsakuzu*, in these terms: "All those born in this society and who eat rice, even if they are not farmers, should nevertheless know the circumstances of rice growing. Those who do not live in the countryside do not see the process from growth to storage. So I have illustrated the peasants' tasks to make them known and to arouse the readers' compassion for the pea- sants, in order that the population will appreciate the value of rice."[255] The agricultural operations can therefore be summarized in these images, reproduced in Figures 1.7–1.13.

Figure 1.1 Rice stacks and racks for drying rice
Source: *Seikei zusetsu*, 1806, book 5, 1933 edition.

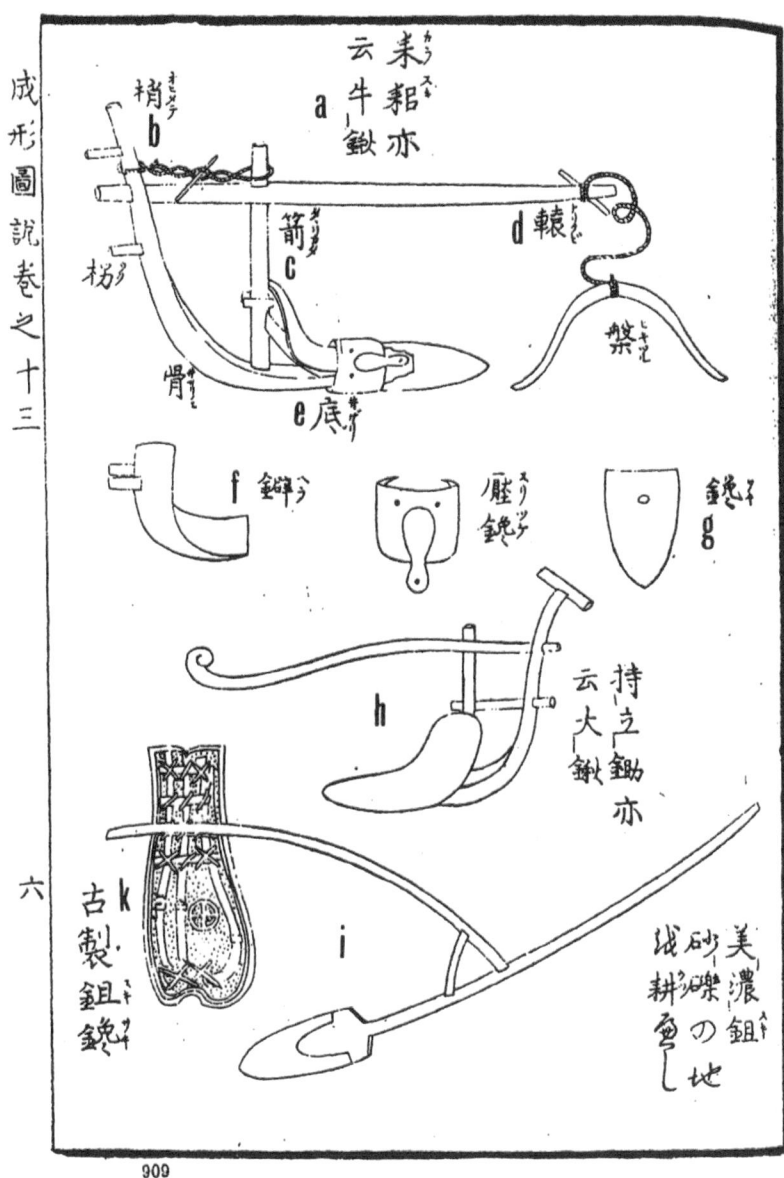

Figure 1.2 Ploughs and scratch-plough (ard)
Notes: (a) *karasuki* or *ushiguwa*, plough, (b) *oitate*, stilt, (c) *tatarikata*, strut, (d) *torikubi*, beam, (e) *izaki*, sole, (f) *hera*, mould-board, (g) *saki*, share, (h) *mottatesuki* or *ōguwa*, scratch-plough, (i) Minō-suki, Minō plough, pulled by a man, (k) *kosei sukisaki*, old-style share?
Source: *Seikei zusetsu*, 1806, book 5, 1933 edition.

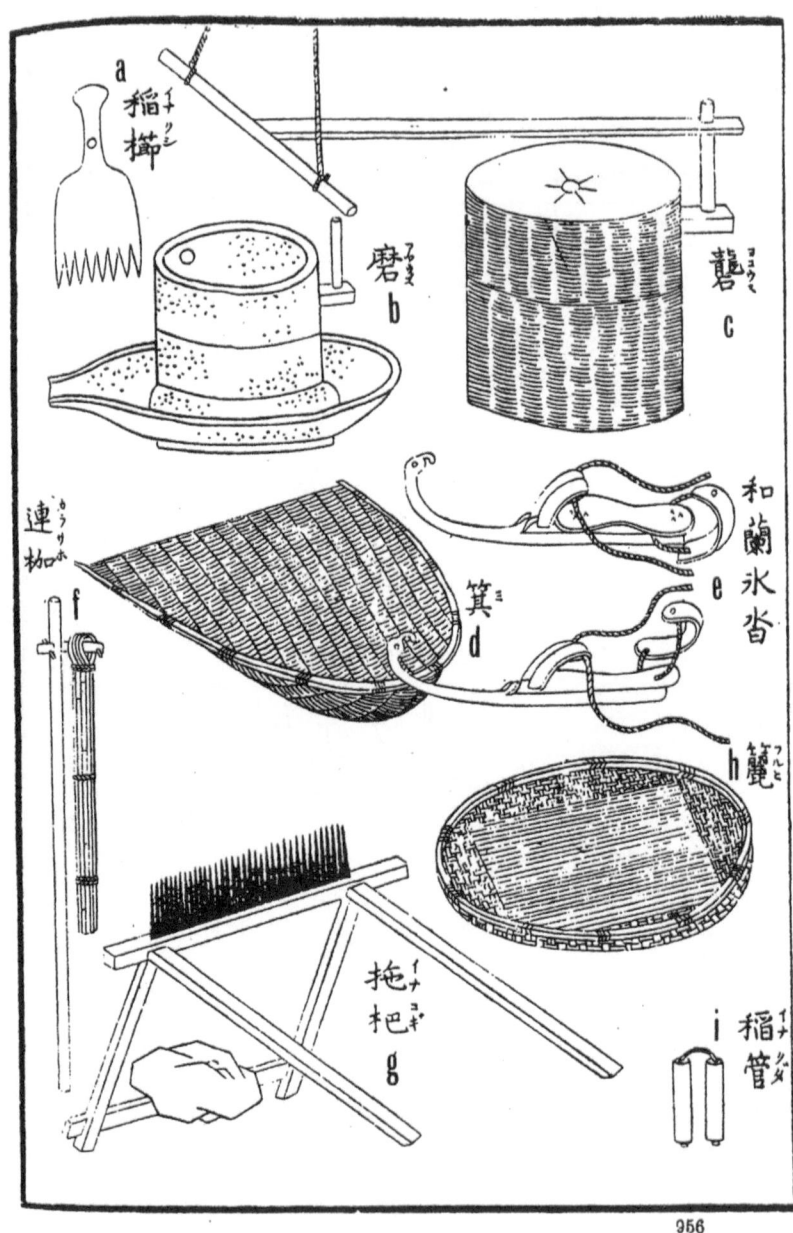

956

Figure 1.3 Implements for processing grains (1)
Note: (a) *inakushi*, comb, (b) *atsuusu* (*ishiusu*), mill, (c) *yokousu* (*suriusu*), hulling-mill, (d) *mi*, winnowing-basket, (e) *oranda kōrikutsu*, ice-skates, (f) *karazao*, flail, (g) *inakogi* (*inakoki*), threshing-comb, (h) *furui*, sieve, (i) *inakuda* (*inekan*), bamboo threshing-sticks.
Source: *Seikei zusetsu*, 1806, book 13, 1933 edition.

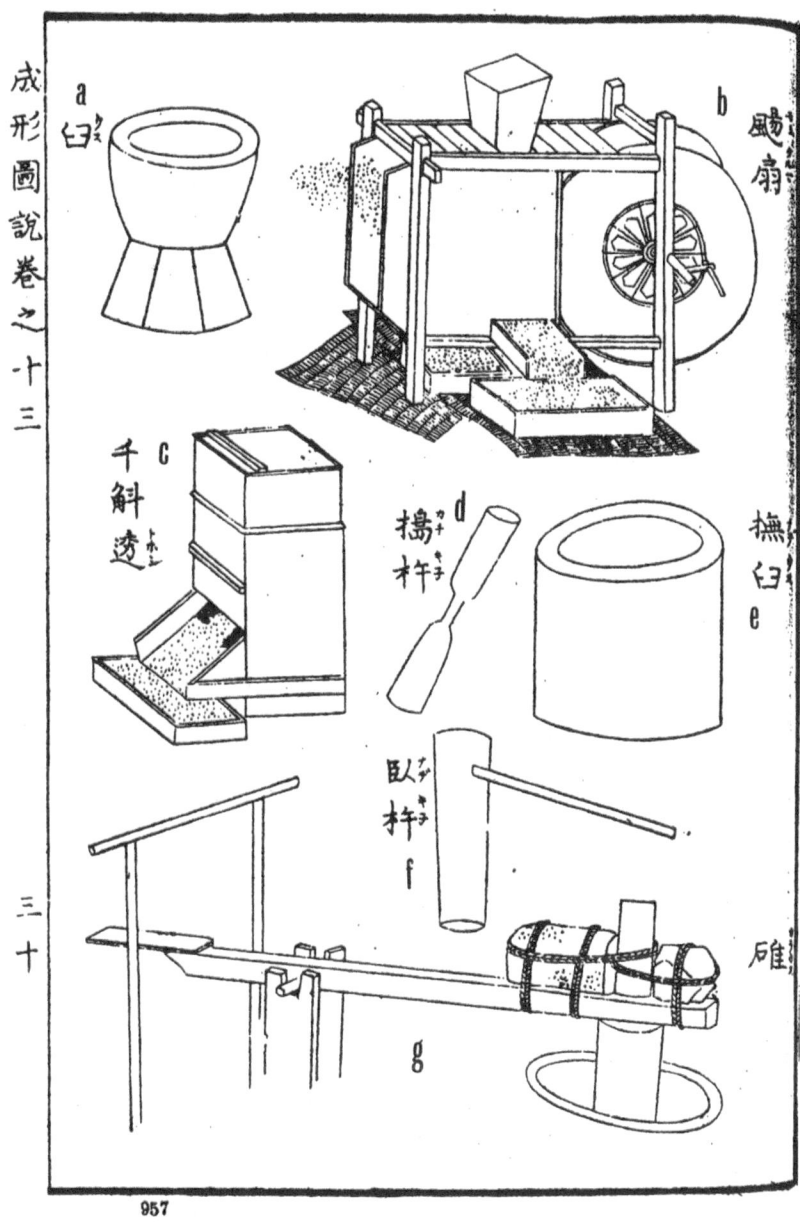

957

Figure 1.4 Implements for processing grains (2)
Note: (a) *usu*, mortar, (b) *momikuruma* (*tōmi*), winnowing-machine, (c) *sengokudōshi*, sieve, (d) *kachikine* (*kine*), pestle, (e) *nadeusu* (*usu*), mortar, (f) *fujikine* (*yokogine*), hammer-pestle, (g) *karausu*, mechanical pestle with lever.
Source: *Seikei zusetsu*, 1806, book 13, 1933 edition.

Figure 1.5 Pounding, winnowing and suspension-drying
Note: Pestle, mortar; winnowing-baskets; rack for drying sheaves.
Source: *Yamato kōsaku eshō*, ca. 1703.

Figure 1.6 Hulling with a hulling-mill
Note: Two iron-tipped hoes; sieve; hulling-mill (*tousu*).
Source: *Yamato kōsaku eshō*, ca. 1703.

Figure 1.7 Soaking the seed (above); removing and drying the seed (below)
Source: *Ehon tsūhōshi*, 1729, book 1.

Figure 1.8 Sowing in the seedbed (above); loud deterrent (small wooden boards) and dummy scarecrow (below)
Source: *Ehon tsūhōshi*, 1729, book 1.

Figure 1.9 Plough tillage of the irrigated rice field (above); harrowing of the irrigated rice field (below)

Source: *Ehon tsūhōshi*, 1806, book 5, 1933 edition.

Figure 1.10 Collecting the seedlings from the seedbed (above); transporting the seedlings and transplanting the rice (below)
Source: *Ehon tsūhōshi*, 1806, book 5, 1933 edition.

Figure 1.11 Weeding and irrigation with a swinging tub *tsurube* (above); irrigation
with a manual "dragon-bone" water-wheel *ryūkotsusha* (below)
Source: *Ehon tsūhōshi*, 1806, book 5, 1933 edition.

Figure 1.12 Harvesting with a sickle and transporting the rice sheaves (above); threshing using threshing-sticks and hulling by beating with a flail (below)

Source: *Ehon tsūhōshi*, 1806, book 5, 1933 edition.

Figure 1.13 Hulling with a clay hulling-mill *tousu* and winnowing (above); filling straw
          sacks and storing them in a granary (below)
Note: Two iron-tipped hoes; sieve; hulling-mill (*tousu*).
Source: *Ehon tsūhōshi*, 1729, book 1.

Notes

1 Bibliography: Furushima Toshio (1949); Kojiruien Sangyōbu 1 (1908, 1984); Naoki Kōjirō (1968); Toyoda Takeshi (1964); Mori Kōichi (1984); Tateno Kazumi (1991); Kimura Shigemitsu (1993); Kimura Shigemitsu (1992); Kinoshita Tadashi (1985); Matsuo Hikaru (1994); Shiraishi Akiomi (1993); Horio Hisashi (1990); Terasawa Kaoru (1986). For Japanese agricultural techniques of the 17th–18th centuries, see Janata *et al.* (1969) and Pauer (1973); for those of China, see Bray (1984); see Mane (1983); for those of medieval Europe, see Comet (1992).

2 Kameda Takashi (1965). See below, note 247, for the water-wheel *suisha*, and "Drying" for the drying-rack *inabata*.

3 RSK 8: 327, 328 (wheat and buckwheat); SN Reiki 1.10.7. 715, Enryaku 8.5.9. 789. See the fragments of almanacs *guchūreki* in DNK 2: 570–574 (746), 3: 347–353 (749), 4: 209–217 (756); these fragments cover only the 1st–4th months of the different years.

4 See the bibliography of Chinese treatises in Bray (1984: 47–80).

5 The *Qiming yaoshu* appears in the 9th-century catalogue *Nihonkoku genzaisho mokuroku*, see GR 30 ge: 42. See the copy of the *Qiming yaoshu* preserved in the Hōsa bunko archive, with an indication of the date of 1166 in the postscript of the tenth book. The Japanese government mentions, for example, the economic treatise from the *Hanshu* (1st century) or the *Houhanshu* (ca. 5th century), cf. RSK 8: 321; it also mentions the calendar *yueling* in the *Liji* written in the 3rd century BC (RSK 8: 328), as well as other classics and early Chinese dictionaries (RSh, p. 347 et seq.).

6 Kōno Michiaki, personal communication; Kōno Michiaki (1996: 72–73). References to flora and fauna are also common in European literature, see Ferdière (1988: 11–17); and in that of China, see Watabe Takeshi (1996); Chiku Katei (1988).

7 *Shinsarugakuki*, pp. 76–78. We have not adopted the readings given in the Heibonsha edition; concerning the irrigation systems, see pp. 32–35.

8 See the *Man'yōshū* (hereafter abbreviated M), M 14: 3561 and 11: 2476 (see also a translation by Péronny (1993: 66). *Kojiki* ge and N Nintoku 30.11. and *Kodai kayōshū*, pp. 75, 76, the poem by Nintoku; *Harima fudoki* (i.e. *Harima no kuni fudoki*; the abbreviated names of the *Fudoki* will be used from now on), p. 351, the poem by Kensō, translation by François Macé, in Rotermund (2000: 50).

9 Kjg, p. 31 and *Toyukegū gishikichō*, p. 61. Nara kokuritsu hakubutsukan ed., *Shōsōin ten*, 1998, p. 48, and 1987, p. 18 (spades). SJ Jōgan 6.2.25. 864 and 8.3. intercal. 1. 866. See above, pp. 27–30. Kuroda Hideo, "Taasobi no seiritsu", in Kuroda (1984: 410) attributes the origin of *dengaku* to the Ise ritual.

10 *Chūyūki* Daiji 2.5.14. 1127.

11 *Fuboku wakashō* 2: 16–12952 (read: "Shinpen kokka taikan", vol. 2, text 16, poem no. 12952), poem by Tsuchimikado Tennō see also *ushi no karasuki* in *Rokujō eisō* (a 19th-century anthology) 9: 21–1489 (unnamed author); *Keiun hōin shū* 4: 24–49 (*ara oda o ara sukikaesu*), poem by Keiun (14th century); *Sōkonshū* 8: 10–1625 (*sukikaesu haru no arata*); and three poems that mention *shirokaki*: *Horikawa-in hyakushu* 4: 26–227, poem by Minamoto no Kunizane (1069?–1111), 4: 26–407, poem by Fujiwara no Nakazane (1057–1118), *Yukimune shū* 7: 43–207, poem by Minamoto no Yukimune (1064–1143), translated below, see note 61. I thank Jacqueline Pigeot for revising the French translations of poems in this chapter.

12 These three poems mention *arata uchikaesu*: *Horikawa-in hyakushu* 4: 26–237, poem by Ryūgen (d. before 1120); *Michinari shū*, poem by Minamoto no Michinari (?–1036), after Ienaga Saburō, 1942, no. 1712; *Shūi wakashū* book 13, 1: 3–812. One finds *uchikaesu* in *Horikawa-in hyakushu* 4: 26–228, poem by

Minamoto no Moroyori (1068–1139); *Kin'yō wakashū* 1: 5-2-74, poem by Tsumori no Kunimoto (1023–1102); and *Dainagon no ie utaawase* of 1029, 5: 61–12, poem by Minamoto no Chikanori (?–1045); and *kaesu* in *Shinsen waka rokujō* 2: 14–634. Furushima Toshio (1949: 160–162), mentions several of these poems, but without giving the references.

13   *Harima fudoki*, pp. 263, 273, 277, 285 et seq., etc. mention the quality of the soils. *Kaden*, p. 883, biography of Fujiwara no Muchimaro.

14   See note 3 for the almanacs. The *byōbu uta* poems were collected by Ienaga Saburō, 1942. We quote the numbers in Ienaga: no. 1328–3 *tagaesu*, poem by Tadakimi, included in 983 in *(Ōnakatomi no) Yoshinobu shū*, Nishihonganjibon; no. 1038 *arata utsu*, poem by Minamoto no Shitagō (911–983) dating from 967, in*Minamoto no Shitagō shū*; sankō no. 71 *arata uchi*, poem in *Sagoromo monogatari* (9th century) "Imahime no kimi"; no. 129 and 183 *tagaesu*, poems by Ki no Tsurayuki (872?–945?) dating from 902 and 906, taken from *Tsurayuki shū* book 2 of the collection "Kasenkeshū" and *Tsurayuki shū* book 1; no. 1473 *tagaesu* in *Kanemori shū*; no. 1195 *arata o uchikaesu* in *Nobuaki shū*; and in no. 829, poem by Fujiwara no Motozane (?–?) dating from 952, taken from *Motozane shū* no. 896 *yamada utsu* in *Shūi wakashū*. *Shinsen waka rokujō* 2: 14–635 *uchiokosu; Horikawa-in hyakushu* 4: 26–228 *uchikaesu*. See Toda Yoshimi, "Jūjūni seiki no nōgyō rōdō to sonraku", in Toda (1991); he also mentions, p. 62, the *Kaidōki* (ca. 1223), according to which fields were tilled *ta o utsu* with spades.

15   ES 39: 878–881; we follow Furushima Toshio (1949: 178–182) on the subject of tillage. See Kōno Michiaki, below, note 21.

16   ES 39: 878–881; we follow Furushima Toshio (1949: 114; 178–182) and Aoba Takashi (1991: 70–77) for the interpretations of terms, except in the cases noted later.

17   Some of these vegetable gardens are glossed with their reading in the Kujōkebon of ES 39: 865, which dates from after 1036. See the localization in WR 6: 1–3. There are two Nara no sono, with different characters, for an obscure reason. There was a garden called Yamashiro no sono, specializing in early melons, set up by Emperor Kanmu (r. 781–806), according to the *Morotō nenjūgyōji*, p. 214, *Moromoto nenjūgyōji*, p. 245 and *Nenjūgyōji shō*, p. 304. This garden may correspond to the Keihoku no sono. In our translation, the notes in the *Engishiki* are in parentheses; our additions and reconstructions are in square brackets.

18   The text indicates the number of men and not the number of days, for management purposes. We calculate the days, though for tillage one working day includes two men. We have given the figures in the text, but because of the day of draught tillage done by two men, the details in each paragraph will show one day less. We use the measures given in *Iwanami Nihonshi jiten*, 1999, p. 1431: 1 *chō* = 1.13 hectares; 1 *tan* = 0.3 acres; 1 *bu* = 3.15 m^2; 1 *koku* = 85 litres; 1 *to* = 8.5 litres; 1 *shō* = 0.85 litres; 1 *gō* = 0.085 litres; 1 *kin* = 670 grammes. See the plant names in the "Catalogue of edible and industrial plants" in the Appendix.

19   For *erabu*, Aoba Takashi (1991) suggests the meaning 'to select'; Furushima Toshio, that of threshing, possibly by rubbing between the hands. *Erabu* appears after the harvest in the case of barley and also taro. This term probably means "select the seedlings" in the case of *mira* (*myōshi o erabu*), which corresponds to *nae o toru*, for *nasubi* and *seri*. *Erabu* refers to the selection of heads for transplanting in the case of *azami*. In other cases, *erabu* has various meanings: 'to choose', e.g. choose a good quality soil (RSK 8: 328); select rushes of a certain length (ES 38: 859); select to remove the millet in the rice field (M 11: 2476, 12: 2999); separate by sieving *furui erabu* (ES 39: 871).

20   *Hashoku san tabi* means 'to transplant' rather than 'to sow'. This explains why there are three *toru*, i.e. two selections and one harvest.

21 This is read *hasu no kō go nin*, five men to guide [the ox?]. Kōno Michiaki (1994: 66–70) thinks that for the water-leek, an aquatic plant, an additional harrowing was required and that this was draught harrowing, but he does not explain the number of men (5). Given that *hasu* is also read *osamu*, this could be an incorrect character for *osamu*, 'to harvest'. See also Takai Yoshihiro (1998).

22 Aoba Takashi (1991: 70–77); see above, pp. 25–27.

23 *Harima fudoki*, p. 347. Another village, called Shikigusa, refers to the seat of a deity, in *Harima fudoki*, p. 321. The meaning of *shiku* is not restricted to green manure. In *Makura no sōshi*, no. 227, it is the sheaves that the harvesters lay out in rows *shiku* in the field.

24 Kjg, p. 3; see the reading *sokusa* instead of *naegusa* in the "Shintō taikei" edition, p. 10, of Kjg. The *Nihon kokugo daijiten* and *Kadokawa kogo daijiten* think that *naegusa* means 'seedlings'. We find the explanations of *naegusa* and *uu* in the *Kadokawa kogo daijiten* unconvincing.

25 Poem attributed to Tamenari (?–?) by Furushima Toshio (1949: 159), and found in *Tango no kami ke hyakushu* 4: 28–103 (an anthology of eight poets compiled by Fujiwara no Tametada, ?–1136). Ten *shiro* (*soshiro*) indicates a small area; 50 *shiro* corresponds to 1 *tan* (0.3 acres). M 9: 1677 mentions the *karishiki* of bamboo leaves collected to make a sleeping space. M 1: 7 and 47, and M 7: 1291 mention the cutting of grass intended for roofing, etc. In the early poetry and ES 8: 176, *karishiku* elsewhere refers to the collecting of wild plants to make a seat, pillow, etc. Green manure is called *karishiki* in *Aizu nōsho*, p. 57, and *Seiryōki*, p. 170; *kusagoe* in *Kōka shunjū*, p. 199.

26 *Mumyōshō*, GR 16: 380, a poetical treatise by Kamo no Chōmei (1155–1216), quoted by Kimura Shigemitsu (1992: 251). One finds *kusakari* for the stables in *Utsuho monogatari* 1: 126, 186, 2: 440, etc. On green manure *karishiki*, see also Morimoto Masahiro (2013: 145–188).

27 *Eishōki, uramonjo*, undated; here *koe* is written like the modern word *koyashi*. Furushima Toshio (1949: 147) and Kuroda Hideo (1984: 68) quote the *Eishōki* for Daiji 4.7. intercal. (1129) *uragaki*; and Kimura Shigemitsu (1992: 127, 147, 251, 271) for Kajō 1.7. intercal. 12. and 13. or 11. and 12. (1106). There is a 7th intercalary month in 1129, but not in 1106. In fact, Kimura (personal communication) refers to the document "Yamada no shō kumon tone shikijira jōan", taken from the *Eishōki, uramonjo*, manuscript Kyōto daigaku eishahon, reproduced between two other documents dating from 1187 and 1204, in Ōsaka shigaku kenkyūkai ed., *Settsu no kuni shiryō shūei*, 1960, p. 15.

28 François Sigaut, personal communication, on the subject of *citimene*. See also Chapter 2.

29 For China we follow the interpretation of Bray (1984: 99, 328, 504). This practice is attested for the region of Hangzhou in 1268, see *Xianchun Lin'anzhi*, by Qian Yueyou, 1268, book 58, p. 2a, or p. 3868, in the collection "Song Yuan fangzhi congkan".

30 *Horikawa-in hyakushu* 4: 26–228.

31 ES 39: 878–881, translated above. ES 39: 879 gives the figures for the quantity of manure and the loads; there one finds 6 round trips and not 6 loads per porter. Given the enormous quantity of manure, it may be asked whether these are not small *kin* of 223 grammes. In this case the weight of the manure amounts to one-third, but a load of 8 kg per person would be too small.

32 *Shasekishū* 5: 7. Kuroda Hideo (1984: 93) thinks this is in fact animal and not human manure. The *Myōgoki* 5: 73 defines *koe* as "black soil and therefore fertile".

33 Furushima Toshio (1949: 108–160, 206–216, 336–340); Naoki Kōjirō (1968: 56); Kuroda Hideo (1984: 68); Kimura Shigemitsu (1992: 128). But Horio Hisashi (1990: 229–230) thinks that green manure was not used until the 14th–15th

centuries. Kinoshita Tadashi (1985: 97–117) bases his hypothesis on the *Kenkyū sannen Kōtaijingū nenjūgyōji* (1192?) and on the commentaries of this text published in 1752 and 1891. Terasawa Kaoru (1986: 335).

34 Furushima Toshio (1949: 206, 207, 274); Naoki (1968: 56); Kuroda (1984: 93, 94). The toponym Kusoda is found in DNK 7: 47.

35 The *Seiryōki*, p. 119, also mentions 105 men for half a day, i.e. 105 half-days of work a year for collecting vegetation. See Furushima Toshio (1949: 340) for the data of 1704 and 1686.

36 See note 3. The *guchūreki* mention the favourable days for cutting grass *kusa kiru* in the 1st, 2nd, 3rd and 4th months. This term must be distinguished from the weeding *kusagiru* of the rice fields that took place in the 7th or 8th month, and was written with a different character. For Aizu, see the *Nōgyō zue*, pp. 60, 62, 80, 90, 103, 139, 147; this was done in the 3rd, 4th, 5th, 6th, 8th and 9th months. This work uses *kusakari* for grass cutting and *kusatori* for weeding. See below, notes 95–96, for weeding *kusagiru*.

37 François Sigaut and Marie Maurin, personal communication. Biodynamic gardeners recommend as fertilizer yellow lupins, sweet clover, white mustard, phacelia, buckwheat and rye; see the review *Germinance* 2001, pp. 31–33.

38 Fertilizer is not so important for rice, according to Furushima Toshio (1949: 234); Kurosaki Tadashi, "Ta ni mizu o hiku", in Tsude Hiroshi (1989: 71–73); Bray (1986: 16, 28, 48–50); and Sasaki Kōmei (1988: 314). On the problem of fertilizer, see Ogawa Naoyuki (1993: 142–175). See the maxim in Oka Mitsuo *et al.* (1990: 120). It seems surprising that in Rome, Pliny maintains that wheat does not need manure: see Ferdière (1988: 45). In the 17th–18th centuries, the *Kōka shunjū* 3 ge and 4: 197–204, *Hyakushō denki* 9: 130–136, and *Aizu nōsho* 1: 45 et seq., 57, 2: 161, recommend fertilizer for rice as well as for dry crops.

39 M 18: 4122.

40 RSK 8: 324–329. According to RSh, pp. 372, 373, the two forms of *uu* are interchangeable; and *hashoku* and *maku* are synonyms.

41 N Jitō 7.3.17. 693. Before that, Yūryaku 16.7. 472, and later, the code RR Denryō 16 stipulate the planting *uu* of mulberry trees. RSK 8: 327 Wadō 6.10.7. 713, and SN Reiki 1.10.7. 715; RSK 8: 327 Yōrō 7.8.28. 723. RSK 8: 328 Kōnin 11.7.9. 820, mentions the *yueling* calendar of the *Liji*. See also RSK 8: 327–328 Jōwa 7.5.2. 840: barnyard millet, common millet, wheat, barley, soybeans, red beans, sesame, and Kōnin 5.10.10. 814 in SY 70: 620. At Ise, in 1118, wheat or barley were sown on the 13th day of the 9th intercalary month: see Kimura Shigemitsu (1992: 193), who quotes Hi 5: 1983. Later, in 1501, wheat or barley were sown, in Izumi, at the beginning of the 10th month: see Kimura Shigemitsu (1993: 120), who quotes a document by Kujō Masamune. See buckwheat in RSK 8: 328 Jōwa 6.7.21. 839. *Mugi* means wheat and barley, but it is wheat in a 12th-century document in Hi 8: 3050: see Kimura Shigemitsu (1992: 192).

42 *Seiryōki*, pp. 24–48: Early foxtail millet is sown in the 2nd month (March) and late millet from the 4th to 6th month (May–July); in Shikoku, the cultivation of millet is especially important. See below, note 123.

43 ES 48: 981.

44 *Ruiju kokushi* 199: 377 Enryaku 19.4.12. 800; see Verschuer (1985: 156, 157, 450).

45 *(Sone no) Yoshitada shū* no. 130, beginning of the 5th month. The *Engishiki* 39: 880 has the early melon being sown in the 2nd month, the late melon at an unspecified time.

46 *Yoshitada shū* no. 206, middle of the 7th month; the editor of the Iwanami edition is, however, thinking of an autumn rice field that had been sown in spring *kami ni*. See *Yoshitada shū* no. 136: hemp sown *makishi asa* in the middle of the 5th month.

47 DNK 2: 43, 49, "Bungo no kuni no shōzeichō" of 737; *Kodai kayōshū*, p. 71, *aona*, and p. 47, *hajikami*.

48 ES 39: 878–881; see our translation above, pp. 27–30. For sowing, we follow the interpretation of Aoba (1991: 70–77). See also the water-leek in M 3: 407, and Péronny (1993: 118); and in M 14: 3415 and 3576. Some plants were "undersown" and the seeds covered with earth; others were "oversown" i.e. on top of the ridge: see Sigaut (1977: 142).

49 See the *guchūreki* in DNK 2: 571, 572, 3: 348, 4: 213, 216.

50 See Furushima Toshio (1949: 422–424); M 20: 4352 mentions *mame* growing by the wayside: see Péronny (1993: 103). M 14: 3451 and Péronny (1993: 27) (the poem translated). See also M 3: 404, 405, 14: 3364 (all *awa maku*), in Chapter 2 in this volume.

51 ES 39: 878 (wheat); DNK 1: 411 (rice, see below, note 57). Trombert (1995: 55) (wheat in the 6th century according to the *Qiming yaoshu* and in medieval Europe). Bray (1984: 287, 379) points out the number of seeds per plant and estimates the proportion of seed to the harvest in China as follows: 1:100 for foxtail millet, and 1:500 for rice in the 20th century; 1:10 for wheat and barley in the 12th century, and 1:6 to 1:20 in the 20th century.

52 To get an idea of the number of poems devoted to this topic, one has only to look at "Shinpen kokka taikan": *Horikawa-in hyakushu* 4: 26–225 to 240 (*nawashiro*), 4: 26–401 to 416 (seedlings *sanae*); *Fuboku wakashō* 2: 16–1865 et seq. (*nawashiro*); *Tango no kami ke hyakushu* 4: 28–99 to 106 (*yamada no nawashiro*), 4: 28–195 to 202 (*kadota no sanae*). The seedbed also appears in *Yoshitada shū*, pp. 52, 58, 62, 71 et seq. But this notion is less common at the time of the *Man'yōshū*: see M 4: 776 (*nawashiro*), 14: 3576 (seedbed of water-leeks), 14: 3418 (*sanae*).

53 *Yoshitada shū* no. 61, end of the 2nd month; *Sanbokuki kashū* 3: 106–162 or *Horikawa-in hyakushu* 4: 26–232 (poem by Toshiyori); *Kyūan rokunen hyakushu* quoted in *Fuboku wakashō* 2: 16–1869. *Muro* are also found in *Tango no kami ke hyakushu* 4: 28–103 (translated above, note 25), *Yukimune shū* 7: 43–207 (translated below, note 61). *Horikawa-in hyakushu* 4: 26–407, 414 and 327; see *Eiga monogatari* 35: 450, p. 23 Neawase.

54 *Horikawa-in hyakushu* 4: 26–227. Other poets mention the *tana-i*: Minamoto no Toshiyori (1055–1129) in *Horikawa-in hyakushu* 4: 26–232, Fujiwara no Tamemori? (d. before 1029) in *Tango no kami ke hyakushu* 4: 28–104; Nijō Tameshige (1334–1385) in *Tameshige shū* 7: 145–26; and Fujiwara no Tomoie (1182–?) in *Shinsen waka rokujō* 2: 14–632. These poems are also quoted in the modern agricultural treatises, cf. *Kojiruien* Sangyōbu 1: 58–61, and by Furushima Toshio (1949: 158–162). See *tanekasu* or *tanenurasu* in *Owari no gebumi* in 988, cf. Hi 2: 474.

55 Furushima (1949: 394, 395). The drying of the seed *tane toriage hoshite* is mentioned in *Akazome Emon shū* 3: 81–180, preface to the poem.

56 SN Yōrō 7.2.14. 723; otherwise there was seed distribution *myōshi* to the provincial officials, according to RR Denryō 34; and of the seeds of dry crops to the militias, according to RR Gunbōryō 62; and in the imperial vegetable gardens. We quote here and later the "Taihō Code" issued in 701, but in fact use the RR (*Ritsuryō*) edition of the "Yōrō Code" issued in 718: the code of 718 corresponds almost exactly to that of 701 and the passages we quote were all found in the 701 code.

57 DNK 1: 397, 399, 400, 411, "Yamato no kuni no shōzeichō" of 730. See Otomasu Shigetaka (1992: 103) for the early medieval and modern practices.

58 RSK 8: 324 and SNK Jōwa 9.3.9. or 11. 842, beginning of the 3rd month, concerning sowing and irrigation; RR Denryō 23.

59 *Chōshū eisō* no. 116, spring.

60   *Yoshitada shū* no. 100, beginning of the 4th month, no. 125, beginning of the 5th
     month. According to the commentary in the Iwanami edition, this refers to the
     intendant of a rice field belonging to a shrine estate.
61   *Goshūi wakashū* 1: 4–205, poem written in 1051 by Fujiwara no Takasuke
     (?–1083); *Yukimune shū* 7: 43–207, poem by Minamoto no Yukimune, of
     uncertain meaning. The subject of the hasty selection of seedlings is also
     found in *Horikawa-in hyakushu* 4: 26–401 to 416, etc.
62   *Horikawa-in hyakushu* 4: 26–413; see another poem about *yui* in *Tametada(kyō)
     senshu* 4: 43–237 by Reizei Tametada (1361–1417).
63   M 15: 3746.
64   *Chōshū eisō* no. 302, yorokobu uta.
65   RR Denryō 34; RSh, p. 375, mentions a decree of Yōrō 8.1.22. 724; RR Denryō
     30: when there are disputes about a rice field, the harvest goes to the person who
     transplanted it.
66   RR Kenyōryō 1 and RSh, p. 945. See also RR Gakuryō 20 and RSh, p. 460:
     music students receive a "field holiday" in the 5th month. The 5th month is
     also mentioned in SN Enryaku 7.5.2. 788, and in *Harima fudoki*, p. 309. In Owari,
     most of the farm work was done in the 4th and 5th months, cf. *Owari no
     gebumi*, Hi 2: 474. See Matsuo (1994: 141–151).
67   NR Tenchō 9.4.14. 832 *ueta no danjo*; and three decrees prohibiting saké in N
     Taika 2.3.22. 646; NK Kōnin 2.5.21. 811, farmers *nōjin*; RSK 19: 625 Enryaku
     9.4.16. 790, men of the fields *denpu*. *Shinsarugakuki*, see above, note 7.
68   *Harima fudoki*, p. 309. See also female transplanters in M 9: 1710.
69   *Eiga monogatari* 19: 110–113 Jian 3.5. 1023, *taue no hi*, with a description;
     Denryaku Eikyū 4.5.16. 1116, *ta uu*; *Hyakurenshō* Daiji 2.5.4. 1127, *taue no kyō*
     (spectacle) and 3.5.11. 1128, *taue*; *Chūyūki* Daiji 2.5.14., *taue no kyō*, with a
     description. *Kyō* (written as in *kyōmi*) means an amusing spectacle, according to
     Makura no sōshi no. 3. *Chōshūki* Daiji 4.5.10. 1179, *ueta*, with a description;
     Gyokuyō Jishō 3.6.6. 1179, *taue*; *Horikawa-in hyakushu* 4: 26–406 mentions
     female transplanters *otome*, 4: 26–401 to 416, "young girls of the fields" *tago*.
     A rustic festival *dengaku* brought out many nobles and minor officials in the
     capital of Heian in 1096; this was the Eichō no dai dengaku, cf. *Chūyūki*
     Eichō 1.6.17. 1096, and the *Rakuyō dengakuki* by Ōe no Masafusa (1041–1111).
     The *dengaku* is preceded in history by *tagaesu no rei* (see above, note 9); and from
     the 4th–5th centuries, by rustic dances *tamai*, cf. N Tenji 10.5.5. 671; RSh, p. 88
     (Gagakuryō); SN Tenpyō 14.1.16. 742, 15.5.5. 743, Hōki 8.5.7. 777 (in 743 and 777
     it is the 5th month); Kjg, p. 31; ES 7: 155 (at the time of the Daijōsai).
70   *Makura no sōshi* no. 226. Transplanting is not limited to rice; see above,
     water-celery and water-leeks.
71   Kjg, p. 31 and *Toyukegū gishikichō*, p. 61 *mita no tane oroshi*; see also "Shintō
     taikei", pp. 142, 265.
72   Furushima Toshio (1949: 109); Naoki Kōjirō (1968: 55), thinks that the sowing
     of rice was widespread at Ise.
73   Kinoshita Tadashi (1985: 130–146) quotes the *Kenkyū sannen Kōtaijingū nenjū-
     gyōji*, p. 380 *hashoku* and *maku*, p. 371 *hanshoku, maku* and *ta o uu asobi*;
     Kinoshita also quotes the *Toyukegū...*, p. 61 *mita o tagaeshi tsukuri uuru*, with-
     out realizing that *uu* here means 'to sow'. Kuroda Hideo (1984: 411) explains
     that this text was copied in 1291, then in 1464, then again with commentaries by
     an anonymous author in 1494; but that, even though the 1494 version is still in
     existence, the passage in question uses Heian-period terminology and is therefore
     a reliable copy of the original text.
74   Kjg, p. 3, and "Shintō taikei", p. 10 (reading *sokusa* instead of *nawagusa*); see
     also sowing and green manure in *Tango no kami ke hyakushu* 4: 28–103,
     translated above, see note 25.

75 M 10: 3603: willow branches may have been used to delimit an area, or driven into reservoir walls or into the ditches next to rice-field inlets for purposes of ritual or exorcism (see M 15: 3603 and 14: 3492). The purified seeds *yudane* correspond, according to some, to seeds soaked in tepid water *yu*, cf. Péronny (1993: 78). Cf. M 12: 2999; M 7: 1110 (translated here): in principle, newly cleared fields are dry. See also *makeru waseda* in M 8: 1624. On this point, see Abe Takeshi (1995: 51, 52).

76 Naoki Kōjirō (1968: 55); Kinoshita (1985: 141); Tateno Kazumi (1991: 175).

77 *Horikawa-in hyakushu* 4: 26–231 (500 *shiro* corresponds to quite a large field, compared to a small field of 10 *shiro*), mentioned by Kimura Shigemitsu (1993: 87). The small field associated with the seedbed *oda no nawashiro* appears in *Chōshū eisō* no. 18, 116; *Horikawa-in hyakushu* 4: 26–230, 229, 234, 238, 240; *Fuboku wakashō* 2: 16–1869, etc. *Kin'yō wakashū* 1: 5–2–74, poem about *nawashiro* that mentions *oda o uchikaeshi tane makite*. But *Shinsen waka rokujō* 2: 14–631, *oyamada no tanemaku* may refer to direct sowing in the rice field.

78 M 10: 2244, and Péronny (1993: 78); we should note that this poem is called "The irrigated rice field" *konata/suiden*. See Satō (2001: 126–134); but these DNA studies have since been called into question.

79 Kinoshita (1985: 140, 141); Takaya Yoshikazu (1990: 181–183); Sutō Isao (1989, 1994: 66–68), *tsumida*, p. 26, a photo of Shizuoka: a distinction was made between wetlands and marshy lands, the latter being suitable for transplanting, which is the case in Shizuoka. See Ogawa Naoyuki (1988: 89–111). Ogawa Naoyuki (1993), with a list of the regions where irrigated rice was sown directly in Japan in the 1900s. Kuroda Hideo (1984: 240–249). See also Kanai Tenbi, *Shitsugen saishi.*

80 See Kameda Takashi (1973); Kelly (1982). Tables of irrigation works are given in Kameda, "Kangai yōsui", in Toyoda Takeshi (1964: 141), and in Kimura Shigemitsu (1992: 130). Matsuo Hikaru (1994: 155, 156) has counted 800 projects in the annals.

81 *Nihon shoki* Jindai jō, 7th episode; there are different versions, summarized in *Kogo shūi*, 4th episode; the legend also notes three other misdeeds of Susanoo, relating to animal sacrifices. See extracts in Kjg, p. 3, and ES 8: 169.

82 *Harima fudoki*, Iibo-gun, p. 307. See the translation by François Macé, in Rotermund (2000: 49).

83 M 8: 1635 (Sahogawa), 3: 241 (The august lord, translation by Sieffert (1997: 245)), 14: 3492 (*ike*), 11: 2721 (*shigarami*), 11: 2833 (canal), 11: 2720 (water-pipe). See also M 4: 776 (*nawashiro mizu*). See Ishino Hironobu, "Man'yōshū to iseki", in Mori Kōichi (1984: 332–335). The same subjects appear in 9th to 12th-century poetry, as for example in *Yoshitada shū* no. 92, 545; *Dainagon Tsunenobu shū* no. 58, 140, etc. See below, note 247, for the noria.

84 See the Tōdaiji documents in DNK 4: 249, 410, 5: 536, 548–550. See other water-pipes in RSK 16: 507.

85 *Shinsen shōjiroku*, Ukyō kōbetsu jō N Buretsu 5.6. 502.

86 DNK Tōdaiji monjo 4: the maps of Minuma (751), Suka (767) and Naruto (767). See *mito* in WR 10: 15; RSK 16: 504, 507; *Nihon ryōiki*, jō no. 3.

87 Kaminogō Toshiaki, "Suiden kaihatsu to Mannō-ike", *Rekishi kaidō* 77, Sept. 1994, pp. 124–131. "Sanuki no kuni Mannō-ike kōhimon", ZGR 33 jō, p. 586, and *Heian ibun* vol. Kinseki mon no. 90; *Konjaku monogatari* 31: 22.

88 For the *guchūreki*, see DNK 2: 570–574, 3: 349–352, 4: 212–216, *ana o fusagu*; DNK 3: 351, 352, *mizo unate o tōsu, teibō o osamu*. Bruno Smolarz, personal communication, for foxes; see Lachiver (1997).

89 M 18: 4122 (an extract). Matsuo (1994: 155) has counted the number of prayer sessions. See other poems about rain in M 6: 1154, 12: 2299 (see above), 11: 2476

(see above), 18: 4123, 4124 (the second translated by Jean-Pierre Berthon, in Rotermund (2000: 88).

90 *Saigūki* 13; *Shingishiki*, pp. 220, 221; see *Heian jidaishi jiten*, art. "Kiu" with a bibliography. Apart from the prayers for rain, the agricultural year was dotted with rituals. Tateno Kazumi (1991), Kimura Shigemitsu (1993), Araki Toshio (1991) and others have attempted to reconstruct the annual celebrations in the rural areas.

91 M 14: 3501, and the translation by Péronny (1993: 160). See the list of weeds from the archaeological excavations in Kuraku Yoshiyuki (1991: 43–45). See barnyard millet in M 12: 2999 and 11: 2476 *hie erabu* (see above), but *kuzu hiku* in M 10: 1942 makes reference to the harvest. *Nōgyō zue*, pp. 129, 243 (for Aizu).

92 *Shinsen waka rokujō* 2: 14–637, poem by Fujiwara no Tameie (1198–1275).

93 *Shinsen waka rokujō* 2: 14–638, poem by Fujiwara no Tomoie. One finds *kusa hiku* in *Yoshitada shū* no. 100 (see above, note 60), and *majiru kusa* in *Gen dainagon ke utaawase* 5: 61–11, poem by Minamoto no Nagahisa dating from 1029, and in *Fuboku wakashō* 2: 16–2587. Poems of the early medieval period mention *tagusa hiku* and *tagusa toru*, see the many references in volumes 8, 9 and 10 of "Shinpen kokka taikan", in the CD-Rom with index published by Kadokawa shoten, a very useful index, though incomplete.

94 Furushima Toshio (1949: 410–411); Horio Hisashi (1990: 215). Reading through the administrative texts, one finds the term "remove the grass" *kusa nozoku/josō* in RSK 16: 501, but it relates to the mowing of grasslands. For China, see Ōsawa Masaaki (1996: 205).

95 See *kusagiru* in *Shinsen jikyō* Tenjibon, p. 465, in *Iroha jiruishō* and *Ruiju myō-gishō*. This term is not in the *Wamyō ruijushō*. The RMS gives a second character (*gei*), glossed *kusagi* in ShJ, p. 425. It is only in 1697, in the *Nōgyō zensho* 1: 5, that weeding is associated with the second tilling, which corresponds to the hoeing after sowing, called *joun/naka uchishi kusagiru*; *Kojiruien* Sangyōbu 1: 85–89. See *kuwa-suki* in ShJ. *Kusa kiru* refers to the cutting of grass; see above, note 36. See DNK 4: 505 (the nine old hoes).

96 ES 39: 878–881, translated above, and note 22. The Kujōkebon of ES glosses *kusagiru* for the character *gei*, but *gei* is read *uu* ('to plant') in SN Tenpyō 17.5.8. 745, and in ShJ Tenjibon: 453. See *kusagiru* written like *gei*, in *Kanke bunsō* no. 28, p. 127.

97 See the code RR Buyakuryō 9, Koryō 45, Kōkaryō 65; RSh, pp. 396–402; SY 60: 482–500; RK 173: 183–189; RSK 15: 429. Typhoons are not mentioned in the code, but for example in SN Taihō 1.8.21. 701, and Keiun 1.8.28. 704.

98 See the reports of locust invasions: SN Taihō 1.8.21. 701, 2.3.5. 702, Keiun 1.8.5. 704, Tenpyō shōhō 1.2.5. 749, Hōki 7.8.15. 706, NK Kōnin 3.6.5. 812, 6.5.14. 815, SJ Jōgan 16.8.13. 874, NR Kannin 1.8. 1017; RK 173: 182, 183. The legend in N Jindai jō 8th episode 6th version; ES 8: 169, 175, only mentions *haumushi no wazawai* and *takatsutori no wazawai*, the latter being interpreted by Torao Toshiya as the bird that flies high, in Torao (2000: vol. 1, p. 479). For locusts and ramparts, see Laurent (1999: 73, 101).

99 *Kogo shūi*, second-last episode "Mitoshi no kami". Insect plagues were the subject of the "Treatise on the fight against locusts" *Jokōroku*, 1826, by Ōkura Nagatsune. Cf. SJ Jōgan 16.8.13. 874, NR Kannin 1.8. 1017, and *Ruiju fusenshō* 3: 59 Kannin 1.8.3.

100 ES 39: 880.

101 M 12: 3000, translation by Jean-Pierre Berthon, in Rotermund (2000: 77); see also M 16: 3848.

102 *Bungo fudoki* p. 373; the editors gloss *ki* for palisade, but the *Man'yōshū* gives the reading *shigarami*; the early dictionaries read it *mazegaki*. See *Rōnō ruigo*, p. 176, for Tsushima.

103 M 10: 2265, 16: 3818, 11: 2649. *Fuboku wakashō* 2: 16–14898, which quotes the *Horikawa-in hyakushu*, poem by Minamoto no Akinaka (1059–1129).

104 Nomoto Kanichi (1984); and Nomoto (1984a: 320–325). The pungent deterrent was widely used for swidden-field crops. At Narada, Yamanashi prefecture, and at Hachiōji, Tōkyō, it was called *kagashi*, see Minzoku bunka eizō kenkyūjo (1987: 22); *Kojiruien* Sangyōbu 1: 172. The etymology proposed by Nomoto is therefore geographically limited.

105 One finds *kayaribi* in *Yoshitada shū* no. 160; *Chōshū eisō* no. 132; *Horikawa-in hyakushu* 4: 26–481 to 496; *Yukimune shū* 7: 43–212; *Fuboku wakashō* 2: 16 book 9 *natsu* 3.

106 M 16: 3856; the poem relates to the Indian monk Bodhisena, see Verschuer, 1985: 104–105.

107 *Chōshū eisō* no. 34 *hita*; and also no. 250 *naruko*. In *Ryōjin hishō* no. 332, and *Horikawa-in hyakushu* 4: 26–1516, the *hita* scares away deer. See *Kojiruien* Sangyōbu 1: 169–172. See Nomoto Kanichi (1993: 256–314).

108 M 7: 1353 and 10: 2219 *shime/nawa haeru*, M 8: 1634 *hikita haeru*. The terms *hita* and *shimenawa* are found together in *Dainagon Tsunenobu shū* no. 102. See a small field with *shime haete* in *Kin'yō wakashū* 1: 5–2–74, and a seedbed with *shime haete* in *Fuboku wakashō* 2: 16–1869 (translated above, note 53); *Horikawa-in hyakushu* 4: 26–226, 238; *Shinsen waka rokujō* 2: 14–633; and a small field with *hiku shimenawa* in *Akishino gesseishū* 3: 130–1381, by Fujiwara no Yoshitsune (1169–1206). *Horikawa-in hyakushu* 4: 26–743 writes *tsuna* instead of *nawa*. See Akimichi Tomoya (1995).

109 See the poems relating to *yamada moru* in M 10: 2156, 2219, 12: 3000, 11: 2649; *Yoshitada shū* no. 125, 159, 207, 220, 395; *Chōshū eisō* no. 250; *Horikawa-in hyakushu* 4: 26–1506, 1510, 1511, 1519; *Fuboku wakashō* 2: 16–4999 to 5025, 14504, 14311; *Shinsen waka rokujō* 2: 14–644, 648, 651.

110 *Yoshitada shū* no. 159, written early in the 6th month; *Kojiki* jō, p. 109.

111 M 10: 2247, 2117 (poem of obscure meaning) and 2133. M 10: 2117: *yukiai* means the meeting of two people like that of two seasons, here summer and autumn, at the beginning of the 7th month; but the lespedeza is an autumn plant. The subject of ears of grain in the autumn fields is also found in M 2: 88, 114, 8: 1624, 1625, 9: 1768, 10: 2246; *Yoshitada shū* no. 201, etc. See the subject of the harvest in M 7: 1275, 8: 1592, 1625, 9: 1635 (see above), 1710, 1758, 10: 2100, 2220, 2251, etc.; *Yoshitada shū* no. 199, middle of the 7th month. The *Man'yōshū* has six poems that associate the wild geese and autumn, and seventy-one poems that associate the lespedeza and autumn; see Péronny (1993: 58).

112 RR Kenyōryō 1 and RSh, p. 945, RR Denryō 2 and RSh, p. 346, RR Buyakuryō 22 and RSh, p. 422, for the times of the year.

113 Hi 3: 878 for Kuroda shō; DNK 5: 243 for Echi. For the property in Ōmi, see an inscribed tablet, Shiga-ken kyōiku iinkai ed., *Kamo iseki*, 1980, and *Mokkan kenkyū* 5, 1983, after Tateno Kazumi (1991: 182). See Kimura Shigemitsu (1992: 352, 353) for a table of documents concerning the harvest in the 11th–12th centuries; Tateno (1991: 181–186), mentions other examples: DNK 2: 62 (harvest of the 1st day of the 9th month on the 9th day of the 12th month! in 737, in Tajima), DNK 4: 507 (a certain Kamo no Umakai wants to begin the harvest on the 27th day of the 8th month in 761, i.e. 1st October).

114 The three poems: *Yoshitada shū* no. 197 and no. 232; *Udaijin Kanezane utaawase* 5: 164–36, dated Angen 1.10.10. 1175, also in *Fuboku wakashō* 2: 16–7242. Poem no. 197 has *nogi wa uchi* or *nogi wa ochi, nogi* being the beard or glume, but the meaning here is unclear. We have already met early rice in M 10: 2117, 8: 1635, *Fuboku wakashō* 2: 16–1869, *Yukimune shū* 7: 43–207, and late rice in *Yukimune shū* 7: 43–315; see here notes 53, 61, 83, 135, 157. We also find early rice in M 8:

1625, 9: 1768, 10: 2220, 2251, 14: 3386, *Horikawa-in hyakushu* 4: 26–235, 407, 414, *Yoshitada shū* no. 201; and late rice in *Shoku kokin wakashū* 1: 11–457 (*oshine no okute*), *Chōshū eisō* no. 127, 219, *Fuboku wakashō* 2: 16–5006, 5030, 14514, *Horikawa-in hyakushu* 4: 26–232, 236, 1511, *Shinsen waka rokujō* 2: 14–645, *Sanbokuki kashū* 3: 106–989, and 162 (see note 53), *Kokin wakashū* 1: 1–842, *Shoku goshūi wakashū* 1: 12–330, *Shoku gosen wakashū* 1: 10–388, *Gyokuyō wakashū* 1: 14–750, *Shin'yō wakashū* 1: 22–348. See the discoveries of inscribed tablets in Hirakawa Minami (1999).

115   See Terasawa (1986, 1991); Sigaut (1978).

116   See above, pp. 53–56. Terasawa Kaoru (1991: 50–69); Furushima (1949: 114); Matsuo (1994: 163) and Shiraishi Akiomi (1993: 83) think that high cutting was widespread between the 8th and 10th century; Terasawa (1986: 330, 331) and Tateno (1991: 183) think that low cutting was the norm at that time. Naoki Kōjirō (1968: 61) and Abe Takeshi (1995: 54) refrain from associating the height of the cut with the method of storage. Terasawa associates low cutting with transplanting that was usual from the 3rd century. See below, note 146, for the pounding of stalks and ears.

117   *Makura no sōshi* no. 227. See above for its account of transplanting at Kamo. The interpretations vary: does this account describe a widespread or an exceptional practice?

118   See the cut ears *kariho* in *Chōshū eisō* no. 308; *Daini Takatō shū* 3: 71–360 (Fujiwara no Takatō, 949–1013); *Fuboku wakashō* 2: 16–5023. *Kariho* is associated with its synonym meaning 'temporary hut' in *Gosen wakashū* 1: 2–295 and 302; *Shinsen waka rokujō* 2: 14–651 to 655. See also *Sōkonshū* (15th century) 8: 10–9802.

119   ES 50: 995; *Nihon ryōiki jō* 33; *Shinsen waka rokujō* 2: 14–646 (translated here); *Fuboku wakashō* 2: 16–14496; see also *Fuboku wakashō* 2: 16–12875, *Tametada senshu* 7: 41–268; see *ochibo* also in *Sōkonshū* 8: 10–3732, 4537, 4541, 9799, 9802 (*kariho*), 9809, 9823, 9827; and other references in "Shinpen kokka taikan", CD-ROM, Kadokawa shoten. See the modern use in *Nōgyō zue*, pp. 160, 227, and *Kōka shunjū*, p. 31. Araki Toshio (1980: 262) and Abe Takeshi (1995: 61) consider that gleaning was a privilege of the poor.

120   See the rituals of the 2nd and 9th months in Kjg, pp. 28, 31 ("Shintō taikei", pp. 130, 143); *Kenkyū sannen Kōtaijungū nenjūgyōji*, p. 438 (14th day of the 9th month); these rituals are omitted in *Toyukegū gishikichō*; there is *nukiho* for these rituals, but harvest *karu* in the 2 *chō* 4 *tan* of shrine rice fields, cf. Kjg, pp. 28, 42. See the Great Thanksgiving in ES 7: 144, 145, 11: 334; *Gishiki*, pp. 81–86; and *Midō kanpakuki* Chōwa 1.8.17. 1012, in F. Hérail, 1988, 2: 614, concerning the messengers *nukiho no tsukai*. There is a shrine called Nukiho jinja, cf. ES 3: 61. See another *nukiho* ritual in *Hyūga fudoki*, p. 523. The legend is found in *Harima fudoki*, p. 271. Furushima Toshio (1949: 110) thinks on the one hand that *nukiho* refers to high cutting with a knife, and on the other hand that this refers to harvesting by hand(?). Matsuo (1994: 163) also confuses *nukiho* with high cutting. For the rice spirit, see Takaya (1990: 166). But Bray (1986: 20, 1984: 328) points out that in Malaysia, the knife, unlike the sickle, did not offend the rice spirit, because it was held in the palm of the hand. M 4: 520 mentions the picking *tekari* of hemp.

121   *Yoshitada shū* no. 135, middle of the 5th month, poem translated below (note 140). Kimura (1992: 193–196, 350, 351) quotes Hi 4: 1402 (Yamato in 1101), Hi 7: 2763 (Yamato in 1170), Hi 8: 3031 (Kii in 1181?), Hi 7: 788–807 (received in 1048–1049). See Sonoda Kōyū, pp. 240–306. Kimura (1992: 350) also mentions a document from Izumi of 1502, 29th day, 4th month, the date on which "people harvest the wheat and transplant the rice fields". *Mugi* means wheat and barley, but wheat when making noodles *muginawa*, Hi 8: 3050.

122 RSK 19: 612, 613, Tenpyō shōhō 3.3.14. 751, Daidō 3.7.13. 808, Kōnin 10.6.2. 819. *Nihon ryōiki* chū 10 (the legend). For the guards, see ES 48: 981.

123 *Masaki monjo* Ōei 34.4.11. 1427, after Kimura (1993: 97) (foxtail and barnyard millet). RSK 8: 328 and SNK Jōwa 6.7.21. 839 (buckwheat). *Nōgyō zue*, pp. 46, 89, 95, 127, 129, 137, 138, 140, 167.

124 Kimura (1992: 350, 351); *Owari no gebumi* in Hi 2: 479; ES 39: 879, 881; *Nōgyō zue*, pp. 67, 174, 175; ES 39: 873 (the imperial kitchens; see Chapter 3 in this volume); *Shisseishoshō*, p. 463; ES 39: 878–881, the imperial vegetable gardens, see our translation above. *Fuboku wakashō* 2: 16–923; see Chapter 2 in this volume.

125 ES 39: 878–881; the text omits the harvesting of the other vegetables.

126 DNK 2: 43, 49; gromwell *murasaki no kusa* was used for dyeing.

127 The *Man'yōshū* has 24 poems about the harvesting of wild rice *makomo*; see Péronny: 94; the *Yoshitada shū* has four: no. 85 (end of the 3rd month), no. 118 (end of the 4th month), no. 126 (beginning of the 5th month), no. 485 (spring). *Chōshū eisō* no. 142 (autumn): in this poem, eulalia is growing thickly in a plain of sweet miscanthus *ogi*; eulalia appears in eight *Man'yōshū* poems, see Péronny (1993: 154). The harvesting of eulalia *karikaya* is the subject of a number of poems in *Horikawa-in hyakushu* 4: 26–641 to 656, and in *Fuboku wakashō* 2: 16–4444 to 4456. See Chapter 3, note 90.

128 *Kōka shunjū*, p. 220 (and p. 29). The *Nōgyō zensho* 2: 91 quotes this passage and also explains how to build drying-racks. The *Seiryōki*, p. 171, gives two methods: drying loose sheaves *kari-boshi* on the ground, and drying bound sheaves *kari tabane* on the ground; elsewhere it mentions that some free-running horses have eaten rice suspended in the field; after Furushima Toshio (1949: 275).

129 *Hyakushō denki* 9: 141. According to Sigaut (1978: 152), the term 'small bundles' is used to refer to the overall number of ears cut and bound. See an example below, note 140.

130 *Nōgyō zue*, pp. 148–158, 252–256; *Seikei zusetsu* book 5; and the iconography of the 18th–19th centuries in Satō Tsuneo *et al.* (1996: 1: 32, 87, 149, 171), these are straw silos; (pp. 30, 84, 170), suspended drying; (pp. 75, 79, 168, 170), rice silos.

131 RSK 8: 324.

132 SY 53: 302.

133 ES 8: 159, 162–166, 171; *Fuboku wakashō* 2: 16–8421.

134 Kōno Michiaki (1997), sees a gradual change from drying in the courtyard to drying in the fields. He bases this on the decrees of 841 and 903, but also quotes the early 12th-century poems that mention the two drying methods. In fact, this change occurs after the disappearance of the high cutting of rice, i.e. after the 12th century (Kōno, personal communication). Hotate Michihisa (1999) quotes *Denryaku* Eikyū 3.9.21. 1115, as an example of a silo of "stacked rice" *ine o tsumu*. He puts forward the hypothesis that silos could also be used for storing rice on the stalk. However, he confuses, as do others, the rice stacked in granaries (in ears) with the rice stacked outside, and quotes the title of the poem in *Chōshū eisō* no. 308 (*ine o tsumu*) and not the text of the poem (*kariho o tsumu*). *Harima fudoki*, p. 291. *Fuboku wakashō* 2: 16–1868, and *Daisaiin saki no gyoshū* 3: 76–188. Orikuchi Shinobu (1953).

135 Inazumi in Ikebe Wataru (1981: 751, 760); and Inazumiyama in *Izumo fudoki*, p. 209. The song, in *Chōshū eisō* no. 285, refers either to *karitsumu*, or *kaketsumu*, depending on the manuscripts, though the second term, "suspend and stack" makes no sense. See *karitsumite* in the poem (translated here) by Ōshikōchi no Mitsune (?–?), dating from 917, in the Kasenkeshūbon version of the *Mitsune shū*, after Ienaga Saburō: 43, but *Mitsune shū* 3: 12–154 Nishihonganjibon (in "Shinpen kokka taikan") writes *karihoshite*. See *karitsumu* in *Fuboku wakashō* 2: 16–14840 and *Kintō shū* 3: 80–237.

136   *Chōshū eisō* no. 308; and Hotate (see above, note 134). See Hozumi in Ikebe Wataru (1981: 239, 269, 306, 430, 600). It is also a family name and a shrine name, in ES 9: 216. The "Izumi gen shōzeichō", DNK 2: 81 et seq., mentions the height *tsumidaka* of the ears (*ei*, not *ine*) stacked in the granaries. Inazumi has the meaning of silo in the early dictionaries IJ and RMS.

137   *Gōke shidai* book 4, p. 128; and the interpretation of this passage in *Kojiruien* Shokubutsuhen 1: 801. But the poem M 16: 3848 mentions that the rice is stored in granaries *ine o kura ni ageru*! Here *ine* seems to be used as a generic term. In the registers of resources, *ine* is a generic term that designates rice in granaries; according to these registers, *ine* includes grains *momi* and ears *ei*, but not rice on the stalk; cf. DNK 1: 397–411, 2: 75–97, etc. Hi 5 no. 1958, etc. also distinguish between ears and stalks.

138   Shibusawa Keizō, *Nihon jōmin seikatsu ebiki*, 1984, 2: 102, 112, 117, 118, 136, 176.

139   *Horikawa-in hyakushu* 4: 26–1518, poem by Higo no naishi (Takashina no Motoko); and 4: 26–1517, by Ryūgen. See also *oshine hosu* in *Shoku gosen wakashū* 1: 10–388; *karihosu ine* in *Fuboku wakashō* 2: 16–14833.

140   *Yoshitada shū* no. 135, middle of the 5th month, with the obscure expression: *kudakete mono o omou koro kana*. The version of this poem in GR Waka 15, p. 292, reads *hata/hada* instead of *hade*. See *hate* in *Wakun no shiori*, chū, p. 812.

141   *Fuboku wakashō* 2: 16–5024, by Minamoto no Tsunenobu (1016–1097), and in *Tsunenobu shū* no. 104. See *inaki* in IJ entry under I zatsu. This term is not found in the other dictionaries WR, RMS and ShJ. See Inaki in WR 6: 14, ES 9: 220, and the family Inaki no mibu of Owari, in Saeki Arikiyo, *Nihon kodai shizoku jiten*. See the images of sheaves suspended from trees in *Kokawadera engi* and *Shigisan engi* (12th century), in Shibusawa Keizō, 1984, 1: 108 and 3: 25. A poem in *Sanbokuki kashū* 3: 106–538 mentions buckwheat suspended from a willow tree in the 9th month, in these words: *somamugi to yu mono o kaketaru*; see also Kimura Shigemitsu, "Waka no naka no minshū seikatsu", p. 2 (an article in a newsletter without references).

142   For all this, see Sigaut (1989 and 2008), Mane (1983), Comet (1992) and Perrine Mane (personal communication) for beating off, Yoshio Abe (personal communication) for the second hulling and Lachiver (1997). For the iconography, see Kōno Michiaki (1996: 21, 25 et seq.); Horio Hisashi (1990: 145, 146); and Yoshio Abe (2007).

143   Leroi-Gourhan (1973: 152).

144   Furushima (1949: 110–115, 167, 318); Naoki Kōjirō (1968: 62); Terasawa Kaoru (1991: 62); Matsuo Hikaru (1994: 168); Horio Hisashi (1990: 247); *Nihon kokugo daijiten*, "inatsuki".

145   Shizuoka-ken maizō bunkazai chōsa kenkyūjo ed. (1994).

146   M 14: 3459; see also pounding in M 14: 3550.

147   RSh, p. 347; DNK 1: 327 fragment of a *momiyone chō* dated by the editors of DNK to Yōrō 7 (723); RSK 8: 349–351 *kachishine* (gloss of the manuscript of 1266); and ES 26: 654, gloss *momi*, in the Kujōkebon copy of ES dating from after 927 (see Shikanai Hirotane).

148   *Harima fudoki*, p. 329.

149   DNK 1: 398, 2: 430–439. Several registers of provincial resources record losses after winnowing/sieving *hifuruu*, *hiru*, *furuu* of rice and foxtail millet *awa*, in DNK 1: 430 et seq., 454 et seq., 2: 12–19, 41, 45, 47, 52, 70, 72, 80–88, 106, 120, 143.

150   See *shōmai* in RGi, pp. 49, 50; ES 26: 655, 23: 584 (list of 22 provinces that deliver it), DNK 1: 398 (*shōmai* of Yamato).

151   The pounding ratio is stipulated in RSh, p. 345, RGi, p. 107; see cases of the application of this ratio (according to our calculations) in DNK 1: 419 *nenryō hakumai*, 1: 607 *nenryō shō hakumai*, 1: 608 *sekimai*, 2: 56 *shō hakumai* and

shōmai, DNK 1: 435 threshed rice. But immature rice gives only 3–4 *shō* of hulled rice per sheaf, in DNK 4: 255.

152 See the reading *hakumai* in IJ, the reading *shirayone* in ShJ; the readings *kokumai* and *kuroyone* are not attested by the texts.

153 Black rice is normally eaten by subordinates: ES 6: 136, 14: 407, Hi 1: 383, DNK 5: 5–33. Since slaves were given either 1.2 *shō* of *hakumai* or 2 *shō* of *kokumai* (DNK 5: 32, 33, 89; Hi 1: 383), it may be supposed that the latter was pounded less. According to ES 23: 585, *hakumai* is sent to the Ōiryō, *kokumai* to the Minbushō and Kuraryō of the Nakatsukasashō. See other deliveries sent to temples, in ES 26: 656–658. For the management of *hakumai*, see RSK 14: 407, RGi, p. 50, RSh, pp. 129, 347, and ES 23: 584, 585.

154 See the costs of husking *shirage-shiro* in ES 26: 655 (2 *soku* paid to obtain 5 *to* of *hakumai*), ES 23: 585, DNK 5: 247, 15: 223, 250, Hi 1: 374, 3: 825, and *Tōdaiji yōroku* 6: 213, 220, 221; and *shiragetaru* in *Shūi wakashū* 1: 3–91, poem of the 4th month by Ōshikōchi no Mitsune, dating from the Engi era (901–923) and *Utsuho monogatari* Fukiage jō 1: 340, 342. Fujii Kazutsugu (1986: 377–378) reproduces the text of the *mokkan* relating to a *shōen* located at Fujiwara-kyō, dating from 810, which gives detailed management figures; he quotes Nara kokuritsu bunkazai kenkyūjo ed., *Fujiwarakyū shutsudo mokkan* 6.

155 ES 35: 803 concerning the Ōiryō with its pounders. I thank Kōno Michiaki for sharing his interpretation of the large tubs. *Yokousu/yokusu* are mentioned in N Ōjin 19.10. For Europe, see Lachiver: 1125. In Izumi, in 738, the ears were pounded until hulled rice was obtained; in the granaries at that time were 130 sheaves of straw without ears, cf. DNK 2: 87. For this reason we think that the ears were separated from the stalks before pounding, and that whole stalks did not go into the mortar, contrary to what Furushima and others think (see note 144). Pounding was also done all at once at Ise: on the 14th day of the 9th month, the collected ears were pounded in preparation for drying, cf. *Toyukegū gishikichō*, pp. 47, 55. But pounding is deferred temporarily in the poem quoted above of the 4th month; and in *Nihon ryōiki* jō 2: the pounders *inatsukihime* hull the rice in the 2nd and 3rd months, even though harvesting had been done in the preceding 8th or 9th month.

156 *Makura no sōshi* no. 99.

157 *Yukimune shū* 7: 43–315 *koku* (Furushima (1949: 168), quotes another version of this poem); *Kokin wakashū* 1: 1–932 (*kokitaru* associated with rice). Modern dictionaries do not mention the *Kokin wakashū* poem, but Furushima (1949: 254) and Naoki (1989: 73) associate it with threshing. See Furushima (1949: 114, 167, 254), on the subject of *koku*. The verb *kokitaru* is also found, with tears, in *Kokin wakashū* 1: 1–639, *Gosen wakashū* 1: 2–874, *Sanbokuki kashū* 3: 106–989 (tears and rice), *Shoku kokin wakashū* 1: 11–456 (dew). See *kokichirasu* in *Kokin wakashū* 1: 1–922, and *kokishikeru* in M 20: 4453 (lespedeza). I thank Kōno Michiaki for providing these references.

158 Furushima (1949: 254–256, 316–319).

159 Yin Shaoting, *Yunnan wuzhi wenhua*, pp. 534–545; and Kōno Michiaki, personal communication.

160 *Hyūga fudoki*, p. 523, after the *Shaku Nihongi* (ca. 1302), book 8. Sigaut (1978: 149).

161 WR 17: 3. See roasted rice as food in ES 39: 868; *Azuma kagami* Genkyū 1.5.8. 1204; *Utsuho monogatari* 3: 230; DNK 6: 84; and in the other early dictionaries: IJ entry under *ya*, section on "food"; RMS, p. 789.

162 Kōno Michiaki (1998); and personal communication. Kōno explains that here "rice on stalks" in fact means 'ears', which leads him to suggest hulling by hand, implements such as threshing-sticks being reserved for stalks. See *koku* in RMS, and *inekoku* in ShJ; see below, note 250.

163   *Makura no sōshi*, no. 99; Kōno Michiaki (1998); Fujii Kazutsugu (1997: 18) (who, however, reads *hikiusu* instead of *suriusu*), quotes DNK 2: 573, 3: 350 (*guchūreki*). Kōno Michiaki, for *tengai*. Abe Takeshi (1998: 133) quotes "Chikaraōmaro tabatake kazai yuzurijō" dating from Kōchō 1 (1261). For the equipment in the 15th century, see Kuroda Hideo (1983: 50); see the poem by Shōtetsu, *alias* Komatsu Masakiyo (1381–1459), in *Sōkonshū* 8: 10–3719.

164   *Hyakushō denki* 9: 143; Furushima Toshio (1949: 315–323, 413–416); Masuda Shōko (1990: 178) (*temomi*); Ferdière (1988: 64); Bray (1984: 382).

165   ES 39: 878–881, which glosses *katsu*, according to a late manuscript; ES 39: 871, 2nd line (imperial kitchens); DNK 13: 277, 305 (pounding *tsuku* of wheat); Nara kokuritsu bunkazai kenkyūjo ed. (1989, 21: 27, 1990, 23: 9). Shizuoka-ken maizō bunkazai chōsa kenkyūjo (1994); Watanabe Makoto (1985).

166   Verschuer (1993: 9, 10, 16, 18, 31, 42, 46).

167   Terasawa Kaoru (1991: 60–69).

168   M 9: 1710. Granaries also appear in M 16: 3848 (translated below, see p. 148; see also p. 141) and M 16: 3832, a difficult poem to interpret.

169   RR Denryō 1; SN Yōrō 3.6.16. 719; ES 26: 643–652 and 26: 671–685 (ears and grains). Matsumura Keiji (1983), for the types of storage mentioned in the registers. Bray (1984: 328, 345, 382); and Sigaut (1981: 166).

170   See the registers in DNK 1: 389–474, 607–628, 2: 9–153. Murao Jirō (1961). The granaries store other foods, such as wheat. See winnowing/sieving *furuu, hifuruu, furuire* in DNK 1: 430 et seq., 454 et seq., 2: 12–19, 41–52, 70–72, 80–84. The regulation tolerated bigger losses *utsukasu/kō* (wastage) for older grains, cf. RR Sōkōryō 7, DNK 1: 398, 430; NK Enryaku 17.9.20; NR Kōnin 11.11.7. Sawada Gōichi (1978: 401), and Murao (1978: 173), are silent on the losses due to sieving. See Sigaut (1981), on the subject of dried rice; and Sigaut, personal communication, for the storage of grains with their husks.

171   See note 170, and Murao (1978: 131 et seq.). ES 26: 684 points out that *hōsō* in the form of *azekura* and *itakura*, as well as *kawarakura* and *tsuchikura* were the norm. See the temple granaries in DNK 2: 651, Hi 1: 41, 170–171, 191. See two *hōsō* in DNK 2: 84, 95, and one *nagakura* in Hi 2: 170; one *hiwadabuki* in DNK 25: 304. See also an analysis of granaries by Takei Noriko, 2015.

172   DNK 2: 75–97 with the granary mentioned, 2: 81. Murao (1978: 169), with a table of all the granaries. The appearance of the opening *soko* is unclear; Murao (1978: 153). Is it a closing cover in the floor?

173   Hi 1: 308–316; and an analysis by Murao (1978: 154–175). One finds *azekura, itakura* and *kuraya* with *soko. Tsutsumi* are used in Iga, cf. DNK 1: 429–439. DNK 25: 304 (*mame*). See above, note 136, for stacked ears.

174   SN Wadō 7.4.26. 714. See two larger granaries in Izumi, DNK 2: 84, 95. ES 22: 559–597 classifies the provinces into *jō, dai, chū* and *ge*. Izumi is *ge*, and Etchū is *dai*.

175   DNK 2: 75–97 (Izumi). In Yamato, they distinguished between stocks of fewer than three years, from three to five years and more than five years, DNK 1: 397–411. In Etchū, cf. Hi 1: 308–316, they recorded in 807 stocks dating from 758 to 766; in 751 stocks dating from 733 to 750; in 757 stocks from 750 to 753; in 771 stocks from 758 to 771; and in 909 stocks from 751, 818 and 830. The old stocks are mentioned in most of the registers in DNK 1 and 2.

176   RR Sōkōryō 6, 7; RGi, p. 268; *Enryaku kōtaishiki*, pp. 5, 6, 9, 10; RSK 8: 349–350. ES 26: 655 stipulates the replacement of damaged rice.

177   Kurosaki Tadashi (1989: 71–73); Bray (1986: 16, 28); Takaya Yoshikazu (1990: 208). See Hattori Hideo (2003); Hotate Michihisa (2004); Sigaut (2008). For the archaeology, see Chapter 4, pp. 208–210 and note 47 in the same chapter.

178   The *Aizu nōsho*, p. 217, recommends *kataarashi* for dry fields and does not mention it for rice fields, pp. 213–216. See the site of Kuroimine, in Noto Takeshi (1991: 94). *Wakun no shiori*, 1898 edition, *jō*, p. 478.

179 RR Denryō 3, 29, and RSK 15: 441; RSh, p. 348; RGi, p. 107.

180 RSK 15: 433, and RK 159: 120, for the decrees of 821 and 827.

181 *Kōninshiki*, p. 18; ES 26: 654, 27: 690, mention that *gegeden* fields carry 150–300 sheaves of rice. During droughts, the government granted tax exemptions, cf. RSK 15: 429, 430. The Japanese legal "Code" *Ritsuryō* was inspired by one of the codes of Tang China (618–907); this code was promulgated when the Chinese capital was in Chang'an (modern Xi'an) located in northern China where millets and other dry crops were widespread.

182 *Shūgyokushū* 3: 131–225, poem by Jien (1155–1225); *Shinsen waka rokujō* 2: 14–543 (and *Fuboku wakashō* 2: 16–10130), poem by Fujiwara no Tomoie. See another poem in *Tamemasa senshu* 4: 43–965; see above, note 108, for the rope.

183 Hi 1: 160–164, 176–178, 207–210, 183–194 et seq., surveys of the Tōdaiji lands in DNK 4 and 5; examples from the 10th–12th centuries in Kimura Shigemitsu (1992), and Kuroda Hideo (1984). The same applied in Europe; see Duby (1977). Kimura Shigemitsu (1996: 111–112) estimates the waste lands at 20% in the 12th century.

184 Toda Yoshimi (1967: 178–184); quoted by Kimura (1992: 12–14, 233); by Kuroda (1984: 66, 67, 121); and by Hattori Hideo (2003), see note 177. *Wakun no shiori* (zen hen 14) is also mistaken about another term, *ta-i*, which does not mean a field hut, but a rice field.

185 Sigaut (1977: 148–164); *Aizu nōsho*, p. 217.

186 Hi 7: 2531; RSK 15: 427. Sigaut (1977: 264); Kuroda (1984: 264), mentions another example of the failure of the irrigation systems.

187 Hi 2: 484.

188 See above, pp. 15–17.

189 Sigaut (1977: 154, 155), for long- and short-term idling; Satō Yōichirō (2001: 126–151); see also the end of Chapter 2 in this volume, pp. 158–159, 168–182.

190 Kimura Shigemitsu (1992: 115–122, 188–198, 256, 257); Kuroda Hideo (1984: 64–70, 261–269); Furushima Toshio (1949: 234–241); Kawane Yoshiyasu (1965), reprinted in Kawane, *Chūsei hōkensei seiritsushi ron* 1972; Ōishi Naomasa (1964: 179–183, 189–191); Isogai Fujio (1988).

191 Hi 2: 478; Kimura (1992: 200), interprets article 15 of this document as showing the harvesting of rice and wheat; he also quotes article 30 (Hi 2: 484) according to which the rice farmers have all absconded and only a small number of dry-field farmers remain.

192 The inhabitants of Owari were no doubt referring not to the second annual harvest, but to the principle of non-imposition for dry crops in general. According to Kimura (1992: 201–205), a tax began to be levied on dry fields from 1009, in the case of an estate of the Eizanji in Yamato, if one examines the exemption certificates for this estate dating from 990 to 1021; cf. Hi 2, no. 341, 359, 443, 451, 471, 478, 484. Kimura (1992: 194, 351) also analyses the tax receipts of another estate on dry crops in 1048–1049, in Hi 3: 788–807; Kimura (1992: 120) notes the *jishi* levied on dry fields in Yamato in 1062, in Hi 3: 1027; and Kimura (1992: 205–214) mentions other examples of *jishi*. The dry-field tax therefore dates at the latest from the 11th century, but an earlier case of *hatake jishi* can be found in Awa in 844, in Hi 1: 67.

193 Hi 5: 1683, see Kimura (1992: 193), who follows the interpretation of Kuroda Hideo (1984). The date of the document is probably read *uruu kugatsu*, 9th intercalary month, 1118.

194 Hi 5: 1727 (in 1123), 5: 1964 (in 1135), 6: 2169–2174 (in 1145), also 5: 1757 (in 1124), 7: 2613 (in 1164); and the interpretations of Kuroda (1984: 263–266) and Takashige Susumu (1975: 17 et seq. and p. 173 et seq.).

195 Hi 7: 2512–2514 (Yuge), 5: 2028, 7: 2501–2502 (Ichii), and Kimura (1992: 188–190); Hi 6: 1464 (Kataoka) dating from 1159; and Kimura (1992: 191).

196 Hi 7: 2742, 8: 3050; *Kamakura ibun* no. 2107, and Kimura (1992: 192, 193), who quotes the *Tsuika hō* art. 45.

197 *Shinpen tsuika*, quoted by Furushima Toshio (1949: 235), and Kimura (1992: 115).

198 Kimura (1992: 256) mentions the gift of a rice field producing two crops at Kōyasan in 1221; p. 350, he notes that "they cut the wheat or barley, then transplant the rice", on the 29th day of the 4th month 1502, on the Hine estate in Izumi; p. 115, he mentions the gift of 2 *tan* (0.6 acres) at Kōyasan in 1276, which give harvests of rice and wheat/barley.

199 RSh, p. 361, commentary of RR Denryō 19.

200 Kawane Yoshiyasu (1965); Kimura (1992: 116).

201 Isogai Fujio (1988), however, is interpreting the decrees relating to dry crops, in RSK 8, as texts that encourage double-cropping in rice fields, although they only concern dry fields; see also note 228 below. Sigaut, personal communication, for Egypt and India. Ōsawa Masaaki, "Ine mugi no nimōsaku ni tsuite", in Ōsawa (1993: 74–81). Kuroda Hideo (1984: 249) mentions the aspects of tillage and manuring.

202 DNK 7: 44–50 (Gufukuji). Takashige Susumu, "Jōrisei shita no kōchi", *Hiroshima daigaku bungakubu kiyō* 12, Sept. 1947; reprinted in Takashige (1975: 7–83); and Ōishi Naomasa, in Toyoda (1964: 182–183), think that crops were rotated, whereas these are converted fields. For field conversions, see Kuroda Hideo (1984: 78–79, 243, 263–265); and Kimura Shigemitsu (1992: 119–122, 139, 311–314).

203 Takaya Yoshikazu (1990: 176–179).

204 *Nōgyō zensho*, chap. "Nōji sōron", kōsaku 1. See Oka Mitsuo *et al.* (1990: 119, 120).

205 Sasaki Kōmei (1971: 260–262).

206 Bray (1986: 117); Takaya (1990: 208).

207 Hi 1: 74 (in 846), Hi 1: 176–178, 207–210 (Kōryūji), Hi 1: 281, 284 (Kanzeonji). One finds *tabatake* with the meaning of rice fields and fields in DNK 3: 46; Ni chū: 412; Hi 1: 115, 324, 358, 359, 378, and many other references in Hi.

208 For the Eizanji, see above, note 192. For Hirose, see Hi 3: 1027, 1028; for Hida, see Hi 7: 2743, 2744; Kimura Shigemitsu (1992: 119–122).

209 Kimura (1992: 120–122, 139), supports his hypothesis with a document relating to the Kuroda estate of the Tōdaiji, dating from 1214, in *Kamakura ibun* no. 2107.

210 See Chapter 2 for land clearing. Hi 1: 322 (in 915).

211 Kuroda (1984: 263–266); see *mizu no tayori*, and other quotations in RSK 15: 427; RR Kyūmokuryō 62; RSh, p. 163; Hi 1: 288, 2: 563, 7: 2531.

212 Hi 1: 332 (in 928); RK 159: 113 (in 830).

213 RSK 8: 328 (in 840). See Tabatake in Hi 1: 324; ES 9: 211; see Hatakeda in Hi 1: 40, 96, 281, 353, 3: 1096, 9: 3628, 10: 3900; ES 5: 116; and in the Gufukuji map in DNK 7: 47, 48, and the Minuma-shō map in DNK Tōdaiji monjo 2: 348.

214 Many land surveys record rice fields, dry fields and uncultivated plots side by side. We wanted to give precise details about waste ground and field conversions to help with the interpretation of these documents, as misunderstandings persist (e.g. Kimura (1992: 120, 122) sees a deliberate alternation of dry and irrigated crops). See Kitamura (2012: 43–48).

215 Farris (2009: 73). Educational textbooks such as *Shōsetsu Nihonshi zuroku*, Yamakawa shuppan (2008: 122) mention draught tillage, intensive fertilization and double-cropping, but we find their explanations inadequate, as we will show later. Takagi Tokurō (2012: 30–33) gives a brief overview of the opinions for and against advances in the late medieval period and deplores the lack of proper agronomic analyses of farming techniques. This is what we aim to do in the following pages.

216 Verschuer (1993: 24–28) and (1998: 428–431); Kimura (1992: 123–126).

217 See an overview of farming implements based on the archaeology and the 1822 treatise in Verschuer (1993: 427–443). Yet during the 1980s it was still thought that the spread of iron tools dated from the late medieval period, see Kuroda (1984: 255).

218 *Konjaku monogatari*, book 26, episode 10; *Uji shūi monogatari*, book 4, episode 4; see Kuroda (1983: 47, 48).

219 In presenting this and the following four types of holdings, we mainly follow the interpretations of Kuroda (1984: 48–54). Amino (1993: 328–339) (a reprint of his article with the same title in *Rekishi chiri kyōiku* 383, June 1985) gives other examples of late medieval landholdings.

220 For late medieval kitchen utensils and garments, see Verschuer, "Daily Life in Medieval Japan According to the 'Scroll Paintings of Ippen's Wanderings' (1299)", *Annuaire de l'École pratique des hautes études 2009–2010*, Sorbonne, 2011.

221 *Jizō bosatsu reigenki*, book 4, episode 12, after Kuroda (1983: 54).

222 Kuroda (1983: 51); Kuroda (1984: 252) ("the plough was absent from the lives of the poorest peasants"); Kimura (1992: 249) (important role of the plough in land clearing); Kōno (1994: 77–84); Kōno (1985) shows the various ploughs in the late medieval iconography.

223 See for example *Echizen kokushi no ge* of 776, in DNK 5: 554–617; *Yamashiro no kuni handen zu*; and Shiryō hensanjo, vols 1 jō, 1 ge and 2.

224 See "Tillage" and "The imperial vegetable gardens" at the beginning of this chapter, pp. 17–21 and "Tillage" at the end of this chapter, p. 82. But Kōno Michiaki suggests that the depth of tillage could be regulated by means of a saddle harness on the animal's back (see Kimura 1992: 249).

225 A comparison with medieval France would probably alter the assessment of livestock rearing and draught tillage in medieval Japan. Scores of different ploughs and devices pulled by animals can be seen in French agricultural museums, whereas the Japanese plough remained unchanged from the 8th to the 18th century. Kuroda (1984: 251) discusses the question of transport.

226 See Isogai (2002: 126) and Kimura (2010: 120–122) for ashes; see Chapter 2 in this volume.

227 Kuroda (1984: 261–268); Isogai (2002: 78–92) has listed the occurrences in the sources.

228 Isogai (2002: 10, 11) (marginality of double-cropping), pp. 111–113 (increase in known cases of double-cropping from the 13th century), pp. 117–123 (decrease in yields), pp. 125–128 (relationship between yields and use of chemical fertilizers). However, Kimura (2010: 89) considers double-cropping to be a widespread practice, but his argument concerns taxation, not techniques. A comparison with European practices could enable a reassessment of the Japanese data. The technical data given above can provide an answer to the question posed by Takagi Tokurō, namely: "Was the late medieval period in fact not a time of agricultural progress?"; see note 215.

229 Kimura (2010: 117). This tale is found in Murai Shōsuke, ed., *Rōshōdō Nihon kōroku*, Iwanami bunko, 1987, p. 144.

230 Discovery of tablets by Hirakawa Minami, mentioned by Kimura (2010: 70–72).

231 For this section concerning varieties, see Kuroda (1983: 73–76).

232 Kuroda (1983: 75); and Kuroda (1984: 70–73).

233 See Chapter 4.

234 Harada (2008: 21–25); Kimura (1996: 110); Verschuer (2009: 354).

235 Kimura (1996: 113–115). However, Kimura (2010: 84, 85) returns to the idea of a wave of agrarian expansion *dai kaikon jidai*.

236 Kimura (1992: 101–102) (a slight expansion); Kimura (2010: 84, 85) (estates set up by the Kamakura military government).

237   See Verschuer (2009c: 349); the figures are given in *Kokushi daijiten*, vol. 7: 811; *Nihon daihyakka zensho*, vol. 12: 507; Kitō Hiroshi, *Jinkō kara yomu Nihon no rekishi*, Kōdansha gakujutsu bunko, 2007: 16 (who gives 4.5 M for the 8th century), and other publications.

238   Kimura (1996: 11, 12); Harada (2008: 25).

239   Kimura (2008: 74).

240   We share the views of Harada (2008: 30–53) ("The diversity of agrarian spaces"); he quotes Ogawa Naoyuki (1995).

241   Tsukuba Hisaharu (1987); Pauer (1973); Hayama Teisaku (1992). See the editions of agricultural treatises in the bibliography.

242   *Seiryōki*, "Nihon nōsho zenshū". The date of this work is unclear, see *Kokushi daijiten* 8: 271.

243   *Hyakushō denki*, "Nihon nōsho zenshū"; see Hayama Teisaku (1992: 186–199).

244   *Aizu nōsho*, "Nihon nōsho zenshū"; Sasaki Takeo (1993: 52–76).

245   Furushima Toshio (1949: 373–376); Tanaka Kōji, in Watabe Tadayo, ed. (1987: 302–312); Tsukamoto Manabu (1984). See above, note 234 and pp. 89–91.

246   Horio Hisashi (1990: 229–236); Kōno Michiaki (1994–1996); Kōno (1994). Haudricourt and Brunhes-Delamarre (1986). See the modern iconography of the implements in Akiyama Takashi *et al.* (1979, 1991).

247   See the water-wheel in the decree of Tenchō 6.5.27. 829, in RSK 8: 323. Kōno Michiaki (1995). On the noria, see Kimura Shigemitsu (1992: 133). See the poem by Toshiyori in *Sanbokuki kashū* 3: 106–1337, and a poem by Minamoto no Nakamasa, dating from 1173, in *Fuboku wakashō* 2: 16–15724; quoted in Furushima Toshio, p. 164; and *Kojiruien* Sangyōbu 1: 266–272. See Furushima (1949: 309–312); *Nōgu benriron* ge, chapter on pumps and hydraulic equipment. *Satsujōshū* in GR 30 ge: 292. *Kōka shunjū* 7: 294: *mizukae oke*; *Hyakushō denki* 5: 217: *tsurube*.

248   Pauer (1973: 140–146); Guyonvarch (1993); *Nōgu benriron*, jō.

249   Furushima Toshio (1949: 506–568), on the subject of fertilizer in each region of Japan; and a table of this in Akiyama Takashi (1979: 235).

250   Horio Hisashi (1990: 243–252); Furushima (1949: 316–319). They think that threshing with chopsticks goes back to the time of Sei Shōnagon. See the *Hyakushō denki* for wheat. Ihara Saikaku, *Nihon eitaikura*, p. 161, mentions the *gokedaoshi*; I thank Abe Yoshio for this reference.

251   Furushima (1949: 318–320, 415); Pauer (1973: 55, 156); *Hyakushō denki*, pp. 144, 145, 185; Miwa Shigeo (1991: 106–124); he thinks, p. 120, that the spread of the *tousu* was limited to certain regions.

252   Pauer (1973: 156–160); Horio Hisashi (1990: 254–257); we quote the *Saizōki* after Furushima (1949: 319); *Hyakushō denki*, p. 185; *Kōka shunjū*, p. 33: this treatise does not mention the winnowing-machine.

253   Miwa Shigeo (1991: 49–78, 139–223); Trombert (1996). For the *daibutsu* in 1195, see *Tōdaiji zōryūkuyōki*, Enkyū 6.3.12., after *Kojiruien* Sangyōbu 1: 104. For Jōjin, see *San Tendai Godaisanki*, Enkyū 4.9.17. 1072. *Hyakushō denki*, pp. 185, 186, 327, 328, for wheat and the *karausu*.

254   Horio Hisashi (1990: 242); Kumazawa Kikuo, quoted in Oka Mitsuo *et al.* (1990: 130).

255   After Kōno Michiaki, in Reizei Tamehito (1996: 104, 105). Kōno (2000). See photocopies of the *Ehon tsūhōshi* from "Shizuoka-ken nōgyō shikenjō", which Kōno Michiaki kindly made available to us (Figures 1.7–1.13).

2   A mountainous environment

Shifting cultivation

Many civilizations in the world have practised shifting cultivation on temporary fields for long periods of their history. In fact:

> cutting down a patch of forest, burning the wood and sowing the ground for one or more years before abandoning it once more to the forest is one of the most widespread practices of early agriculture. It is especially in tropical countries, where this system was still [in 1975] the basis of agricultural production in many regions, that it has been most often studied and described; but it was no less important in several parts of Europe, where it continued up to the 19th century and sometimes even until the early 20th century. Moreover, north-eastern America was covered by forests when European settlers arrived; in order to live, they had to become land clearers, thereby imitating the Indians they supplanted.[1]

Various types of shifting cultivation have characterized the history of civilizations throughout the world. In this chapter, we discuss our hypothesis according to which Japan is no exception. We think in fact that shifting crops represented, in early medieval Japan, "another agriculture" practised in parallel with permanent irrigated and dry crops. However, as this is an area of the private (individual) economy that falls outside the public administration, shifting agriculture is very poorly documented by the administrative and legislative texts. We therefore adopt an approach based essentially on the early medieval poetry, lexicography and toponymy. Before starting our detailed analysis, we need to devote a few pages to introducing the general context and the basic comparative data, in order to show that shifting agriculture is not a marginal phenomenon.

What is shifting agriculture?

The most usual type of shifting agriculture is represented by swidden-field crops grown on mountain clearings, i.e. plots cleaned by clearing or felling trees. Swidden fields have been defined by Yoshio Abe as follows:

The plot is generally obtained by clearing the side of a forested hill. The trees are felled at the end of the dry season. The felled trees are burned. Sowing and planting are therefore done on the ashes. The plot is neither irrigated nor tilled and is fertilized only by ashes. It is worked for just a few years (and sometimes a single year) and is then left resting three to four times longer or even much longer in some cases than the cultivation period, before being worked again. One of the characteristics of swidden farming is complantation. Cereals of other types [than rice], tubers, pulses and vegetables are grown at the same time.[2]

After having worked a patch of ground, the farmer clears another. It is thus a form of cultivation that is both temporary and shifting. The plot is called a swidden field, and the technique is known as slash-and-burn. Below we will use 'swidden farming' and 'slash-and-burn' as well as the older term 'assartage' (which strictly speaking only refers to the felling of trees before setting fire to the land), all with the same meaning.

Fire techniques in the European traditions

François Sigaut has examined the archival sources relating to swidden or forest agriculture in Europe from the 18th–20th century. We have taken the following remarks from his book *L'agriculture et le feu* (1975). Swidden fields made possible two to four or five years of harvests. Two or three cereals (wheat or rye) were usually grown; these crops were sometimes preceded by a summer crop (buckwheat, turnips), followed by one or two harvests of hay and several years of pasturage (pp. 122–124). It is difficult to compare the yield from forest and permanent agricultures, but slash-and-burn had the advantage of reducing the work of fertilizing, tilling and weeding the fields. Contrary to what may be thought *a priori*, forest agriculture (on swidden fields) is not so much a necessity imposed by the shortage of available land in certain regions, as a conscious choice made by farmers because of the many advantages they derive from it (p. 168).

However, forest fields could only be cultivated for a limited period, the exhaustion of the mineral nutrients in the soil and the profusion of weeds forcing the farmer to abandon his field. The cultivation period and the forestation period varied according to the regions. In Europe, the former seldom exceeded a maximum of five to six years, while the period of natural reafforestation lasted between ten and fifty years, with the average being around twenty years. The forestation period could be shorter – from eight to twelve years – in the scrub of southern France. In certain regions (Black Forest, northern Russia), the new undergrowth was cut again in some cases after eight to twelve years (pp. 121–123).

The investment in labour was far less for swidden-field crops than for permanent agriculture, since time could be saved on tillage, fertilizing and weeding. The most important task, in short, was the actual assartage, i.e. tree felling. A man needed thirty to sixty days on average to cut down a hectare of forest (p. 167). For example, in 1800, in New Hampshire (America), one man took a week to fell the trees on an acre (a quarter of a hectare) of forest (p. 18). In Europe, in the Ardennes, in 1881, nineteen days per hectare were needed for felling, spreading the trees around the site and burning them (p. 152).

François Sigaut also mentions other fire techniques used on flat ground. Slash-and-burn, widespread in France in the 18th century, consisted of burning squares of turf removed from a moor or grassland to fertilize and prepare it for a temporary crop (pp. 11–14). The bonfire allowed a field to be fertilized by bringing vegetation from a moor and setting fire to it (p. 40). Burning with running fire was found everywhere in the African bush, the Eurasian steppe, the European moors and grasslands (and in our view, Japan). This technique enabled lands to be cleared, animal pastures to be improved (pp. 37, 38) and, generally speaking, promoted the self-sown regrowth of vegetation.

For Germany, Wilhelm Abel mentions various farming practices attested since the end of the medieval period in his work on the history of German agriculture.[3] On the one hand there is forest assartage, and on the other a form of agri-sylviculture that alternates six to eight years of shrub growing with one year of cereal cultivation. It is interesting to find that in Germany these plots were given different names depending on the region. Various terms have in fact been noted, including "waste land" – Wild – and Scheffelland, "overgrown field" – Wildfelder, "mountain land" – Bergland, "moorland" – Heideland, "idle field" – wüste Felder. In the sandy regions in western Prussia, according to a forester in 1820, permanent fields were cultivated near dwellings at the same time as cleared plots based on three years of cultivation for a period of forestation from twelve to twenty years. The land maps of this region distinguish, for certain villages, permanent fields, plots cultivated temporarily and the forest proper. For nine villages, the maps show that only four-tenths of the farmed areas were permanently cultivated. This data reminds us that in the 8th century only one-third or four-tenths of the lands on some estates belonging to the Tōdaiji temple were farmed.

Swidden-field crops in Asia

For China, Francesca Bray thinks that swidden farming preceded permanent agriculture in Zhou times (1134–250 BC), but in an uneven manner, depending on the regions and periods. It predominated in southern China and among the non-Han populations. The Yao cultivated in particular non-irrigated taro on swidden fields up to the 20th century. There is little written Chinese evidence, but references to this practice are found in Tang (618–907) and Song (960–1279) poetry. Ōsawa Masaaki has analysed these written references and has drawn up a geographical list of swidden fields in China.[4]

More so than history, ethnography provides a great deal of information on shifting cultivation. Research has been especially rewarding in tropical countries. We give here a short summary of the investigations carried out by Georges Condominas in 1948/1949 in Vietnam with the Mnong Gar tribe in the village of Sar Luk and published in his book *Nous avons mangé la forêt* (1957). This report can be used as a reference for reconstructing the lives of individual clearers in a tropical climatic zone.

The Mnong Gar did not practise permanent agriculture, but lived solely by swidden farming, gathering, fishing, and, to a lesser extent, bovine rearing and hunting. At Sar Luk, the land-clearing season was between mid-January and mid-March. Each household in the village, which had 146 inhabitants, felled its patch of forest every year, the forest fields only able to be cultivated in principle for one year. The work of felling took eighteen to twenty-nine days for a household. Then, it seems, they set aside only a few days for drying the wood, before setting fire to the cleared patch. After burning, they formed work groups to cut the partially burned trees for reburning. At this time, the clearers built field huts on posts, in two days. They then went ahead with sowing, choosing the moment when the first storms of the monsoon season hit, namely 5th May at Sar Luk in 1948.

The Mnong Gar practised intercropping, like most clearers in tropical regions. In the same field they sowed rice and about thirty other plants including maize, tobacco, cotton, indigo plants and two poisonous plants. The "sowing season" lasted fifteen to twenty days and was followed by the "weeding season". Under the tropical rains, from early May to the end of September, they had to dig up the weeds, which could easily strangle the growing rice, using a small hoe with a short handle. The village was almost deserted while the rice was growing and the maize was ripening, as it was necessary to stay near the fields in order to protect the crops, by day from birds, and at night from herbivorous animals. They used traps, xylophones and noisy devices. Maize was harvested in the second half of August and the second weeding was completed in mid-September. Around the beginning of October, everyone began harvesting the early rice. Three different varieties were harvested successively up to the beginning of November. In November, the families built temporary granaries near their field at the entrance to the track leading to their village. The autumn rice had to be harvested and brought in very quickly to avoid unseasonable rains. To do this, the villagers formed groups who worked for sixteen days. They harvested by day and pounded the rice at night. After that they dug up the cassava and picked the cotton and beans. The harvest ended with the rice being transported from the temporary granaries to the granaries owned by each family in the village. The men and women went back and forth for two days, laden with heavy baskets filled with hulled rice. At the end of the agricultural cycle, rituals were celebrated, from 5 December in 1948; but the entire year had been dotted with rituals that accompanied each farming operation. Moreover, throughout the year the villagers went fishing and foraging and did handicrafts: basket-work, weaving and pottery. At the end of the year, the fields were again left resting for a period of eighteen to twenty years.

As regards areas, the poorest family, a couple with small children, normally cleared and cultivated a plot of 1.03–1.25 ha. However, being unable to turn regularly to groups who had to be paid in work or money, this couple abandoned 0.22 ha because of a shortage of labour. This shows that it was easier to clear the ground than to cultivate it. By contrast, the wealthiest family, that of the village headman, cultivated 3.5 ha thanks to the abundant labour at its disposal.

There are several differences between the systems in Europe and the Vietnamese village. At Sar Luk, the time required for felling was longer than in Europe, but the time for drying the wood was very short. At Sar Luk, weeding had to be done several times, whereas there was no weeding on cleared plots in Europe. At Sar Luk, the cultivation period lasted only one year, with mixed crops on a single plot, whereas in Europe the same field was cultivated for four or five years with successive different crops. Swidden farming therefore takes various forms depending on the geographical and climatic conditions. Ethnographers who study the countries of South-east Asia think that the yield from swidden-field agriculture is comparable to that of permanent crops and that the plots allow a greater number of edible plant species to be grown on the same arable land.

Swidden farming across the world

There are many other ethnographical reports on this practice. Yin Shaoting, Watabe Takeshi and Christian Daniels have noted the methods used in the province 1of Yunnan in China. Inoue Makoto has studied those of the Indonesian island of Kalimantan. Olivier Ducourtieux has done a thorough ethnographical survey of swidden farming in Laos and Harada Nobuo has outlined the practices in Japan.[5] In Europe, Axel Steensberg has directed experimental slash-and-burn projects and he presents the situation worldwide in his book *Fire-Clearance Husbandry: Traditional Techniques Throughout the World* (1993). According to the FAO, the population of clearers was 200 million in 1950 and the area of swidden fields 36 million square kilometres. For Asia, J. E. Spencer noted in 1960 some 50 million clearers and areas of 10–18 million hectares. The National Ethnographical Museum of Japan in 1990 published a study of the ways of life of the (minority) ethnic groups of South-East Asia carried out by a research group led by Ōbayashi Taryō. This study showed 31 ethnic groups of rice farmers, 108 ethnic groups of clearers and 18 ethnic groups who practised rice growing and swidden farming together in the 1980s.[6] In all cases these were tropical regions. Yet we may ask whether it might not also be possible to envisage for Japan not an ethnic diversity, but a regional one and the coexistence of methods of permanent and itinerant farming, with a predominance of one or the other depending on the geographical situation. Ethnology has in fact shown the existence of villages of clearers in 20th-century Japan. We present below the results of Japanese ethnographical research.

The ethnology of swidden-field crops in Japan

Sasaki Kōmei carried out studies on swidden agrieulture in Japan and Nepal in the 1950s and published the results in two works: *Nihon no yakihata* ("Swidden-Field Crops in Japan", 1972) and *Inasaku izen* ("Before Rice Cultivation", 1971).[7] His investigations subsequently influenced the research of the post-war generation and his publications are standard references even today. Sasaki's first "terrain" was a village in Kyūshū: between 1958 and 1960, he spent several periods in the hamlet of Kajiwara that forms part of the village of Itsuki in the mountains of Kumamoto. Each of the fifteen families worked an area of 1.8 ha on average, on slopes of 20–30°, located more than an hour's walk from the hamlet. They did not practise irrigated rice growing, but lived a self-sufficient existence by means of swidden farming, gathering and a few permanent dry crops.

The inhabitants of Kajiwara practised alternating four-yearly crop rotation on plots of 0.8–1 acre. Each family simultaneously farmed plots that it had cleared one, two, three and four years before, and in two different parallel cycles. The first cycle included successively: buckwheat *soba*, barnyard millet *hie*, red beans *azuki* and foxtail millet *awa*, with a small quantity of taro *satoimo*; and if the soil allowed it, tubers were also grown in the fifth year: a variety of sweet potato *satsumaimo* (Ipomoea Batatas Lam. var. edulis Makino) and *satoimo*. The other cycle provided for successive crops of wheat *komugi* or barnyard millet, foxtail millet, red beans with soybeans *daizu*, and two tubers *satoimo* and *satsumaimo*. On the new lands intended firstly for growing buckwheat, the trees were felled in July or early August. A group of four to five men could fell 0.26 acres of trees in a day. The plot was immediately burned and prepared for the first sowing. For crops of wheat or barnyard millet, felling was done between mid-September and mid-October and burning before sowing, either in October, or in April of the following year. In this case, the wood was left to dry all winter.

Two weedings were needed every year, from mid-June to mid-July and from early August to early September. This task, which was by far the most time-consuming of the farming work for the entire year, occupied eight to ten people for a month on an area of 0.26 acres. Despite this, the arable area declined every year on the edge of plots invaded by weeds, so that sometimes only half the field remained after four years. For most crops, the harvesting season was in October and November. From March until the harvest, the villagers lived mainly in their field huts, but they sometimes returned to the village to tend the vegetable gardens and small permanent fields near their houses.

At Kajiwara, the first plant cultivated on cleared plots was foxtail millet, the second barnyard millet and the third red beans. These three species were part of the two crop cycles. The last-mentioned ended with tubers, when the areas were reduced in size by the advance of self-sown vegetation bordering the fields. Rice was entirely absent from forest agriculture. After cropping, the field was abandoned for a forestation period of twenty to thirty years.

Swidden farming across Japan

Outside Kyūshū, Sasaki Kōmei found the same swidden-field crops in Shikoku, the difference being that here they added the paper mulberry *kōzō* and *mitsumata* (Edgeworthia chrysantha Lindl.), two plants used for making paper. Sasaki thinks that the crops listed above were grown throughout the laurel-forest zone, i.e. forests of evergreen trees covering the south-western half of Japan, a zone that mainly has a hot-temperate climate.

Sasaki identified another cycle of crops in the cold-temperate zone that extends from central Japan to Hokkaidō. He developed a prototype based on the observations he made in 1965 in the village of Shiramine in Ishikawa prefecture, where permanent and shifting cultivation were practised together.

Near Shiramine, in permanent fields, the farmers grew barnyard millet, *shikokubie* (see below), maize, sweet potatoes *satsumaimo*, cowpeas *sasage* and various other leguminous plants. On the cleared mountain plots, the villagers mainly felled alder *hannoki* and oak *konara*. On these plots worked for four years, they successively cultivated barnyard millet, foxtail millet, soybeans, and again in the fourth year foxtail or barnyard millet with red beans and soybeans, and sometimes, in the fifth year, did intercropping with perilla *egoma* as well. In this region, three weedings were done. One sees the preponderance of foxtail millet, as in the south of the country, and the absence of wheat and tubers. Rice was not grown either in the permanent fields or in the cleared plots, which means that part of the mountain population of Japan did not eat rice until the 1950s.

In total, Sasaki identified in Japan five zones characterized by a tradition of swidden farming. From north to south (see Map 2.1), they are:

- The mountains of Kitakami (Iwate) in north-eastern Honshū, where large quantities of beans were grown, a zone where the word *araki* is commonly used in the dialects to refer to swidden fields.
- The mountain chain of Mutsu and Dewa in the central-north (Akita-Iwate) up to that of Jōetsu and Kubiki (Yamagata, Fukushima, Niigata), where they practised among other things felling and burning in summer, followed by a first crop of buckwheat in autumn; in this region, the dialectal word *kanno* is used for swidden fields.
- In central Honshū, the eastern region of Shizuoka and the mountains of Toyama, Ishikawa and Fukui on the Japan Sea coast, a region where the swidden field was called *nagihata* and where barnyard millet, foxtail millet and beans predominated; here they felled trees in autumn, burned and sowed in spring, or else felled trees in summer, followed by burning and the sowing of wheat in autumn.

- The islands of Shikoku (Kōchi, Ehime) and Kyūshū (Kumamoto and Miyazaki), with the more complex crops mentioned above; in this region, they used the word *koba* for swidden field.
- The subtropical zone of the southern islands, Hachijōjima and Okinawa, where mainly tubers were cultivated.

In other words, swidden farming was most widespread on the Japan Sea coast, in Shikoku and Kyūshū. The crops identified by Sasaki in 72 villages of

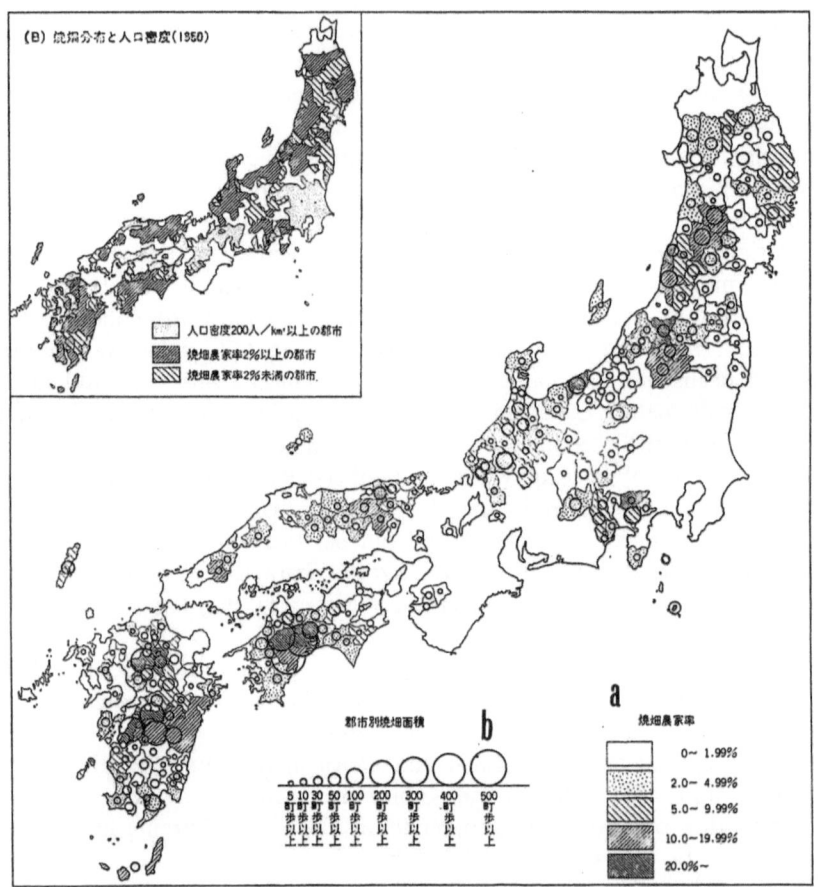

Map 2.1 Swidden-farming regions in 1950 – distribution of swidden-field areas and the population of clearers
Notes: a: Percentage of clearers among the farming population: white, less than 2%, progressive increase up to black, more than 19%. b: Area of swidden fields by district or town: large circle, 500 ha, progressive reduction down to the smallest circle, less than 10 ha.
Source: Sasaki Kōmei, *Nihon no yakihata*, Kokon shoin (1972: 23).

Japan included, before 1945, as a percentage of the number of villages that grew these crops, foxtail millet (83.8%), buckwheat (80.5%), soybeans (76.4%), red beans (73%), barnyard millet (56.9%), wheat (20.8%), taro (19.4%), sweet potatoes *satsumaimo* (15.3%) and *mitsumata* (15.3%).[8] This means that millets and buckwheat were the first swidden-field cereals in Japan.

Today, on the trail of swidden farming

The research of Sasaki Kōmei drew the attention of other ethnologists to the importance of swidden-field crops in Japan and gave an impetus to field work. A number of ethnologists have carried out studies and published monographs on a region or village. Certain places have become known to the public to the point where today they attract tourists who seek a return to rural traditions. The following examples from different regions of Japan may be mentioned.

The village Akiyama-gō straddles the prefectures of Niigata (Nakauonuma-gun, Tsunan-chō) and Nagano (Shimominochi-gun, Sakae-mura), four hours' drive north of Tōkyō. It is a group of twelve hamlets situated 800 m above sea level, in a depression carved out by the Nakatsugawa river in Mount Torikabuto (2,023 m). Every year, these hamlets host several hundred visitors including many agronomy students, on the first weekend of August; they come to see the burning of a mountain plot where the local council has crops grown on swidden fields. The vegetation cut down beforehand burns fiercely, making a loud crackling noise. Only the very large trees remain standing. The land is cleaned and strewn with ashes in three hours. The local council also offers a series of lectures and a museum visit that makes it possible to reconstruct the life of the mountain clearers of Japan, the last of whom disappeared in the 1990s.

In this village, the ethnologist Ichikawa Takeo conducted inquiries with some of the inhabitants and in the local archives, and in 1961 published a book called "The Valley of the Heike: Akiyama-gō, Secret Land of Shin'etsu". Ichikawa situates this village in a deciduous forest zone. In his previous publications, he called "Fagus zone" *buna-tai* the zone covered by fagaceae of the species beech *buna* and oak *konara*, which occupy the entire north-eastern half of the archipelago.[9] Akiyama-gō lived mainly by shifting cultivation, bear hunting, fishing and foraging. The first plants grown, in 1897, were foxtail millet and perilla, then barnyard millet, then buckwheat and soybeans. In 1951, it was half foxtail millet, a quarter soybeans and for the rest the other plants. From 1891, a few rice fields were noted, but the staple food was foxtail millet. Today, the village inns offer visitors traditional specialities, namely vegetables picked in spring and macerated in salt or soy, such as bracken *warabi*, royal fern *zenmai*, bamboo shoots and stems of horseradish *wasabi* as well as cakes made partly with horse chestnut *tochi* flour. Horse chestnuts, which are considered inedible in Europe, in Japan had their tannin removed by a complicated process of drying and repeated washings and were eaten either mixed with millets or in the form of dumplings. At Akiyama-gō, one has only to admire the landscape marked by very steep slopes that climb skywards on all sides of the village to realize that irrigated rice cultivation, even on terraces, was difficult and that the inhabitants of the high mountains led a different life from the populations concentrated on the plains of the country.

The island of Shikoku is the "terrain" of the ethnologist Fukui Katsuyoshi. In 1970–1971, Fukui identified seven hundred villages practising swidden farming in Shikoku and he chose for his inquiry Tsubayama at Ikegawa-chō, Agawa-gun, in Kōchi prefecture, a village 50 km west of Kōchi city. In his report, "The Village of Swidden Fields" (1974),[10] Fukui describes the annual cycle of work in the fields; and the summary of his observations can be seen in a video preserved in the National Museum of Ethnology library in Ōsaka. The first crop that fed Tsubayama up to the early 20th century was barnyard millet. *Mitsumata*, a kind of pappiferous daphne, was added as a commercial crop in the 1880s. Fukui stresses the usefulness of the self-sown vegetation growing on cleared plots that are abandoned after four years of cultivation. At Tsubayama, the land was invaded by pueraria *kuzu*. The roots were dug up at the end of autumn, a starch was extracted to make flour (*katakuri*), the leaves were fed to the livestock or used as fertilizer and ropes were woven with the stems. For Fukui, swidden-field crops provided the villagers with a range of products that enhanced their daily lives. Cereal crops on swidden fields and the dietary customs of Shikoku are also the subject of a richly documented collection of photos published by Kondō Hideo (1999), an amateur ethnologist from Ehime.

Nomoto Kanichi published, in 1984, the results of his research conducted over twenty years in Shizuoka prefecture, in the mountain villages north-west of Shizuoka city, near the Ōigawa river.[11] He studied in detail the technical aspects of swidden-field crops, the dietary customs and the other activities involved in obtaining food, as well as the oral and folkloric traditions, toponyms, the dialectal agrarian vocabulary, the rituals and beliefs of this region. This book, with its detailed index, is a reference work on agricultural traditions.

Tachibana Reikichi carried out work of the same scope and minute detail over many years in another region, Ishikawa prefecture. He conducted his research in the village of Shiramine, Ishikawa-gun, located on Mount Hakusan whose summit Shakadake is 2,053 m above sea level. Shiramine is 40 km south of Kanazawa city.[12] With the construction of a dam for the production of electricity in 1972–1973, five hamlets belonging to Shiramine were destroyed and submerged in the reservoir. Tachibana Reikichi was tasked with an ethnographical study before construction work began and had several houses from the affected hamlets as well as their traditional farming and everyday equipment moved to the village ethnographical museum. Swidden-field crops (noted by Sasaki Kōmei, see above) have gradually disappeared since the 1950s but, in 2001, three-quarters of an hour's drive from the village, one still came across in the mountains "amateur" growers of the Japanese radish *daikon* on swidden fields who lived in temporary huts during the farming season. The ethnographical museum cultivates in its garden *shikokubie*, a graminae that is gradually disappearing in Japan. *Shikokubie*, Eleusine indica Gaertn. var. coracana

Makino, was a permanent field crop at Shiramine. The very small grains, pounded and mixed with hot water, give a soft, sweet, chocolate-coloured dough with a very pleasant taste; but the low yield and almost constant attention required by this plant have led to the abandonment of this crop. The ethnographical studies referred to above and the archives of the Edo period for the villages mentioned are a rich source of material that can be used by those interested in the popular traditions and material culture of the mountain-dwelling populations of Japan.

Another village, this one in Kyūshū, was partially drowned at the bottom of a dam: Shiiba-mura at Higashi-usuki-gun in Miyazaki prefecture, high up in the Kyūshū-sanchi mountains, 60 km east of Hyūga city. On the other side of the same mountain chain is the village of Itsuki in Kumamoto, studied by Sasaki Kōmei in the late 1950s. Shiiba-mura is dealt with in one of the first works by Yanagita Kunio (1875–1962), *Go karikotoba no ki* (1909), which describes the customs relating to hunting in this village. According to an old land survey, in 1838 there were 5,208 inhabitants who farmed 0.6 acres of rice fields, 50 ha of dry cereal fields and 500 ha of cleared mountain plots. Apart from cereals for food, the villagers produced tea, mulberry bark, hemp, tobacco, mushrooms, vegetables from collecting and deer and serow *kamoshika* skins. In the 1930s, they cultivated two types of cleared plots: swidden fields of barnyard millet at 500–900 m above sea level, cleared in autumn, then burned and sown in spring; and swidden fields of buckwheat at 300–800 m above sea level, cleared in summer and immediately burned and sown. Besides the two staple foods, the inhabitants grew in these same fields millet, beans and some-times taro. The village has an ethnographical museum and the implements and objects for daily use relating to swidden-field crops, obtained from the local people, are displayed in another museum attached to the school in the village of Osaki, situated beyond the dam that is 16 km long. In 1997, Shiiba Hideyuki and Kuniko were the last farmers practising slash-and-burn cultivation in Japan. They still farmed 7.8 acres of swidden fields on the area of 1.3 ha that they had worked since the 1950s, and ran a small inn "Minshuku Yakihata" located a short distance from the centre of the village. The ethnographer Sasaki Akira of Miyazaki University recorded during 1995 the remarks of Shiiba Kuniko and published them in a book, "Diary of an Old Woman from a Mountain Village". Shiiba Kuniko retraces in her account the calendar of farm work and rituals, as well as the dietary customs, the use of implements, the beliefs and customs of daily life. In a work edited by another ethnologist, Shiiba Kuniko mentions her knowledge of the use of cultivated and wild plants.[13]

Each of the villages mentioned above, as well as other mountain commu-nities,[14] is a kind of repository of local traditions. The practices here are still partly alive, partly reconstructed for the purpose of rural tourism. Visitors can taste the fruits of gathering and hunting and local specialities derived from non-rice-growing cereal traditions, take note of the importance of popular beliefs and watch traditional arts, such as ritual Kagura dances. They can realize that swidden is not merely a farming technique, but is set within a natural environment that has given rise among the mountain populations to a way of life different from that of the inhabitants of the plains and coastal areas of the Japanese archipelago. It is a way of life governed by a landscape where villages are set deep in the valleys of the mountain chains, surrounded by steep wooded slopes and overhung by jagged ridges on all sides. The access routes are narrow roads that follow the contours of the steep slopes. But this very narrow landscape of wooded mountains is not exceptional, since it occupies more than two-thirds of the total land area of Japan.

Some figures

We know some figures, thanks to studies published by the Japanese Ministry of Agriculture. In 1936, the total area of swidden fields was 76,795 ha, or 15% of the arable acreage of the country; there were 152,000 households practising assartage, or 21% of farmers. In 1950, there remained only 9,457 ha recorded as swidden fields, but a relatively high number of 110,500 households of clearers. However, Sasaki Kōmei thinks that the areas are under-estimated. For example, he noted 400–500 ha of swidden fields at Itsuki in Kyūshū, whereas the land survey of this community recorded just 77.6 ha. For this reason, Sasaki estimates the total area of shifting cultivation in Japan in 1950 at around 50,000–60,000 ha.[15]

In fact, the areas had already decreased markedly in 1936. In that year, the Japanese government issued a decree aimed at encouraging shifting cultivation, in order to guard against food shortages. But since 1888, previous governments had prohibited swidden-field crops and had banned them by law in 1897 because of their perceived negative effects. Mentioned at the time were the risk of forest fires, landslides and damage to hydraulic installations. The governmental measures attracted the attention of researchers. Thus Ono Takeo, Yamaguchi Sadao, Yamaguchi Yaichirō[16] and Furushima Toshio devoted studies to the practice of shifting cultivation in the 1930s and 1940s, a generation before Sasaki Kōmei. These scholars, especially Ono Takeo, retraced the traditions that were current in the Edo period.

Swidden farming in the 18th–19th centuries

For the Edo period, we have no figures, but there are accounts in the local monographs and agricultural treatises. The "New Geography of Musashi Province" *Shinpen Musashi fudoki kō*, written by Mamiya Kotonobu between 1810 and 1828 for the military government of Edo, provides information on the swidden-farming practices in the village of Ōtaki at Chichibu in present-day Saitama prefecture. Here is an extract:

> The old village of Ōtaki is surrounded by wooded mountains. There are no rice fields. Only two-tenths of the land has permanent dry fields

honbata 本畑, the rest being given over to crops on swidden fields *yaki-hata* in the mountains. The soil quality is half average, half poor(?). In this region, they burn the vegetation *kusagi*, spread the ashes as fertilizer and grow foxtail millet, barnyard millet, soybeans, red beans and buckwheat. This is what they call "swidden fields" *yakihata* or *sasu*. All the villagers live off these crops. But as the lands suitable for cultivation, in the mountains and valleys, have always been insufficient, the harvests only provide subsistence for six to seven months of the year. This is how the inhabitants grow crops. From the end of spring (3rd month, April) until the start of winter (10th month, November), they go far away, one or two leagues (4–8 km) and build a field hut *iori* 庵 at the foot or top of the mountain. Father, mother, children and grandmother live there to do the sowing *tane o maku* 播種. Until the fruits of the field ripen, the field must be protected from monkeys by day and from deer at night. To do this, husband and wife cross the mountain or the valley here and there, to spend the night apart (in turns) in a temporary shelter *kari no oya* 仮小屋. Night after night, they strike small wooden boards *bangi o utsu* 打板木 or call out, to frighten the wild boar and deer, undeterred by rain or storms. We must feel for their suffering.[17]

The agronomy manual *Jikata hanreiroku* (ca. 1794) states that people waited for rain after having spread the ashes, before going ahead with sowing. According to this work, swidden fields "are not found near dwellings but in the mountains; they are especially numerous in the region of Shinshū" (Nagano).[18]

Several agricultural treatises of the 17th–18th centuries also mention cleaned fields. The *Aizu nōsho* (1684) recommends sowing buckwheat immediately after burning the trees to stop the ashes being carried away by the rain or wind. This treatise discusses the Aizu region, i.e. present-day Fukushima prefecture. It suggests successive crops of buckwheat, foxtail millet and soybeans on swidden fields called *kano koba* and recommends lands with a south-easterly exposure.[19]

Two other treatises from the Kaga region (Ishikawa prefecture) where the village of Shiramine is also located (see above) mention shifting cultivation. The *Nōgyō zue* (1717) shows two polychrome images of setting fire to a mountain in the 4th month (May) and the 6th month (July), titled "cleaned and burned fields *nagihata* in the distant mountain of spring/of autumn". According to the *Kōka shunjū* (1707), areas of 500, 1,000 or even 1,500 *bu* (0.4 *tan* or 1.3 acres) were cleared *ki shiba kaya kozue o kiru* in the 4th or 5th month (May–June) and burned in the 6th or 7th month (July–August), starting at the top of the slope. Next day, any charred remains were collected and light raking was done with a hoe while spreading the ashes, then sowing took place: buckwheat in the first year, foxtail millet in the second year, soybeans in the third year and red beans in the fourth year. Weeding was done in the 6th and 7th months. At the end of four years, the vegetation was again allowed to take over the land.[20]

The swidden fields of the Kaga region were exempt from taxes, but yields were only one-seventh of those for irrigated rice fields. On the island of

Tsushima, there was no rice growing. Here, the struggle against wild boar was a major task on swidden fields *koba*, according to the *Rōnō ruigo* (1722). A corvée labour system of young men had been put in place for the construction and upkeep of the palisades around the swidden fields. The second danger that was especially feared on Tsushima was the forest fire.[21] Ono Takeo examined the archives of the Edo period in his "History of Swidden-Field Crops".[22] He found many references to techniques and even many cases of their use in the land surveys of several regions. It is clear that the farming calendar varied from one region to another, that different methods were used requiring weedings and other more or less constant attention, and that crop rotation showed specific regional characteristics. But overall, the accounts noted by Ono Takeo give the impression that swidden fields were quite widespread in Edo-period Japan.

Land clearing

The technique of swidden was not limited to forest agriculture. It was in fact the method most often used for land clearing. The expression "to open up a clearing" was used when virgin land was to be brought under cultivation. Miyazaki Yasusada explains this in his treatise *Nōgyō zensho* (1697) in these words:

> Not only in the mountains and wilderness areas, but also when one wants to open up new land for the first time, the land is burned in spring, then the burned area is hoed and a [permanent] field is established. Spring is the best time for land clearing, according to Miyazaki Yasusada, because the soil is friable and the vegetation abundant, so that there is a considerable energy saving compared with tillage done in autumn or winter. Moreover, the summer sun aids the decomposition of plants and the fertilization of the soil.[23]

The encyclopedia *Wakansansaizue* (1712) explains the stages in bringing new soils under cultivation and gives the terms, quoting a Chinese dictionary of the Yuan dynasty (13th–14th centuries): "A newly formed field is called in the first year *arakihari* 菑; the grass is [dug in] by turning over [the soil]. In the second year, the field is called *kochita* 畲; the soil becomes friable. In the third year it is a "new field" *arata* 新田; i.e. a field that is still recent. In the fourth year it is a [true] "field" *ta* 田".[24] The same four-year cycle can be observed as for shifting crops. In other words, after burning the land, the farmer did the same thing for four years, whether to grow cereals for a limited time or to clear the land in order to farm it over a long period. After four years, he abandoned the land to the forest in the first instance and turned it into a permanent field in the second instance, by bringing tillage and fertilizers to it. Later, this field could also be turned into a rice field by the installation of a water-supply system.

Therefore, both land clearing and forest assartage involved the technique of fire and the first swidden-field crops.

Swidden fields: the poor relation of historiography

Specialist historians of the early medieval period have shown little interest in fire techniques, no doubt mainly because of the scarcity of sources. Hatai Hiromu was the first Japanese scholar to show the importance of assartage in early medieval Japan, but his article "Agriculture on swidden fields in the Nara-Heian period" published in 1976 hardly caught the attention of historians.[25] It is true that it contains some errors of interpretation. There is, in our view, confusion between toponyms and generic terms taken from the early medieval land surveys, as well as some misreading of certain decrees relating to the management of wilderness areas. These errors may have contributed to the scholarly isolation of this publication, though the author also collected many interesting references relating to shifting agriculture in the early medieval texts. The study of the history of non-rice-growing agriculture thus got off to a very difficult start.

Nonetheless, in his work on land clearing in the medieval period published in 1984, Kuroda Hideo considers the possible existence of swidden fields, citing Hatai Hiromu with some reservations. Kuroda thinks that future research should evaluate their place in history. Wayne Farris, in 1985, speaks of "slash-and-burn agriculture" in his book on the agrarian system and demography. In his view, the practice was fairly widespread around the year 900, basing his argument on the low population density. Farris quotes an article on swidden fields, published in 1953 in the United States by Robert Hall and Toshio Noh. It is in fact an ethnologist, Nomoto Kanichi, who since 1984 has most vigorously defended the idea of the existence of the practice of slash-and-burn in early medieval Japan. Kimura Shigemitsu mentions "the theory of swidden fields" in his work on the history of dry cereals, published in 1992. One also finds a few references to assartage or scattered remarks in the writings of some other historians such as Kodama Kōta, Morita Tei and Abe Takeshi; yet the general scholarly opinion remains unchanged, namely that "rice cultivation was the basis of agriculture" in early medieval Japan (Yoshida Takashi). Thus, swidden-field crops were all but absent from historical studies on the bringing under cultivation of agrarian spaces. However, things changed after the establishment in 2001 of the Institute for Humanity and Nature in Kyōto, where a programme has been devoted to ecohistory. Within its framework, a team led by Harada Nobuo, a medieval historian, has undertaken research into the 'agriculture of fire' *ka-kō*. Among the Institute's publications may be mentioned "Japan Archipelago 3500 Years of Environmental History Series" *Shiriizu Nihon rettō no sanman gosen nen hito to shizen no kankyōshi* (6 vols, 2011), with the work "Environmental History of Mountains and Forests" and Harada Nobuo *et al.*, eds, "Environmental Studies of Swidden Farming" (2011).[26]

Thanks to these studies, there has been a growing awareness of the importance of fire techniques in Japanese history and ethnology. As regards the examination of the early medieval Japanese sources, this is the subject of the sections in this chapter. We first of all present the legislative and administrative documents, which reflect the point of view of the government in 8th–10th-century Japan concerning this practice.

Swidden fields in the 8th–10th-century administrative texts

The government of early medieval Japan does not explicitly deal with swidden in any text, but it implicitly acknowledged its existence in the administrative regulations issued in 701. The penal code curbed forest fires by sentencing to fifty strokes of the cane those who "lose fire" and those who "burn the fields and virgin lands (*no*) outside the authorized period" from the 30th day of the 10th month (end of November) to the 1st day of the 3rd month (beginning of April). Thus the penal code implicitly authorized the burning of fields and virgin lands during certain periods of the year. The local militia regulations warned soldiers against the wilful misuse of fire signals, as they could be mistaken for the fires of clearers. Moreover, the regulations concerning pastures recommended burning grasslands at the start of the year, except those "in places where fire was prohibited", namely certain mountain forests and bamboo groves. The burning of pastures seems to have been practised up to the Edo period.[27]

The government had prohibited fires in some mountains of the central region from 677. These appear to have been sacred sites occupied by shrines or imperial tombs, or places where access was restricted to the imperial family. In 841, the court forbad the felling of trees and the gathering of brushwood in the mountains of Kasuga near Nara, on the land of the Kasuga Shrine, citing the displeasure of the deities. It issued another edict in 867 prohibiting "the farmers of the province of Yamato from burning the sacred site of Mount Isonokami to sow foxtail millet and beans". In fact, according to the Regulations of the Engi Era (927), trees could not be cut down on shrine lands. It was also necessary to avoid any risk of fire and the defilement that followed from it. The regulations imposed a prohibition of seven days for shrine officials who had passed near a forest fire.[28]

However, the government also introduced an exception to these interdictions: in 814, it authorized the Heian palace guards to grow foxtail millet each year in the mountains and on the Kitano moor, a sacred site north of the imperial palace. The term "each year" probably refers to the idea of moving around: each year, the soldiers burned and cultivated a different patch of moorland.[29]

Some accounts of fires, dating from the 10th–12th centuries, also refer to the practice of burning wilderness areas. The fires of clearers at various times destroyed a plantation, a row of trees near a temple and even a temple building. On occasion, several oxen died in the flames.[30] But in these cases burning was

Table 2.1 Terminology of the administrative texts

	Japanese transcription/characters	*English*
Cleaning by fire	*noyaki* 野焼	burning wilderness areas/virgin lands or grasslands
	nobi 野火	fire in wilderness areas or on grasslands
	ta-ya o yaku 焼田野	to burn fields and virgin lands
	bokuchi...yōyaku yake, kusa ofuru ni itarite amanekarashimeyo 牧地...漸焼、至草生使遍	to completely burn pasture..., to maximize the spread of grass regrowth
	rei ni yorite no o yakiharau 依例焼掃野	to burn grasslands as is customary
	mainen no haru nobi o hanatsu 毎年之春放野火	to burn grasslands every spring
	shikka 失火	to lose fire, accidental fire
	hito no shataku oyobi zaimotsu no enshōsu. 延焼人舎宅及財物	fire damages dwellings and people's possessions
	kudan no hayashi o enshōshi sudeni owannu 延焼件林既畢	fire has damaged this wood
Cultivation on swidden fields	*sanya... midari ni yakikiru koto nakare* 山野...莫妄焼折	it is forbidden to clear and burn mountain and virgin lands at will
	Isonokami no shinzan o yaki, awa mame o hashi suru ... kinshi su 禁止...焼石上神山播蒔禾豆	it is forbidden to burn the sacred mountain of Isonokami and to sow millet and beans
	awa-batake o tsukuri; mainen chi o ukite tsukurashimu 営粟畠,毎年授地令営	to cultivate a field of millet; to receive a plot every year to cultivate

Sources: The texts mentioned in this section (see notes 27–30).

not necessarily followed by cultivation. It is in fact important to distinguish two different techniques.

Two fire techniques

In the quotations given above, one can identify two practices. In 677, 814 and 867 there is the burning of a patch of mountain or forest *yama o yaku*, or wilderness area *san-ya (yama-no) o yaku*. The fire was followed by a crop of millet and/or beans. These were therefore swidden-field crops.

The other references mention the burning of virgin lands or grasslands *no o yaku* or *noyaki* and the burning of pastures *maki*. This is the technique of cleaning by running fire intended to promote the regrowth of self-sown vegetation through the fertilization of the land with ashes. The term *noyaki* does not, *a priori*, exclude the idea of cultivation, but the notion of "spring burning" refers simply to cleaning by fire. In the reports relating to fires on temple lands, one finds the word *nobi* (*no hi*), a fire on virgin land, which also refers to cleaning by fire. The Japanese term "*no*" is often translated "moor" in poetry, but we prefer "wilderness area" or "grassland", as moor, strictly speaking, implies a specific type of wild vegetation.

We are therefore in the presence of two techniques: on the one hand, the clearing of trees and the burning of a patch of mountain, followed by sowing; on the other hand, the cutting of brushwood in a mountainous or lowland area that is burned and left uncultivated. Assartage (or swidden farming) required a single burning at the start of the cultivation period; it allowed virgin lands to be used for food crops. Cleaning by running fire involved annual burning; it facilitated the gathering of edible plants and the harvesting of industrial stems.

The references to burning in the administrative texts of the early medieval period were noted by Ono Takeo from 1942, taken up by Furushima Toshio in 1947 and analysed in detail by Hatai Hiromu in 1976.[31] However, judging by the corpus of archives of the 8th–12th centuries, these references are very brief compared with the wealth of information about rice cultivation. This dearth of sources may have led historians to regard swidden-field crops as a marginal phenomenon. As we are dealing with lands not recorded in the early medieval surveys and a method of farming that falls outside the tax system and land management, it would be a mistake to expect any more information from the administrative texts. We therefore propose to study the fire practices outside these texts and to extend our inquiry to notions other than those of burning per se. It is important to draw attention to the terminology. Today, the Japanese words *hata* and *hatake* mean a non-irrigated permanent field as opposed to the rice field *suiden*, but the meaning was not the same in the 8th–12th centuries. In fact, in the early medieval period, *hata* was reserved for a swidden field on a burned clearing.

Hata – *the swidden field;* hatake – *the permanent field*

The original meaning of the word "*hata*" can be found in the dictionary *Wamyō ruijushō*, compiled by Minamoto no Shitagō (911–983) during the 930s. Here is the definition:

> *Ryū* 疁 that is, a swidden field *yaihata* 火田: it is a field sown [straight after] burning without tillage. The *Tangyin* [Chinese dictionary, 7th–9th centuries] and the *Kan[go]shō* [Japanese dictionary, 717–724] call it *yaihata*. Consider the "Tale of the Old Peasant" *Yarōden* (*Yelao?*) which

says: 'They call "cut mountain" *kirihata* 截幡 a field established after cutting the side of a mountain [into terraces], and "burned field" *yakihata* 焼幡 a field burned then tilled.' But if they say "field", why is it not tilled? The explanation in the *Kan[go]shō* [saying that tillage follows burning] and that of the *Tangyin* [saying that there is no tillage] are contradictory.[32]

In this dictionary, *hata* has the meaning of swidden or temporary field; it occurs in the three compounds *yaihata, yakihata* and *kirihata* and is written with the Chinese character for flag *hata* in the two latter cases. However, the *Wakansansaizue* (1712) explains that *kirihata* is a terraced permanent field bounded by a ditch and that the character for flag *hata* is explained by the appearance of this field seen from above. According to this encyclopedia, the *Tangyin* rightly notes that [only] *yakihata* is a [temporary] field burned not tilled.[33] It is clear that the 18th-century compiler gives *hata* the two meanings of permanent field and temporary field, whereas the 10th-century writer assigns to *hata* only the meaning of swidden field.

Hata has in fact changed its meaning over the centuries. Today *hata* means 'permanent field' and is a synonym of *hatake*, as in the 18th century. For the two words, there are two interchangeable characters: 畠 and 畑. This can be verified in all modern dictionaries. Both terms are still found in many toponyms throughout Japan and suggest non-rice-growing cereal fields. Kuroda Hideo retraced the evolution of the word *hata*. According to him, the documents "Katsuragawa myōōin shiryō" no. 34, 42, 68 and 155 dating from the 13th and 14th centuries differentiated *hata* and *hatake*. Then they became confused at the end of the Tenshō era (1573–1592).[34] Before that, one finds the two readings used for a single character in the dictionary *Hyōta jiruishō*, compiled around 1299. It is even more remarkable that the poetic anthology *Fuboku wakashō* from the early 14th century makes the distinction between *hata* and *hatake*.[35] In the Edo period, the agricultural treatises, apart from the *Kōka shunjū* (1707), confuse *hata* and *hatake*, whereas Arai Hakuseki (1657–1725) distinguishes the two notions based on the 10th-century *Wamyō ruijushō*.[36] The treatise *Nōgu zoroe* (1856) explains this dual tradition as follows:

> *Ta-hata* 田圃: originally, *ta* meant 'rice field' and *hata* was used for the "fire field" (*hi-ta*) denoting the burned clearing *yakihata* in the mountains. For this reason, they created [in Japan] the character *hata* composed of the fire radical *hi* and the field *ta*. However, this character became widespread in our province (Hida, present-day Gifu), and [today] it refers here to the permanent dry field.[37]

Thus the linguistic factor may have contributed to swidden fields being forgotten about until Kuroda Hideo, in 1980, reminded historians of the earlier meaning of *hata* in Japanese. *Hata* therefore meant 'swidden field' until the 15th century, i.e. a field prepared by burning and cultivated for three to five years. It was a temporary field and this was a shifting form of agriculture, since a new field was established periodically on another patch of forest on the same mountain. Other early terms, now forgotten, also refer to swidden fields. We will now identify them in the early medieval dictionaries.

Swidden farming in the early medieval lexicography

The *Wamyō ruijushō* (ca. 930) assimilates the Chinese character *liu* 嶚 to the compound *hida/kaden* 火田 and glosses it by *yaihata*. It also gives the compound *yakihata* 燒幡 corresponding to the modern *yakihata* 燒畑. In the 12th century, the dictionary *Iroha jiruishō* (1144, completed ca. 1200) glosses the compound *kaden* by *yakihata*. This dictionary and the *Ruiju myōgishō* (ca. 1100, re-edited in the early 13th century) gloss the Chinese character *liu* by both *yakihata* and *hata*.[38]

As for the single word *hata*, the *Iroha jiruishō* writes it with one character 畠, ancestor of the form 畑. The character *hata* 畑, unknown in China, was created in Japan. It is a native character *kokuji*. Its origin is unknown, but it may have appeared in the first glossary of "New Characters", *Niina*, ordered by Emperor Tenmu (r. 673–686). According to Itō Toshikazu, the first occurrence of the Japanese character *hata* is found in a document from Mount Kōya dating from 1094.[39]

It may be noted in passing that the character *hatake* 畠 meaning 'permanent dry field' was also created in Japan. It is found in its separated form *haku-den* 白田 in the early medieval documents. The three dictionaries mentioned above give three ways of writing *hatake*: 畠, 白田, 陸田. The reading *hatake* is attested by the *Man'yōshū*. *Hatake* is also the reading given for the characters 陸田 by the Kujōkebon manuscript (ca. 1036) of the *Engishiki*.[40]

The early medieval dictionaries also mention the characteristics of swidden fields, namely the fact of sowing on ashes without tilling the land. The *Ruiju myōgishō* notes the Chinese word *ti* 稿 (not glossed) and defines it as "to sow without tilling", after the early medieval Chinese dictionaries. This character had been mentioned previously by the *Shinsen jikyō* (898–901), compiled by the monk Shōjū (?–?), which partially follows a tradition different from the three other dictionaries. The *Shinsen jikyō* glosses the character *ti* by *aramaki*, "sowing on virgin land". This dictionary also gives the native Japanese character 秇 and gives two readings for it: *aramaki* and *yakimaki*, "sowing on burned land, that is the sown swidden field", explaining that the swidden field was not tilled. It uses the character "to sow" *maku* 蒔 in its Japanese sense, this character meaning "to transplant" in Chinese. The other Japanese dictionaries use the word *ueru* 種 meaning 'to sow' and 'to plant' in Japanese and Chinese. These details indicate that the *Shinsen jikyō* was relatively close to the Japanese traditions of the 9th century.

Besides *hata*, *yakihata*, *yaihata*, *aramaki* and *yakimaki*, one finds other terms relating to assartage. The two 12th-century Japanese dictionaries give the Chinese character *zhi* 畲 or 畬. The *Iroha jiruishō* assimilates this term to the unglossed Japanese character 曙 and defines it as *arata* 荒田 "wilderness field". The *Ruiju myōgishō* glosses the Chinese character *zhi* in three ways: *une*, the ridge; *arata-haru*, to clear a "wilderness field" or to clear a still wilderness field; and *arata*. This dictionary defines *arata* as a field of the first year after clearing, copying the definitions of the early medieval Chinese lexicons. Thus the word *arata* here takes on the meaning of first-year swidden field.

The *Ruiju myōgishō* also takes the Chinese character *ji* 畸, without glossing it, giving it the two senses of *arata* and *kikori no ta, kikoru* meaning 'to cut down trees'. *Arata* was indeed a field prepared after having felled trees, as shown in an early poem that mentions sowing in a newly cleared field *arata hirakamu*. But poetry knows a second meaning of *arata*, that of idle land or an abandoned field, like the "wilderness field *arata* where the grass and flowers grow thickly", which is therefore not brought under cultivation. The Chinese character *arata* 曖 in the *Ruiju myōgishō* and the Japanese character *arata* 稴 in the *Shinsen jikyō* mean "wilderness field", i.e. abandoned or temporarily uncultivated.[41]

Two early Japanese dictionaries give another Chinese term, *she* 畬. The *Iroha jiruishō* glosses this character by *arata* and *arakihari*, a word close to *arataharu* 畬, and defines it as a "field of the third year" [since the clearing of virgin land]. This corresponds to the definition of *she* as "a new field *arata* 新田 of the third year", in the 18th-century *Wakansansaizue*. *Arakihari* can probably be transcribed 新伐墾 and means 'newly cleared land'. According to the *Ruiju myōgishō, she* is read *konata tsukuri*, "a prepared rice field", and corresponds to a "field of the second (not the third) year". These two Japanese dictionaries precisely reflect the contradictions of the Chinese works they quote. They also note for *she* the meaning of swidden field. The *Ruiju myōgishō* gives for *arakihari* the character 畾, which is thought to be a variant of the Chinese character *she*.

It may be said that the words *arata, arataharu* and *arakihari* reflect the early medieval techniques, since land clearing was done by assartage followed by some years of swidden-field crops. The semantic element *araki* is also found in literature, written *araki* 荒木 and *arakida* 新城田 in the *Man'yōshū*, and *araki* 開墾 ("to clear") in the *Bungo fudoki*. *Araki*, in the 20th century, is a dialectal word denoting swidden fields throughout north-eastern Honshū and parts of Kantō. According to Yanagita Kunio, the modern toponym Araki, found nearly everywhere in Japan, also reflects the tradition of swidden fields.[42] After having taken note of the earlier terminology, we will look at swidden-field practices in early and late medieval literature.

Swidden fields in the medieval period

The medieval historian Itō Toshikazu was inspired by the terminology of the fields *hata* and *hatake* and became interested in the role of wilderness areas *no*

Table 2.2 Lexicography of the 10th and 12th centuries

	Wamyō ruijushō	Shinsen jikyō	Iroha jiruishō	Ruiju myōgishō
swidden field *hata*			畷・畚・(畲)	畷
burned field *yaihata*	火田・畷			
burned field *yakihata*	燒幡		火田・畷	畷
sown waste ground *aramaki*		秔・穡		(穡)
sown swidden field *yakimaki*		秔		
field prepared after felling trees *kikori no ta*				畸
wilderness field, of the first year *arata*	荒田		(曙)・荒田・畚	甾・畸
new field, of the third year *arata*			畲	(菑)
idle land, uncultivated field *arata*	荒田	荒田・穡		畦
second-year field *konata tsukuri*				畲
to clean or clear *arataharu*				甾・畲
newly cleared plot *arakihari*			畲・畚・菑・載	

Notes: *Bolded: Japanese character *kokuji*.
*In brackets: unglossed characters, with no reading.

and wooded mountains *yama* in agriculture. He examined the medieval documents and noted certain facts that are summarized below.

In the medieval period, references appear in estate land documents to areas called "field on virgin land" *no-batake* 野畠, i.e. a field established in a wilderness area. Itō found eighty-five cases in documents from the 12th–16th centuries. He thinks these were swidden fields (that he calls "unstable" *fuantei*, even though they fit into a clearly defined cycle), despite the character used being that of the permanent field *hatake*. These fields were difficult to record in the land surveys because of their shifting nature, but they are nonetheless recorded with their measurements in the registers. Estate owners, from the 12th century, levied on these swidden fields a modest tax, which led Itō to think that yields were very low (compared to irrigated rice?). Fields of barnyard millet and taro or yams *imo* were, however, tax exempt. A fairly complete register of the Kengunsha Shrine estate in Bungo (Ōita prefecture), dating from 1353, shows that half the land consisted of irrigated rice fields, 40% of swidden fields and 10% of permanent dry fields. Yet half the "fields on virgin lands" were in fact lying idle; the other half was given over to crops of foxtail millet (67.9%), dry rice *no-ine* (17.1%), soybeans (14.7%) and red beans (0.3%).[43]

Itō Toshikazu also found instances of swidden fields in the administrative documents. He noted eight cases of swidden fields *hata* 畑 from 1094 to 1221 and fifty-five cases of "mountain swidden fields" *yamahata* 山畑 from 1223 to 1333. The register of an estate in Kii (Wakayama prefecture) recorded, in 1201, on the one hand permanent fields, dry and irrigated, with their areas and location, and on the other "mountain swidden fields" without specifying their area and with the simple mention "a place". These shifting swidden fields were therefore likely to have been concentrated in one part of the estate, but their exact location was difficult to specify. However, the register of another estate in the same region gives more precise figures for 1272. Here one finds: a plot of 17.5 *chō* (over 17 ha), including 2.3 *chō* of dry fields *hatake*, taxable at 1.84 *koku* (1 *koku* = ca. 85 litres); and 3.9 *chō* of "mountain swidden fields", taxable at 1.96 *koku*. The owner therefore taxed even the swidden fields, but the rate of tax for these fields was far lower than that for permanent fields. The document states "that mountain swidden fields must be inspected every year", no doubt because their location changed. It also notes that there were swidden fields intended for a summer harvest and those intended for an autumn harvest. The former were burned in autumn, sown with a winter crop such as wheat and harvested in summer (in the 4th month); the latter were burned in spring, grew summer crops such as foxtail millet and dry rice and were harvested in autumn.[44]

Itō paid even greater attention to the Kawakami estate of the Tōdaiji temple in the Yamato region. There remain seven documents from the 14th–15th centuries relating to this estate. In 1373, the tenant farmers were exempted from tax for "a new plot" *shinbata* 新畑 in a mountain area, which they "had cleared" *kaihatsu*. They explain that they have cleared uncultivated waste land *kōya/ arano o hiraku* 開荒野 for sowing *sakutsuke* 作毛. In 1377, there is also mention of a "new swidden field" *shinbata*, "recently cleared" *shinkai* 新開, which this time is subject to a tax. The terminology used in these documents is interesting: the notion of "clearing" is literally expressed by "to open up" *hiraku/kai* a plot. The cleared area is virgin land *no* that is uncultivated *ara/kō*, a semantic element also found in the early medieval lexicography. The land clearing was intended not for permanent but for shifting cultivation, since a swidden field and not a permanent field was established there; however, in 1419, some of the swidden fields were turned into a tea plantation.[45]

Takemoto Toyoshige and Itō Masatoshi drew attention to the presence of swidden fields on two other medieval estates and were supported by the historian Ishii Susumu. Itō noted a "mountain intended for swidden fields" *yaki no yama* in the documents of the Kada estate, at the western tip of the Kii peninsula. Takemoto discovered that the Niimi estate of the Tōdaiji, in Bitchū (in present-day Niimi city, Okayama), managed "mountain fields" *yama-batake* 山畠 from the early 13th century at the latest. At one time, an intendant, a certain Narimatsu, responsible for a third of the lands on the estate, managed 47 *chō* of "permanent dry fields" *sato-batake* 里畠, located near the farmers' dwellings, and 170 *chō* of "mountain fields". The areas of swidden fields thus far exceeded those of permanent fields. According to Takemoto, "mountain fields", though written with the character *hatake* of the permanent field, were swidden fields. He localized them in the present-day landscape and found that the lands in question were on a slope with a gradient of 30°. Based on his farming experience in the 1950s, Takemoto states that swidden fields were cultivated in this village on slopes of 30°, permanent fields on lands sloping 12–13° and rice fields on flat land.[46]

It is therefore clear that shifting agriculture played quite an important role on temple estates in the medieval period. This is confirmed by the fact that swidden fields were mentioned in the tax documents of the time. In fact, "mountain swidden fields" and "wilderness fields" first appeared in public administration in the early 13th century. In 1232 and 1235, the Kamakura military government began levying taxes on both kinds of temporary fields and granted taxation rights to the estate intendants *jitō* of the lands under its control. Needless to say, the government was aware of the extent to which swidden farming was practised on estates at that period.[47]

The examples mentioned above suggest that medieval holdings were very diverse. They were not restricted to rice fields side by side in regular squares, but included other agrarian spaces, indicated in the texts by: flat fields *hira-batake* or fields near the village *sato-batake*, fields on virgin lands *no-batake*, mountain fields *yama-batake* and cleared mountain plots *yamahata*.[48] Even if owners had a preference for rice fields, the morphology of the terrain often obliged farmers to use other cultivation techniques. It was, in short, the altitude and land gradient that dictated the farming methods on the medieval estates.

Swidden fields in the 8th–12th centuries

As the taxation of shifting crops only began officially in the 13th century, it would be pointless to look in the administrative documents of the 8th–12th centuries for references other than the few brief remarks given above ("Swidden fields in the 8th to 10th-century administrative texts"). We must therefore turn to a different kind of documentation. The analysis of the dictionary terminology of the 10th–12th centuries has already given us some important information. We now propose an approach using the early medieval poetry. We are interested first of all in swidden fields *hata*.

Poetry and swidden fields

Here is a poem by Fujiwara no Nagayoshi (949–1009):

> On the mountain side *katayama*, someone sets fire to a swidden field *hata yaku*; set fire to your clearing, but spare from the fire the cherry trees deep in the mountain.

Nagayoshi was then in Tanba with his father Tomoyasu (?–977), governor of the province from 968 until 971. He explains the context in his introduction to the poem:

> In Tanba, I happened to fall ill and stayed there for a long time without returning to the capital. I have collected a few poems composed during this idle time about what I saw in the vicinity. This poem is set in spring. Seeing a peasant set fire *hi tsuketaru* to a hillside *kataoka*, I asked him: 'What are you doing?', whereupon he replied: 'We call "setting fire to a swidden field" *hata yaku* what I am doing now'.[49]

Fujiwara no Nagayoshi was thus able to witness the burning of a patch of mountain. Nagayoshi's poem is also found in the *Shūi wakashū*. Itō Toshikazu sees here the first appearance of the Japanese character *hata* 畑 of swidden field, but in fact this word is written in *hiragana* in the earliest copies of the *Nagayoshi shū* and in the *Shūi wakashū* version.[50] Here, we have a word in the spoken language that did not yet have its own character and was still rendered using the "borrowed" character for flag *hata* in the *Wamyō ruijushō*. This is why the *Nihon shoki* refers to the "fruits of swidden fields" *hata-tsumono* in the legend about the birth of cereals, and the 10th-century commentary also transcribes this using phonetic script.[51]

Another poem mentions the swidden field in phonetic transcription, this time in *man'yōgana*. It is found in the *Man'yōshū*:

> The road travelled/ both morning and evening/ by the men of the [swidden fields]/ the rest of us have made it/ the palace road.[52]

Here, Kakinomoto no Hitomaro (?–?), in 689, is on his way to the funeral of Prince Kusakabe. He refers to the "men of the swidden fields" *hata kora* 八多籠良, a term not explained in the "Nihon koten bungaku taikei" edition, but that Sano Masami and Nakanishi Susumu interpret as *hata kora* 畑子等.[53]

There are some other poems of a later date. Fujiwara no Nobuzane (1176?–1265?) composed this:

The village at the foot of the mountain where they burn the cleared plots *hata yaku*, seems from afar in the twilight to be but a ribbon of smoke.

Another 13th-century poet describes the landscape in these words:

It has become a village built by people to live in, the swidden field of the lowly peasant, at the end of the mountain path![54]

Several other poems have cleared mountain plots as their subject. In them one finds the following varied vocabulary:[55]

- *hata yaku*, set fire to a swidden field,
- *yamahata*, mountain swidden field,
- *katayama/ no hata*, swidden field/on a mountain side,
- *yakehata*, swidden field set on fire,
- *hata no yakeyama*, swidden field prepared on a burned mountain.

However, there are few references to swidden fields in the early medieval texts. It is by mistake that this word occurs in the *Heian ibun*. In a letter of 1183, a monk refers to the "rice fields and swidden fields" *ta-hata* of land in Kii, but this is quoted from an Edo-period monograph on Kii and is a late transcription of the expression *ta-batake*. We should remember that the distinction between *hata* and *hatake* disappears in the modern period. The character *hata* is also found in a Yamashiro land survey of 864 in the *Heian ibun*, which quotes the documents from the Ninnaji temple, but the facsimile manuscript of this text shows the character *hatake* and not *hata*. It is even more interesting to see a "wilderness swidden field" in a document of the 1180s, written *nohata no hatake* 野はたの畠. The swidden field *hata* is therefore attested at this time, in its phonetic form with no character.[56]

Wilderness fields

In the document of 1180 is another example of the notion of wilderness that recalls the fields *nobatake* of the medieval estates. In fact, a few earlier scattered references to these fields can be found in documents of the Heian period. The "Map of household fields for the district of Kadono in the province of Yamashiro", dating from the first third of the 9th century, records each plot of land and mentions in four places *nobatake* 野畠 and a further *nobatake* 野陸田. A register of the Niijima estate in Awa, dating from 850, notes for two of the plots that there is a "wilderness field" *nobatake* 野圃. Elsewhere, one finds an area *nobatake* of 2 ha, recorded in 1042, on land situated at the edge of the capital of Heian and belonging to the Jakurakuji temple, and in 1177, several small plots of *nobatake* owned by the Chōfukuji on the bank of the Katsuragawa river.[57] Given that these plots pay a tax, albeit half that of permanent fields, one may ask what in fact is the precise nature of these "fields on virgin lands".

Nevertheless, such information allows us to put forward a hypothesis regarding the Kuwabara estate of the Tōdaiji temple in the province of Echizen (Fukui prefecture). In 757, the administration of Echizen calculates the irrigation works for the rice fields on the estate and notes: "The digging of irrigation canals allows good rice fields to be established; the virgin land outside must be cleared [and brought under cultivation] for one or two years *nokoru no, ichi ni kanen hiraku beshi*." Fujii Kazutsugu and another historian think the virgin lands must also have been turned into rice fields and that one to two years' preparation were generally allowed for this. However, we saw above that in 1373 the Kawakami estate "opened up a plot" *hiraku* to sow crops (not irrigated rice) and that swidden fields were not uncommon on temple estates. We therefore think that the Tōdaiji also farmed areas unsuitable for rice growing as shifting swidden fields[58] and possibly grew foxtail millet on them.

The cultivation of millet

We now return to poetry to obtain information on the types of crops and the methods used. Two poems in the *Man'yōshū* mention sowing on mountain land:

> Even if my beloved's horse nibbles the millet that I sow on Sanatsura hill, I will not tell him to go away.

And again:

> I sowed millet on Mount Hakone in the Ashigara range; though it has ripened, I cannot meet with you.[59]

Both poems are about the cultivation of foxtail millet *awa*. The millet is sown not in a field but directly on a hill and in the mountains. This implies the preparation of a patch of mountain by burning, as the crop could not ripen without fertilization by ashes. Two other poems also mention sowing on a mountain:

> If the shrine of the fierce deities were not there, I could sow millet on Kasuga moor.

And again:

> After sowing millet on Kasuga moor, I wait there for the doe [that grazes on it]; while I continue to come and go, you [the doe] avoid the Kasuga Shrine.[60]

Kasuga is the usual name for the mountain ranges east of Nara that inspired many poets of the 7th–8th centuries. Here, we have adopted the conventional translation of "Kasuga moor (*no*)". In the mountains of Kasuga, one particular slope was revered as the seat of the deities and considered a sacred site. Swidden was prohibited on this part of the mountain, as on all shrine lands, but elsewhere the burning of a patch of land seems to have been commonplace in the mountains, including "Kasuga moor". The last poem mentions "waiting for the doe", yet (paradoxically) one would try to frighten this doe, as the poem alludes to the damage caused by deer coming to graze on the ripening cereal crops. Here is another poem that mentions the damage caused by animals:

> The rice of the clearing *arakida* destroyed by the deer and wild boar *shishida*, I stored it [long ago]; like this rice, our love has aged.[61]

The term *arakida* (in phonetic transcription in the poem) recalls the expression *araki hari* in the early medieval dictionaries, meaning "a newly cleared plot". Here this seems to refer to dry rice cultivation, as irrigated rice growing is possible only after the first dry crops and terracing and irrigation works. Another poem contains a similar expression:

> I have gone to look for a small field to clear *araki no oda* to sow the purified seed, left my home, my feet wetted by the ford.[62]

The ravages of deer and wild boar in swidden fields were a disaster that all clearers had to contend with. There are several accounts on this subject from the Edo period. The passage translated earlier, relating to the village of Ōtaki in Saitama, informs us that the cultivators of swidden fields built a hut *iori* in the mountains and lived there from spring until autumn, i.e. from sowing until harvesting. It was necessary to stay near the plots to protect them from animals. Moreover, if one plot was located away from the others, someone had to spend the night in another makeshift shelter near the swidden field and strike small wooden boards *bangi* during the night to frighten the deer and wild boar. These practices known in the modern period were the same in the early medieval period, as can be seen in these lines:

> With a light heart, in every mountain field shelter, in autumn, can be heard the loud noise of the small boards that frighten the deer; like the sound of the mallet beating hempen cloth.[63]

There were several ways of frightening animals to "protect the mountain fields" *yamada moru*, while staying in a summer hut. Far from being limited to swidden fields, these methods were in fact the same for permanent fields. If an irrigated rice field or a dry field was located away from dwellings, it was necessary to take special care of it, using various methods that we listed in

"The struggle against disasters" in Chapter 1. One poem alludes to the distress of farmers, speaking of "the small primitive shelter *oya* that I would like to burn";[64] and another poem mentions the summer hut:

> The makeshift hut *kari ho* for harvesting the autumn field has not yet been destroyed, but already one hears the calls of the wild geese [come from the north] heralding the cold, and already the frost has settled [all around].[65]

The former may well refer to the "small shelter" in which a person tasked with scaring away animals from an isolated clearing stays, and the latter to the seasonal hut in which the whole family stays while working in the fields. Several poems in the *Man'yōshū* speak of the "makeshift huts for harvesting the autumn field" *aki ta karu kari ho*. In principle, this applied equally to rice growing and the other crops, but the two poems quoted above mention the destruction of the small shelter and the hut after the harvest. Permanent crops were grown every year in the same fields, which meant that the summer hut was probably reused, whereas shifting agriculture involved mobility and the destruction of the hut after four or five years of cultivation. The references given above attest to the practice of swidden-field crops. Another practice is documented far more often in early Japanese poetry: the cleaning by fire of virgin lands, intended not for cultivation but to promote the regrowth of the self-sown fruits of nature.

The poetry of the spring fires

The spring fires have been a favourite topic of poets since the 8th century. The fires marked (the end?) of spring and also provided an occasion to express amorous longings. Here is an example:

> Could it be that the peasants, who after wintering, in spring burn *yaku* the large clearing *ōno*, have not burned enough [land]? They make even my heart burn.

And another:

> This clearing *no* so fine, do not let it burn *yaku*; mixed in with withered grasses, young grasses already grow thickly there.[66]

According to these poems, burning *yaku* was done in a wilderness area *no*, which we here translate by 'clearing', rather than by the usual, more poetic term "moor". The second verse mentions *omoshiroki no* (written phonetically), an adjective with two meanings: *omoshiroki*, full of interest or memories, and *omoshiroki*, with a clear and open surface. The second meaning relates to the clearing. The "fire of the clearings" was called *nobi* (*nohi*) in poetry, as in the administrative documents.[67] It was in fact in sunny places like clearings that the plants often associated with the space *no*, namely bracken *warabi* and eulalia *susuki*, grew.

This is the technique called in Europe "burning with running fire", used for cleaning an uncultivated area to promote the regrowth of wild plants. In early medieval Japan, the main purpose of this practice was the gathering of edible plants. It is found in many poems, such as the following:

> And you! The guardian of the fields responsible for the flying fire *tobuhi* [of Tobuhino moor] at Kasuga, go and see if in a few days we will be able to pick the young shoots *wakana tsumu*.[68]

Tobuhi, "flying fire", meant both the signalling fire of the militias and the running fire of burning. This word also gave its name to the peak Tobuhiga-take in the mountains of Kasuga, where a guard-post was established in 712. In the poetry, we keep the usual translation of "moor" for toponyms such as Kasugano and Tobuhino. The subject here is the gathering of young grasses or young shoots *wakana*, and the next poem gives a concrete example:

> On Kasuga moor, one sees smoke rising *keburi tatsu*: some young girls are no doubt cooking the asters *uhagi* they have picked in the spring clearing *haruno*.[69]

White-flowered asters *uhagi, yomena* in modern Japanese, grew in rather damp mountain soils. In spring, the young shoots and leaves of the aster were boiled and probably added to cereal gruels. The aster is a vegetable high in vitamins and calcium.[70] Besides clearings, people also set fire to lowlands where they wanted to gather plants:

> Apart from you, who can I invite today, to beat a path with me, into the burned clearing *yakehara* of sweet miscanthus *ogi*, to pick young shoots *wakana*?[71]

Burning was also done in areas overgrown with lespedeza, which is mentioned in this poem:

> If I am still of this world, I will go to see them all at Yoshino, in the flowering season, the lespedezas *hagi* of the burned field *yakefu*.[72]

The writer Sone no Yoshitada (?–ca. 1003) observes, late in the 2nd month, the burning of a "field" *fu* that is not, however, to be understood here in the strict sense of 'cultivated field'. The lespedeza, which flowers in autumn, is also found in poetry as a sign presaging this season. It was not an edible plant, but in spring the root-stock of the previous year was burned in order to promote the regrowth of other plants.

There was a great diversity of collected plants, of which we give an overview in Chapter 3, but the poetry mentions above all bracken as a plant gathered in burned areas. It is prominent in this next poem:

> Though one does not see its leaves burn *moyu/* or shoot *moyu* or smoke *keburi tachi*, who then called this plant bracken *warabi/*fire of straw *wara-bi?*[73]

Here the writer indulges in two word-plays: he likens the name of the bracken to the fire of straw, and he mentions burning *moyu* while describing germination *moyu*. Straw was in fact used to make a fire by the peasants and by the guards who sent light signals. It is not known whether wilderness areas were set alight with loose sheaves of straw or with torches made of straw, but a 17th-century image shows burning on a hill using torches.[74] Bracken is also the subject of the following poem:

> The fire of straw/the bracken *sa-warabi* that grows *moyu* in the clearing at the foot of the mountain in spring *haruyama*, is its smoke the mist on the peak?[75]

The stem and crosier of the bracken shoots were used. This plant could be kept by macerating it, but several washes were needed to get rid of its bitter taste. From a certain point in time (unknown to us), the roots were also used to extract a very nutritious starch. It is known only that bracken roots were dug up in the 11th century. But even the stem had a high calorific value.[76] Bracken is in fact the collected plant most often mentioned in poetry. Here is another poem:

> The fire of straw that burns itself out/ the bracken that grows *moewataru*, despite the falling rain, no doubt shows the site of the clearing that they burn *yakuno* in spring.[77]

The last two poems come from the poetic anthology of the time of Emperor Horikawa (r. 1086–1107), *Horikawa-in hyakushu*; the first version was presented to the sovereign in 1105 by Ōe no Masafusa (1041–1111). This collection has sixteen poems about "young bracken", most of them associating the plant with burning. The number of poems on this subject is much higher still in the *Fuboku wakashō* (14th century). Burning also occurs in the *Kokin waka rokujō* (10th century), the first poetic anthology classified thematically. It contains a section on "spring grasses", often gathered in a burned area. Inspired by this collection, the *Shinsen waka rokujō* (13th century) has sections on "spring grasses" and "the spring clearing", most of the poems relating to burning.[78]

Thus, entire sections of poetry collections are about the burning of clearings. One also finds references in several other anthologies, such as the *Goshūi*

Table 2.3 Poetic terminology of the spring fires

Item	Words appearing in the poems
places	Kasugano, Tobuhino, Sagano, Musashino, Yoshino
clearings/grasslands	*no, nobe, katano, harunono, susono no hara, ōno, hara*
mountains	*yamabe, haruyama no susono, toyama, yamano, katayama* 片山
season	spring *haru*
vegetation	young grasses/shoots *wakana, wakakusa,* grasses *kusa, fukakusa,* aster *uhagi,* bracken *warabi, sa-warabi, uchi-warabi,* polygonum *itadori,* lespedeza *hagi,* miscanthus *ogi*
burning	*moyu* もゆ, *yaku* 焼く, *tobuhi* 飛火, *keburi tatsu* 煙立, *nobi* 野火, *noyaki* 野焼
burned areas	*yakehara* 焼原, *yakeno* 焼野, *yakefu* 焼生
blackened clearings	*suguro* すぐろ (末黒), eulalia: *susuki**

Note: * Words are often written in *hiragana* in poems.
Source: The poems mentioned in this chapter (see notes 59–87).

wakashū, compiled in 1086 by Fujiwara no Michitoshi (1047–1099).[79] The poetry shows a range of vocabulary relating to fire and open spaces. See Table 2.3 above.

Running fire in clearings and other grasslands may be regarded as a common practice in early medieval Japan. If there are many more references in poetry to this practice than to swidden-field crops, this may be attributed to the appeal of the subject: the swidden field tended to recall the hard work of tillage, whereas the spring clearing and fire provided material for associations relating to amorous encounters, a topic dear to poets. But the evocation of spring fires and the gathering of bracken also reflect the everyday seasonal life of the Japanese of the medieval period. Ethnologists have noted that fire contributed beneficially to an increase in the quantity of bracken and pueraria *kuzu*, two vegetables with a high calorific value.[80] The early poets also speak of the blackened appearance of clearings after fire.

The poetry of the blackened clearing

Fujiwara no Nobuzane composed the following on the topic of the spring clearing:

> Under the spring shower that bathes the [charred] plants of the blackened clearing *suguro*, making tiny suckers at their feet, clumps of eulalia *susuki* rise up in the burned clearing *yakeno*.[81]

"The blackened tips" *suguro* 末黒 refer to the burned appearance characteristic of an area set on fire *yakeno*. In this poem, one sees the young shoots of new vegetation coming out on a plain or slope overgrown with eulalia. Plants were gathered in the blackened clearings:

> On the young shoots that we pick while making our way among the [charred] plants of the small blackened clearing in front of the spring mountain, light snow is falling.[82]

The blackened clearing is in fact associated with young grasses or young shoots *wakana* and also with bracken,[83] but it is characterized above all by a plant cover of graminae such as eulalia *susuki*. This is expressed, among others, by Jōen (1016–1074) in a poem composed in 1041 on the topic of the horse in spring:

> When on the Awazu moor the eulalia puts forth its young leaves, the horses neigh, still numb from winter.[84]

Eulalia or Chinese silver grass, Miscanthus sinensis, was not a collected plant, but its stems were highly prized for handicrafts and as a roofing material.[85] The poetry speaks of the gathering of eulalia *karukaya* in autumn, although this was sometimes done in summer in burned clearings, according to this poem by Sone no Yoshitada:

> While even yesterday the spring mist stretched out, today – from what hour? – it seems they are harvesting *karu* the charred plants *suguro* from the mountain side *yamabe*?[86]

The poem does not specify whether this is eulalia (Miscanthus sinensis); and in fact the poetry also mentions "burned plains *yakehara* of sweet miscanthus *ogi*" (Miscanthus sacchariflorus Benth. et Hook.).[87] The two graminae are very similar, but sweet miscanthus grows in damp places and near water, whereas eulalia likes sunny places.

Today, eulalia still characterizes certain landscapes in Japan. The Sengokuhara plain in the mountains of Hakone is renowned for the beauty of the eulalia whose silver colour shines under the October sun over broad areas. The annual burning at the end of spring helps to ensure gleaming ears and uniformity of the stems. This tradition was revived in 1989, after the Sengokuhara plain was abandoned because the practice of cleaning by running fire was halted in 1970. Today, as in the early medieval period, industrial graminae are of very high quality if they are regularly burned.[88] After having examined the poetry in order to identify certain farming practices of early medieval Japan, we now turn to toponymy.

Toponymy

Toponyms are a legacy of both the oral and written traditions. Most of those found in the written sources come originally from regional uses that have been transmitted orally. In modern Japan, their study has interested philologists since the Edo period and ethnologists since the Meiji period. A census taken by the government in 1884 gave Yanagita Kunio the chance to become interested in this research. As secretary for the census, Yanagita at that time managed a "List of toponyms of small localities", *Azana shū*, which included the "little places" *azana* within each village and hamlet. While the official administrative units up to that time included the names of the prefectures, districts, villages and hamlets, these "nicknames" concerned the areas around villages, names known and used by the local inhabitants, but not necessarily recorded in the official land surveys. Although the *Azana shū* was lost at the time of the 1923 earthquake, Yanagita continued to note down toponyms during his many field trips throughout Japan.[89]

In his study of place names, Yanagita discussed swidden-field crops and put forward hypotheses that may be summarized as follows. Different families of toponyms can be distinguished, according to the dialectal words referring to the swidden field. Names relating to *koba tsukuri* and *koba kiri* (swidden field) are common in Kyūshū and Shikoku, but absent from the east of the country. In the mountainous regions of Kantō one finds names containing the word *sasu* (swidden field) especially in Musashi and Sagami, and also the word *sori*. *Sori* is a verb meaning *sorasu*, 'allow to become waste land'. This may have to do with the cycle of the temporary bringing under cultivation *sasu*, followed by the abandonment of the patch of forest *sorasu*. Be that as it may, there are in fact places where swidden farming is practised in Yamanashi that are called XY-sōri 草里, as well as in Shizuoka where they are called XY-sōre. On maps, these are often transcribed -sōren 蔵連, which is simply a phonetic way of writing, *ateji*. This kind of toponym is also found in the mountains of Aichi and therefore, it seems, throughout eastern Japan. Sometimes *sori* is found in place names in the abbreviated form -*so*.

The suffix -*kusa* (grass) indicates other toponyms that probably relate to swidden fields followed by the planting of trees *kirikae hata*. This name characterizes places in the same mountains (of Aichi) with fertile soil and an easterly exposure. As they are suitable for growing crops, they have attracted populations practising permanent agriculture, so much so that the present-day inhabitants no longer know the original meaning of the name with the suffix -*kusa* of their village. Also worth mentioning is the name Natsuake 夏明 meaning *natsu yake* ('to burn in summer'). It denotes patches of forest cleared in autumn that can be burned in summer (and not immediately in spring), thanks to a favourable aspect and the fertility of the soil. These are places with dwellings of clearers that have been turned into permanent houses. One also comes across the toponym Yaku ('to burn') even in places where today there are permanent fields.

The name Araku or Arako (swidden field, *araku*) characterizes the villages of the Aichi region. However, in Kantō, Araku is common, while *araku okoshi* in the regional dialect means 'to clear and establish permanent fields'. This too may be a case of a gradual change from swidden to permanent agriculture. In Aichi, the name Hirako, normally meaning 'to clear', is also very common. This may refer to the preparation of rice fields, like Hokku in the Kantō region. It would be necessary to verify on-site the altitude and morphology of the places called Hirako in the Aichi region to ascertain the meaning. As regards the preparation of rice fields, in Kantō there are toponyms relating to *-arata* (new rice field), in the Kyōto-Nara region the names Shin-yashiki and Ima-zaike (new residence), and in Aichi the name Arai 荒居 formerly known as Niii or Niinoi.[90]

For the north of the country, Yamaguchi Yaichirō published, in 1944, a study of the dialects and toponyms relating to swidden farming. He listed the dialectal terms for swidden field and in addition distinguished the words and names relating to farming practices, i.e. the felling of trees and the cutting of vegetation, as well as burning and the stay in a summer or makeshift hut.[91] For Shizuoka, Yamanashi and Nagano prefectures, Nomoto Kanichi extended his inquiry to other related techniques. The toponyms he noted fill twenty-five pages of his book. He tells us that the same swidden field had a different name for each of the four years of cultivation. For example, at Tashiro, near Shizuoka city, the swidden field was called *araku* in the first year, *kawashi* in the second year, *kuna* in the third year and *ekkoji* or *shigguna* in the fourth year (all in *katakana*).[92]

Thus, toponymy is heavily influenced by dialects. The six volumes of the "Linguistic Atlas of Japan" *Nihon gengo chizu* show that each word had many variants and diverse pronunciations in each region of Japan, even in the 20th century. When studying toponyms, it is also necessary to take account of semantic shifts, i.e. a possible change in the meaning of the same term from one region to another. The examples given by Yanagita Kunio also show phonetic changes, such as *araku – arako, sōri – sōre*. And when it came to writing down place names, phonetic script was sometimes used that took no account of the meaning, such as the names Natsuake and Sōren. It is therefore by studies in the field that ethnologists have been able to verify the meanings of toponyms in modern Japan. The Institute for Toponymic Research, Nihon chimei kenkyūjo in Kawasaki, run by Ichikawa Kenichi, today collects toponyms throughout the country with the help of members of the society affiliated with this Institute, and compiles and classifies them. Yet, when one consults the large number of geographical dictionaries, it becomes apparent that there are various explanations for each name. For the historian Ishii Susumu (personal communication), it is difficult for history to make use of such material. Nonetheless, we will try an approach using the toponymy of the early medieval period.

The toponyms of early medieval Japan

Toponyms constitute a very rich cultural heritage for the history of everyday life. In most countries, this heritage dates from the modern period or at best from the medieval period, and a process of linguistic reconstruction is used for the semantic study of place names. The reconstruction of early forms is in fact an important task of present-day toponymy. However, Japan and Denmark are among the few countries to have preserved even a small fraction of their ancient and medieval toponymic heritage.[93]

For early medieval Japan, we have the names of the administrative units, i.e. provinces *kuni*, districts *gun*, villages *sato* (*gō*) and shrines for the country as a whole, thanks to the lists in the *Wamyō ruijushō* encyclopedia (930s) and the *Engishiki* regulations (927). Also known are several hundred "little names" *azana* of smaller localities, thanks to the regional land surveys and temple inventories of the 8th–9th centuries.[94] In 1942, Ono Takeo drew attention to the early medieval toponyms "Fire field" *Hida*, and "Field sown after fire" *Yakimakida* that relate to swidden fields. It is with this in mind and by reference to modern research that we have undertaken a careful study of the early medieval documents. Given that the difficulties remain the same as for modern toponymy, namely phonetic and writing changes and semantic development, our suggestions relating to early medieval place names and their meanings can only be hypothetical, all the more so as there are many homonyms and the dialects of the early medieval period are unknown.[95]

For the meanings of toponyms, we have relied on the terms noted above in the early dictionaries and poetry, as they seem to us to be reliable indicators of the terminology of the medieval period. Sometimes we have gone to the legends in the local monographs, *Fudoki*, of the 8th century for the etymology. Lastly, we have been to the sites: we have visited the places with names that could relate to swidden fields, such as Hata, to verify the altitude and morphology of the terrain in order to find out whether the place in question was, at least in theory, suitable for swidden-field crops.

Swidden fields, cleared plots and land clearing

A number of toponyms recorded in the *Wamyō ruijushō*, *Engishiki* and *Fudoki* reproduce terms given by the early dictionaries or used in the early medieval poetry. They are: *Wilderness field* Arata, *Swidden field* Hata, *Large swidden field* Ōhata, *Small swidden field* Ohata and *Cleared plot* Araki. The name *Mountain field* Yamada is extremely common across the country. It recalls the mountain fields mentioned in poetry. It may refer to rice or other fields, but there are two examples of villages called *Mountain field* located respectively in the districts *Cleared plot* Araki and *Swidden field* Hata.[96]

The characters vary for the same names. Hata was written in nine different ways and Araki in five ways (see Table 2.4). In the early medieval period, district officials did not always know the standard characters, that is, those

Table 2.4 Toponyms possibly associated with swidden farming: names of districts (*gun*), villages (*sato*) and public post stations (*eki*)

Japanese transcription/characters		English
Arata* (5)	荒田	Wilderness field
Araki* (11)	荒木・荒城・ 原木・新木・新城	Cleared plot
Hata* (9)	幡・幡多・播多	Swidden field
Ōhata* (2)	大幡	Large swidden field
Ohata* (4)	小幡・少幡	Small swidden field
Yamada* (40)	山田	Mountain field
Hikita* (2)	曳田	Field with shaken wooden boards
Niida/Nitta/Niuda (14)	新田	New field, of the third year
Niihari (4)	新治	Newly prepared field
Hida* (4)	斐田	Fire field
Atsuta* (1)	熱田	Burning field
Yaku (2)	夜久	Burning
Yaki* (1)	養宜・八木	Burning
Yata* (5)	矢田	= Hata, Swidden field?
Yata (11)	八田	= Hata, Swidden field?
Ita* (1)	猪田	Field with wild boar
Karuta? (2)	鹿田	Field with deer, Harvested field
Taka* (5)	田鹿・多可・田可	Field with deer
Karunoda (1)	苅野田	Cleared wilderness field
Karuno (1)	鹿野	Cleared waste land
Shishiya? (1)	鹿屋	Shelter for scaring away deer
Noda (4)	野田	Wilderness field
Irino/Ihino/Nyūnoya* (5)	入野	Entering virgin land
Hatano (1)	波多野	Wilderness with swidden fields
Hatahara (1)	秦原	Plain with swidden field
Takahata (1)	高幡	High swidden field
Hatanoya (1)	幡屋	Shelter near the swidden field
Mahata (1)	真幡	Swidden field...
Ōara (1), Oara (1)	大荒・小荒	Large, Small wilderness plot
Kamutsuara? (1)	上荒	Upper wilderness plot
Arayama (1)	荒山	Wild mountain

Note: * Also a shrine name.
? Reading uncertain.
( ) Number of occurrences in the sources.
Source: The texts of the 8th–10th centuries (see notes 96–100).

adopted by the court, and they sometimes used a phonetic transcription, i.e. 波多 instead of 畑. Thus, the single village of Hata in the district Takechi-gun in Yamato is written in three different ways in the early medieval texts.

There were also vernacular words with no fixed way of writing. We know that the character for *hata* appeared at the earliest in the 11th century. Conversely, the same characters were sometimes pronounced differently according to the region. The toponym 治田 was read Hata by some, Harita by others and Haruta by still others.[97] In this instance, the meaning possibly shifted from one aspect to another of the same technique, from burning to clearing *haru*: to clear it was necessary to burn. One may ask whether the names *Newly cleared land/field* Niihari and Niida/Niuda/Nitta relate to assartage or rather, more accurately, to land clearing, as Niida/Nitta is written like the new field or that of the third year *arata*, intended for permanent cultivation according to the definition in the early medieval dictionaries. However, the name Araki and some compounds of *ara* (wild) perhaps more accurately referred to a cleared plot. Other toponyms were related to assartage. These are the villages *Fire field* Hida, *Burning field* Atsuta, *Wilderness with swidden fields* Hatano and other compounds of *hata*.

The wilderness, moor and clearing: *no*

The word *no* is found in many toponyms. A number of scholars have expressed an opinion on the etymology and meaning of *no*, and the dictionaries give various definitions. Today, these "neither cultivated nor wooded expanses" would include the lower parts of mountains or less rugged slopes than the high mountains.[98] In this chapter, the reader will have noticed that the meaning of *no* changes according to the context and that we have used different translations. In the early medieval administrative texts, *no* refers to wilderness or virgin lands situated outside the administrative units or else on estates. In the former case they are uncultivated lands, and in the latter case lands unsuitable for irrigated rice growing. The land surveys of the medieval period distinguish between "(shifting?) fields established in a wilderness area" *nobatake* and "mountain swidden fields" *yamahata*. In fact, *yama* and *no* are both wilderness areas. There may have been a distinction according to the altitude or morphology of the terrain, or merely a psychological difference: in the eyes of villagers, the wilderness *no* may have seemed less hostile and inaccessible than the mountain *yama*. In poetry, *yama* is sometimes associated with swidden-field crops, whereas *no* is burned every year to promote the regrowth of useful plants. As the plants are often (not always) those that like the sun, the reference in these cases is to clearings. There is also frequent mention in poetry of mountains with names ending in -no, such as Yoshino and Kasugano. We have translated them by "moor" for reasons of convention and convenience.

Among the names of early medieval districts and administrative villages are compounds of *no*. Noda, *Wilderness field*, and Irino/Ihino/Nyūnoya, *Entering virgin land*, could refer to the bringing under cultivation of waste ground. On the Yamana estate in the province of Owari, Irino refers to the part where dry fields have been established.[99] One finds other compounds of *no*, such as Karunoda written like *Field on virgin land (ravaged by) deer*, and also like *Wilderness field where the vegetation has been cut (karu)*, possibly associated with clearing before burning. The name Karunoda recurs, in the 12th century, in Hyūga (Kyūshū). In modern Japan, it is a toponym of Mutsu, in eastern Honshū. According to an agricultural treatise of the Edo period, *karuno* was read *kano* and in dialect meant 'swidden field' *kanohata*.[100] Also, names such as *Deer shelter* Shishiya and *Wild boar field* Ita possibly allude to the struggle against the animals that destroy the mountain crops. They recall the "fields of deer and wild boar" *shishida* in the early poetry.

The toponyms of small units

The toponyms mentioned above concern the administrative units, namely the official districts and villages, and do not necessarily reflect the actual situation. They were probably fixed when the first land surveys were done in the 7th century, based on the regional oral traditions and the language used by officials. By contrast, the names of smaller units, those of the block (*ri*, about 40 ha) and the plot (*tsubo*, just over 1 ha) were able to be determined later, as the need arose or as lands were developed. They are found in the administrative documents containing the regional land surveys and land maps of the 8th–10th centuries (Table 2.5).

The following examples come from the province of Echizen that corresponds to present-day Fukui prefecture. In the district of Asuwa was the Chimori estate belonging to the Tōdaiji temple in Nara. This locality corresponds to the western suburbs of present-day Fukui city, i.e. the Wakasugi-chō district in the fork at the confluence of the Asuwagawa and Hinogawa rivers. On this estate were at least five plots called *Field of hot ashes* Atsuhaida. According to the 8th-century map, they were located at the foot of a mountain, close to other plots called *Mountain field* Yamada. Many of the adjacent plots were still "virgin" (*no*). Others, under cultivation, were called *Wilderness field* Noda, a toponym also designating three villages in the same province. One farmer recorded in the land surveys for the estate was called Ikagabe no Noyaki 伊何我部野焼, *Burner of wilderness*, which proves that swidden fields were used on this temple estate. Other names of farmers in the same land survey, written phonetically, can be interpreted as follows: one man had Hita, *Fire field*, as his given name and Noo, *On virgin land* 野於斐太 as his family name. A public slave had exactly the same name in 743. Another farmer from Chimori was called Tamaro with the family name Hatasaki, *In front of the swidden field* 秦前田麿. Yet, visiting the site of the old Chimori estate in Fukui, it is difficult to imagine forest agriculture, as the land is completely flat and very built up. Even outside mountainous regions, swidden farming could be practised on gentle slopes at low altitude. The presence of a mountain on the 8th-century map and the absence of the slightest relief in the present-day terrain had put us on our guard

Table 2.5 Toponyms possibly associated with swidden farming: names of small units: blocks (*ri*) and plots (*tsubo*)

Japanese transcriptions/characters		English
Arata*	荒田	Wilderness field
Araki*	荒木	Cleared plot
Arakida	荒木田	Field on a cleared plot
Ōarakida	大荒木田	Large field on a cleared plot
Tera-oarakida	寺小荒木田	Small cleared temple plot
Aramaki	荒蒔	Sown waste land
Aramakida	荒蒔田	Sown wilderness field
Yakimakida	焼蒔田	Field sown on a swidden field
Araha(ri)ta	荒治田	Field prepared on waste land
Toshi-ha(ri)ta	年治田	One-year field
Ara-tsukurita	荒作田	Field cultivated on waste land
Atsuhaida	熱灰田	Field of hot ashes
Yaefuda	夜恵父田	Dry field set on fire (Yakifuda? 焼圃田)
Yakida	焼田	Burned field
Yamada*	山田	Mountain field
Noda*	野田	Field on waste land
Noyorita	野依田	Field next to a wilderness
Nonakada	野中田	Field in a wilderness
Nobatake	野畠	Dry field on waste land
Hata*	幡	Swidden field
Hatasaki takamine	八多前高峯	Summit with swidden fields in front
Hatayama	波多山	Mountain with swidden fields
Kori	口利	Cleared
Korita	口利田	Cleared field
Ikakida	猪垣田	Field fenced against wild boar
Iorita	伊保理田	Field with a hut

Note: * These are also the names of a district, village or post station.
Source: The documents of the 8th–10th centuries (see notes 101–108).

against the misinterpretation of toponyms. However, an old aerial photo showed that this locality still had a high wooded hill in 1948, which it seems was subsequently levelled during the construction of housing developments in what has become a suburban area.[101]

Elsewhere, the toponyms Noda, Yamada and Hata (*Swidden field*) denote several blocks of the Ōkuni estate of the Tōji temple in the province of Ise (Mie prefecture), taking in the districts Iino-gun and Take-gun, a locality corresponding to present-day Taki-chō south of Matsuzaka city. This estate straddled the Kushidagawa river, with its right bank entirely on the flat and its left bank running along the mountains. There were also compounds of *hata*. For example Hatasaki(saka) takamine, *Summit with swidden fields in front*(?), is the name of several blocks on an estate in Iga (Mie) and Hatayama, *Mountain with swidden fields*, is found in Izumi (Ōsaka).[102]

In the district of Sakai, adjoining that of Asuwa in Echizen (Fukui), the toponym *Field sown on a swidden field* Yakimakida designates nine plots. On one side, these are next to plots called *Field on a reed plain* Ashiharada. Nine other plots in this district are called *Field on a cleared plot* Arakida. In fact the *Clearer* Araki family lived in the locality. Plots with the name *Field on a cleared plot* are found, too, in the provinces of Kii, Ōmi, Tanba and Yamashiro. There were also compounds such as *Large field on a cleared plot* Ōarakida at Sakai in Echizen. The name *Small field on a cleared temple plot* Tera-oarakida denotes a plot actually belonging to a temple; it appears on the map "Nukatadera garan narabini jōri zu" of the Nukatadera in Yamato. However this temple, the present-day Kakuanji, now stands on low, marshy ground south of Nara, at Yamatokōriyama-shi, Nukatabekita-chō.[103]

At Sakai, near the Arakida plots, there are three other plots called *Field set on fire* Yaefuda, and next to them several plots and several blocks called Kori, *Cleared*(?), and Korita, *Cleared field*. Though written phonetically, this name is suggestive of the verb *koru*, 'to clear/to fell', and perhaps relates to the "field prepared after having felled the trees" *kikori no ta*, which is found in the 12th-century *Ruiju myōgishō*. However, the old district of Sakai is on a plain between Kanazu-chō and Harue-chō north of Fukui city, and, at least today, does not seem at all suitable for forest agriculture. Was this region wooded in the early medieval period? If so, agri-sylviculture (see below) may have been practised. The name Korita also appears in a plot located at Kuse in Yamashiro.[104]

For Yamashiro, there still exists a 9th-century map of land in the district Kadono-gun, a locality in the present-day districts of Saga and Arashiyama in Kyōto, on either side of the Katsuragawa. The map records a plot called *Sown wilderness field* Aramakida and notes that this land is virgin/wild *arashi*. The land also includes several toponyms meaning *Field in/next to a wilderness area*: Noda, Noyorita and Nonakada. In five plots, the fields are marked as: "wilderness field" *nobatake*, which could also be a toponym.[105] Close by was the Kōryūji temple, in the block *Sown waste land* Aramaki. Today, the grounds of this temple are in the Uzumasa district of Kyōto, very built up and flat, but with a hillock that may be the remains of a hill of the early medieval period. The Kōryūji estate included blocks and plots called *Sown wilderness field* Aramakida, *Field on a cleared plot* Arakida, *Field prepared on waste land* Araharita/Arahata and *Field with a hut* Iorita. This last name recalls the farmer staying in a makeshift hut to drive away the wild animals. This land also includes toponyms possibly relating to agri-sylviculture.[106]

The region of Uzumasa in Kyōto was inhabited in the early medieval period by the Hata 秦, an old immigrant family from the continent. According to the ethnologist Miyamoto Tsuneichi, the name Hata suggests the idea of the swidden field *hata*. But historians are divided as to its origin. They think that Hata could go back to the Korean word *pada* meaning the sea or else to the Japanese word *hata*, the loom. Historical studies on the Hata family have not taken account of the meaning 'swidden field' for *hata*. Perhaps scholars were unaware of this until the publication of the etymological study of *hata* by Kuroda Hideo in 1984.[107]

The names mentioned above mainly relate to certain lands in Echizen and Yamashiro. For the other provinces, there is far less information on the toponyms for small units, but we can give a few examples. In Iga, there was a plot called *Field prepared on waste land* Araha(ri)ta. A land map for Ōmi (Shiga prefecture) includes the name *One-year field* Toshi-harita/Toshi-hata. Another map showing the districts north of the palace at Nara has three plots called *Field cultivated on waste land* Ara-tsukurita, next to a plot that "had just been cleared" *aratani hiraku*. On an estate belonging to the Ninnaji temple in Kai (Yamanashi prefecture) was a block called *Burned field* Yakida.[108]

Our toponyms of small localities are limited geographically to around fifteen lands situated mainly in central Japan, owing to the lack of sources. We have land surveys and deeds of property transactions with the names of blocks and plots for only eight central provinces, those of Yamato, Yamashiro, Kii, Ōmi, Iga, Ise, Settsu, Izumi, and for three others on the Sea of Japan, those of Echizen, Echigo and Inaba. Since these eleven provinces represent one-sixth of the country, a quantitative evaluation for the country as a whole should multiply the number of toponyms by six. While only twenty-five names of small localities have been identified in the Japan of the 8th–10th centuries (Table 2.5), it must be remembered that there could be six times as many. The interpretations put forward above are based on the physical environment, that is, the natural setting and the utilization of space, but sometimes there is also a link with family names and mythology. This is the case with shrine names.

Shrine names

The toponyms of the official units recorded for the entire country, that is, the provinces, districts and villages, often overlap with shrine and family names. The land surveys record the family names Arata, Araki, Arakida, Hata, Ōhata, Ohata, Yamada, Hikita, Noda and Hatano.[109]

Among the 2,861 shrines in the country, recorded in books 9 and 10 of the *Engishiki*, are first listed family names, spread across the provinces: four Araki no jinja (*Cleared plot*) including one in the village of Araki in Hida (*Fire field*, Gifu prefecture), two Arata jinja (*Wilderness field*) including one in the village of Arata in Harima (Hyōgo prefecture), five Hata jinja written with different characters, including one in the village of Hata in Yamato, and three Yamada no jinja (*Mountain field*), but no Noda jinja (*Field in a wilderness area*). Moreover, shrines often have the same names as districts and administrative villages: Hida, Atsuta, Yata (or Hata?), Ōhata, Ohata, Irino and Yaki[110] (Table 2.6). But there were also a great many compounds, such as Hohata.

Shrine names can be significant when they refer to the local toponymy and the physical environment. In this case, they may well reflect the customs of the local inhabitants. In order to verify the possibilities associated with swidden farming, we visited the sites of a number of shrines in Nara prefecture.

In the early medieval period, the shrine *Swidden field on fire* Hohata 火幡 was located, according to the *Engishiki*, in the district of Katsuraginoshimo in Yamato (Nara prefecture), and today it is at Ōji, 15 km south-east of Nara, in the hamlet of Hatakeda. As with all the other shrines, the buildings date from the modern period. In the *Engishiki*, the 12th-century Kongōjibon copy glosses this name by Hohata (swidden field on fire) no jinja, and the Kujōkebon copy, dating from after 1078, glosses it by Hota no jinja. Today, this shrine is not situated at altitude, but in a thicket on a hillock, in undulating surrounds, the steep areas being occupied by houses and the flat areas by rice fields. While the shrine name certainly suggests swidden farming, the panel above the entrance gives another interpretation: Hobata jinja, founded in 806, was dedicated to the deity Honokohatahime 火之戸幡姫, whose name is indeed written with the character for fire *ho* and the old phonetic transcription for swidden field *hata*, but who nonetheless remains the tutelary deity of weaving *hata-ori* 布織.

A research group conducted a study of the *Engishiki* shrines and published the results in 1982 in the editions of the Kōgakukan University, attached to the Ise Shrine. This report quotes the early and modern sources. The shrine Hohata jinja was allocated twenty households in 806 and raised to the 5th rank in 859. The name was read Hohata or Hota 火田. The Edo-period register of shrines *Jinja kakuroku*, anonymous and undated, notes that it would be necessary to visit the site to see whether there were swidden fields *hata* or permanent fields *hatakeda*, and suggests that, being in the hamlet Ōhata(ke)-mura, its origin could go back to Ōhata. Ban Nobutomo (1773–1846) writes in 1813 in his work "Reflections on the *Engishiki* (927) register of shrines" (*Jinmeichō kōshō*) that the local inhabitants called the shrine Yahata or Hachiman. The god Amenokoyane, the goddess Amaterasu, the spirit of Jingu kōgō and two other deities were worshipped here. Only a shrine chronicle, which may date from the Meiji period and has been preserved by the shrine, states that the goddess of weaving was worshipped here before the other deities.[111]

It is therefore *a posteriori* that worship was reshaped. One can accept a possible link between the Hohata shrine name and swidden farming, without being influenced by the modern legend. The example of Hohata jinja tells us that the shrine legends are sometimes, and in fact nearly always, later than their founding. Some legends date from the Meiji period. However, care is needed with homonyms like *hata* – "swidden field" and *hata* – "loom", since the names, written phonetically, do not reveal their meaning by their written form. This example also shows us that shrines can change their name. Here Hohata becomes Yahata with its first character changed. Another shrine changed its protector. Hatamikai jinja 波多甕井, which can be translated as *Swidden field-offering vessel*, was situated in the early medieval period in the village of Hata (*Swidden field*), in the district Takechi-gun in Yamato. Today it is in Takatori-chō, at the foot of the mountains south of Asuka. It is undulating and wooded with a few terraced rice fields here and there. The shrine was dedicated to Amaterasu according to the Edo-period monograph *Yamatoshi*, and to the god Nigihayabi (with the reading *nigi* for the character *mika*), according to a Meiji-period register. It seems that in the Meiji period, they wished to explain the name by a mythological reference, but the geomorphological situation and the localization of the shrine in the village of Hata in the early medieval period enable a possible link to be established with swidden-field practices.[112]

Another shrine seems to us to have an especially convincing name, that of Yakifu-yamakuchi jinja 夜支布山口. Located at Yagyū-chō, in the mountains north-east of Nara, its name translates as *Entrance to the mountain with burned fields* yakifu 焼圃, recalling the "burned areas" *yakefu* (*yakifu*) evoked by the poet Sone no Yoshitada. From early times, this shrine was also called Yagifu 養父, which today is read *yabu*, the thicket, and in addition means swidden field in some dialects. A modern register designates the god Susanoo as the tutelary deity of this shrine, but Ban Nobutomo states that here they worshipped the mountain god Ōyamatsumi, who can be associated with swidden farming.[113]

After having mentioned compounds of *hata*, here is a shrine called simply Hata jinja 波多 in the old village of Hata, in the district Takechi-gun in Yamato. Its present-day site is either near Hatamikai jinja in the same village, or in the mountains east of Asuka, where there is still a Hata jinja near Danzan jinja. The latter place is 600 m above sea level in steep terrain unsuitable for rice growing. This shrine was dedicated from the early medieval period to the ancestors of the Hata 波多 family, distinct from the Hata family of Yamashiro (Kyōto).[114]

The *Engishiki* lists 221 shrines in the province of Yamato. One can point to names "suggestive of burning" other than those mentioned above, but they are more difficult to interpret. Unehi-yamanokuchi jinja 畝火山口, i.e. *Ridge fire at the mountain entrance*, was moved in 1940 from the top to the foot of Mount Unebi. The Meiji government had founded there the shrine of Kashihara honouring "Emperor" Jinmu. Mount Unebi is not normally associated with swidden farming, but the etymology of its name remains

Table 2.6 Shrine names possibly associated with swidden farming

Japanese transcriptions/characters		English
Hida*	非多・日田・斐太	Field on fire
Hota/Hohata	火幡	Swidden field on fire
Atsuta*	熱田	Burning field
Unehi-yamanokuchi	畝火山口	Ridge fire at the foot of the mountain
Hibashiri	火走	Running fire
Hichi	火走	Land to be set(?) on fire
Yahata/Hachiman	八幡	Swidden field
Yata*	矢田	= Hata, Swidden field?
Yaki	夜疑・養基	Burned
Yakifu-yamakuchi	夜支布山口	Burned field at the foot of the mountain
Yakitsu	焼津	Swidden field trap/port
Hata/Haruta*	治田・波多	Swidden/Cleared field
Hatai	幡井	Swidden field
Kohata	許波多・巨幡	Swidden field with trees
Ohata*	小幡	Small swidden field
Ōhata*	大幡	Large swidden field
Kamuhata	神幡多	Upper swidden field
Hatamikai	波多甌井	Swidden field-offering vessel
Awawa	阿波々	Abundant millet
Awano	粟野	Wilderness area with millet
Asahi-hataka	朝日波加	Swidden field facing the sun...
Achiha/Ehata	謁播	Swidden field...
Himukai	日向	Facing the sun
Hatahisa	幡日佐	Swidden-field sun...
Ihata	伊波多	Swidden field with wild boar
Itsuhata	五幡	Five swidden fields
Hatasaka/Hatanari*	幡生	Swidden field-dry field (hatafu 幡圃)
Hataki	波太伎	Swidden field...
Mahatasu	真幡寸	Swidden field...

Japanese transcriptions/characters		English
Irino*	入野	Entering a wilderness area
Katano	片野	Alternating wilderness area
Karino	苅野	Cut-down wilderness area
Kusakari/Kusanagi	草薙	Grass-cutter/Cleared plot of grass
Kusanagi	久佐奈崎・草名伎	Cleared plot of grass
Arata*	荒田	Wilderness field
Araki*	荒木・荒城	Cleared plot
Suchi-araki	須智荒木	Cleared plot...
Arami	荒見	Wild...
Iotachi	五百立	Built hut
Ikami	猪上	Wild boar up high
Ita*	猪田	Field with wild boar
Hikitabe	引田部	Shaken wooden board...
Hikita*	曳田	Field with shaken wooden boards
Yamada*	山田	Mountain field

Note: * These are also district or village names.
Source: *Engishiki* (927), books 9 and 10 (see notes 109–116).

unexplained. The name of another shrine, Hikita jinja 曳田, could have been interpreted as *Shaken wooden boards* (*hikita* 引板) used to frighten animals in the mountain fields, yet this is not so. Hikita jinja is dedicated to the ancestor of the Hikita family. Ikami jinja or Ikamu jinja 猪上 could also suggest the wild animals that ravage the fields. This shrine, which has disappeared, formerly housed the tutelary deity of Mount Shigisan, a place today called Shigisan-hata, but its origin may go back to the ancestral cult of the Ikamube 猪甘部 family. The Ihata 伊波多 Shrine could mean *Swidden field invaded by wild boar*, but the early dictionary *Wamyō ruijushō* gives for another shrine of the same name in Ise the characters Iwata 石田, *Rocky field*. Ihata jinja was also called Yahata (Hachiman) in the 19th century. Haruta jinja 治田, *Cleared field*, is today at Asuka next to the Okadera temple. According to Edo-period writers, they commemorated here the clearing and establishment of rice fields carried out in mythological times, in the place where the Oharida palace (603–630) of Empress Suiko later stood. Although clearing is done using the technique of swidden, there is a strong rice-growing connotation here. We should, however, note that Haruta is read Hata in the *Engishiki*, in the Kujōkebon manuscript of ca. 1078, and that the link with Oharida was established *a posteriori* by the philology of the Edo period. Moreover, the name Araki jinja 荒木 could be associated with swidden fields, because *araki* meant 'a cleared plot' in the early dictionaries and today denotes swidden fields in some dialects; but in the specific case of the Araki Shrine in Yamato, it seems that there is a link with the Arasaka pass and with the deity Ōaraki no mikoto.[115]

Care is needed with these examples. We have referred to the report of the inquiry into shrines published in 1982, but as this report nowhere mentions assartage, we have added our own interpretations. Swidden fields were not taken into consideration in studies of the 1970s and the etymology of *hata* was hardly known at that time. We have therefore not taken out of Table 2.6 the names of shrines of uncertain meaning, so as not to exclude them from a possible revision of their interpretations.

Nor is the name Kamuhata jinja 神波多 associated with assartage by specialists, but it could be translated as *Upper* (kami-) *swidden field*. This shrine is in the mountains 30 km east of Nara, in the village of Hatano, near the summit Hatayokoyama (two toponyms containing the word *hata*). In the early medieval period, this place was on the route to Ise. Today, the shrine still preserves two stone lanterns with inscriptions dating from 1316 and 1346. In the Edo period, a shrine panel mentioned the name of the god Gozu Tennō. The 1982 report makes no mention of any link with swidden fields, but notes, quoting the monograph *Yamabe gunshi* of 1916, that the local inhabitants also called the shrine "Hata no tennō san", a reference to Gozu Tennō, who is none other than the "*Heavenly king of swidden fields*". Given that the 1982 report does not actually suggest another interpretation, it is permissible to include Kamuhata jinja among the shrines with names that recall the swidden field.[116]

In short, present-day scholars emphasize the mythological references and ancestor worship, like the philologists of the Edo period. However, the toponyms of small units and the *Fudoki* legends tell us that some place names relate to the physical environment. Be that as it may, in both cases, we remain dependent on conjecture because of the unknowns relating to the pronunciation of the characters, as well as the dialects and homonyms of the early medieval period.

Homonyms and dialects

Above, we made choices according to the context. We opted for swidden farming based on semantic interpretations and also after considering certain associations of several toponyms of similar meaning in the same locality. For example, grouped on a single estate are blocks called Noda, Yamada and Hata, and in another locality the plots Arakida, Yaefuda and Korita. However, it is necessary to take account of the existence of homonyms.

Hata means not only 'swidden field', but also 'flag' and 'banner' *hata*. This latter meaning occurs in poetry, and one finds in the *Izumo fudoki* "eulalia waving (in the wind) like a large flag" *hatasusuki*, an expression very reminiscent of the burned areas of eulalia, likewise known in poetry.[117] *Hata* also means 'loom' *hata* and 'woven cloth'. This is its meaning in the following legend.

> Seven leagues (*ri*) east of the district headquarters [of Kuji] (Hitachi and Hitachi-ōta in Ibaraki prefecture), is the shrine of Nagahatabe 長幡部 in the village of Ōta. According to the old men, when the god Sumemima wished to come down to earth, the goddess Kamuhatahime followed him in order to weave his garments. They arrived at the summit of Futakami at Himuka on the island of Tsukushi (Kyūshū), then on the hill of Hikitsune in Mino (Okayama prefecture). Later, under the reign of Emperor Mimaki (Sūjin Tennō), the ancestor of the Nagahatabe, the god Tate, left Mino and settled in the district of Kuji. There he set up a weaving bureau *hatadono* 機殿 and introduced weaving [to this district]. The woven fabrics became garments all by themselves, without being sown, and were called 'full garments' *utsuhata* 内幡.[118]

The name of the shrine Nagahatabe therefore means 'long material', *hata* signifying 'loom' and *utsuhata* 'cloth', with the character for *hata* being that of swidden field *hata* in the *Wamyō ruijushō*. *Hata* also means 'fish fins' *hata* and denotes the offerings of fish in rituals. *Hata* further means 'extremity' *hada* (mod. *hashi*) and lastly 'skin' *hada*. All these meanings are attested by the early medieval literature.[119]

The word *araki* is likewise widely used in the early language: in poetry, it not only means 'a cleared plot', i.e. 'a newly cleared field' *araki(da)*, but also 'undressed timber' *araki*, 'anger' *ara-ki*, and 'a funerary pavilion' *araki (mogari) no miya*.[120] We have seen that *arata* is defined in the early dictionaries as a "wilderness field" cultivated for the first time after clearing, as a "new field" cultivated in the third year, as well as an abandoned, uncultivated field, i.e. the opposite. It would be a mistake to interpret the name of the shrine Sue no arata 陶荒田 as *Last wilderness field* sue no arata 末荒田: this shrine was in the district Ōtori-gun in the province of Izumi, between the cities of Sakai and Takaishi (Ōsaka), a place reputed for the production of pottery *sueki* at high temperature, where many remains of kilns from the 6th–9th centuries have been discovered. Here, *arata* is therefore associated with pottery *sue* and must have a different meaning.[121] Still other words with multiple meanings could be listed, such as *hi* (fire, sun, ice, supernatural power, as in Takamusubi, the Japanese cypress *hinoki*).

The word *hi*, 'fire', was probably pronounced *fi* and was also read *ho* in the early medieval period. In the 8th century, there still existed three vowels that subsequently disappeared (ï, ë and ö) and a larger range of consonants. Thus philologists distinguish, in this language, pronunciations known as *kō* and *otsu*. While these rules impose limits on semantic reconstructions, they also allow phonetic associations. A link can no doubt be established between the name Yakifu (*Burned field*), the word *yakefu* and the name Yaefuda, resulting from a contraction of *yakefu* to *yaefu*. One may also ask whether the toponym Yata was a phonetic variant of Hata, as the character *hachi* (eight) of this name was pronounced both *ha* and *ya*. A contraction in the direction *yakihata – yaihata – yata* could have played a role.[122]

Miyamoto Tsuneichi has drawn attention to the possible link between the toponym Yahata and swidden farming. Yahata 八幡 gave its name to the Hachiman 八幡 shrine at Usa in the old province of Buzen, now at Usa in Ōita prefecture (Kyūshū), and to the other shrines with this name dotted throughout Japan. The origin of the Hachiman Shrine is unclear, as is the time when the reading shifted from Yahata to that of Hachiman. The *Engishiki* mentions the Usa Shrine without glossing its name. The classical literature sometimes reads Yahata, but Hachiman also seems to be an early reading. The dictionary *Iroha jiruishō* gives the two readings Yahata and Hatsuman.[123] It was noted above that some shrines with names containing the semantic element *hata*, such as Hohata and Ihata, were called Yahata or Hachiman in the modern period. Perhaps there is a tradition at Yahata relating to assartage. An early poem seems in fact to allude to such a tradition. In it we read: "Beneath the moon adorned with a halo, I sow the seeds in the swidden field *yahata*; come, I shall clear this wilderness field *arata*."[124] The writer of the poem has compared the swidden field *yahata*, written like Hachiman, and the field *arata* to be cleared.

Generally speaking, one encounters in toponyms a vocabulary belonging to different registers. There are on the one hand common words and semantic elements, known from the early dictionaries (*hata, no, ara, hi*), and on the other hand less usual words, such as *yaefu, aramaki* and *yakimaki*. In fact, there was almost certainly a vernacular language that is unknown to us. Often, it is impossible to know what is hidden behind toponyms. How can one guess, for example, that the name Awawa refers to swidden farming? It is a village in the district Kako-gun in Harima, in present-day Kakogawa city near Himeji. The *Harima fudoki* says this about it: "The village of Awawa 鴨波: soil of average quality. Long ago, Korime 古里売 (Clearing woman *kori-me* 伐女?), ancestor of the family Ōtomo no miyatsuko, tilled the waste land (*no o tagaeshi*) and sowed a lot of foxtail millet. This is why they call the village Awawa 粟々." Awawa therefore meant the cultivation of millet in a wilderness field. While the characters 鴨波 are incomprehensible, the latter 粟々 reveal part of the meaning. The *Engishiki* also records an Awawa shrine in the province of Tōtōmi and another shrine called *Wilderness area of millet* Awano in Shinano.[125]

The early medieval vocabulary could be very varied, judging from modern dialects. In the 20th century, swidden fields were designated by about thirty different words according to the region. One finds for example in the northeast of the country *araki* 荒起, in Tōhoku *kannō* 刈野, in Kantō *sasu/sashi* 刺, in Chūbu and the west coast *nagihata* 薙畑 and *yamahata* 山畑, in the central-south (Chūgoku) *karihata* 苅畑 and *kariyama* 苅山, in Shikoku *kirihata* 切畑 and *yabu* 薮, in Kyūshū *koba* 木場, etc. These are vernacular words written phonetically that had no fixed way of writing.[126]

Among the early medieval toponyms listed above, some recall these dialectal expressions, such as Araki and Karunoda (written like *kannō*).

However, one could probably add other compounds, such as Kusanagi Shrine (containing the semantic element *nagi* of *nagihata*), and toponyms containing *sashi, yabu* and *koba* with its variant *kuwa*. Moreover, certain techniques could provide other names, including *shishi* that refers to the struggle against mountain animals, burning *yaki* (in the toponym Yaki and the shrine name Yakitsu), the easterly exposure *himuka* that is indispensable for mountain agriculture (in the toponym Hyūga or Himukai and the shrine name Himukai), etc. This means that, on the one hand, our tables may include names that will turn out to be unrelated to swidden farming, while on the other hand omitting names that future research may add. Toponymy remains "a field to be cleared". Meanwhile, we point out some toponyms possibly relating to another fire technique belonging to the system of agri-sylviculture.

Agri-sylviculture: the traditional forest management

The historian Hatai Hiromu drew our attention to the toponym *Wooded field* Hayashida. In the 8th century, it designated three villages located respectively in the provinces of Harima (Hyōgo), Mimasaka (Okayama) and Sanuki (Kagawa in Shikoku). In the two latter cases, it was pronounced Haida.[127]

The name Hayashida also appears, in the 9th century, in some small localities: there were two plots called Hayashida north-west of the palace at Nara and another south of Nara. Two other plots with the same name were in the district Nagusa-gun in Kii, in the region of the Kinugawa delta and present-day Wakayama city. The two plots contained irrigated rice fields and adjoined a wood *hayashi*. Six *Wooded field* plots, adjacent to each other, and a seventh were in the block *In front of the chestnut trees* Kurisaki, in the village of Ōkuni or Yabu (*Thicket*), district Echi-gun in Ōmi, a place corresponding to the northern part of the Ōmi Plain that runs along the eastern shore of Lake Biwa.[128]

The name *Wooded field* Hayashida suggests fields associated with a wood. One can point to a few cases of fields and woods side by side in 9th-century deeds of property transactions. An aristocratic residence of 1 *chō* (1.13 ha) in Yamashiro near Heian-kyō consists half of cultivated fields and half of a wood *hayashi*. Another aristocratic residence in Yamato owns fields on 3 *tan* (0.78 acres) and a grove of chestnut trees on 1.5 *tan* (0.39 acres).[129] On an estate of the Kanshinji temple in Kawachi (Ōsaka), two areas each of more than 5 ha contain respectively rice fields, dry fields and woods. And in the 770s, a landowner from Ōmi (Shiga) exchanges his idle or waste ground for land with a wood and a dry field.[130]

In these examples, the field and wood form a pair. It is possible that this type of land was exploited alternately with cereal cultivation and arboriculture, i.e. with plantations. There is a form of agri-sylviculture that combines the planting of useful trees, providing timber, industrial fibres, fuel, edible fruits or charcoal, with cereal intercropping. This practice differs from swidden fields where trees are allowed to regrow naturally. With intercropping, there is active reafforestation and the recovery of trees, implying selective burning when the land is cleaned after the forested period: the trees are felled, but only the branches and leaves are burned and additional fuel is brought in.

Certain forms of forest agriculture existed in Germany in the medieval period. They knew on the one hand sylviculture alternating with one or two years of cereal cultivation, resulting in "wilderness fields" (Wildfelder), and on the other hand a system of planting fruit trees with cereal catch crops between the trees. In South China, a system of agri-sylviculture, known as *taungya*, was practised up to the 17th century: these were plantations of valuable woods such as the conifer *shamu*, Cunninghamia lanceolata (Lamb.) Hook., with inter-calary cereal catch crops and even shrub crops that need shade, such as the Camellia oleosa (Chin. *cha you*) and tung-tree (Chin. *you tong*, Aleurites fordii Hemsl.).[131]

In Meiji-period Japan, *kirikaehata* referred to a form of shifting forest agriculture, which differed from swidden fields *yakihata* by an active reaf-forestation. It was therefore an alternating and not an intercropping agri-sylviculture. They planted oaks *kunugi* and *nara*, cryptomeria *sugi*, Japanese cypress *hinoki* and chestnut trees *kuri*. This system was known in the Edo period by the names "mountain swidden fields" *yamahata* or "wood swidden fields" *hayashihata*. Some *yamahata* mentioned in the medieval documents may have referred to this form of agri-sylviculture.[132] We think that it was practised from the early medieval period. Since they did not wait for natural regrowth, but planted trees after the crop cycle, this system could also have been used on lowland areas. Perhaps *Wooded field* Hayashida meant agri-sylviculture on lowlands, as opposed to that in the mountains. In fact, the plots called Hayashida in Kii and Ōmi were situated on lowlands.

As regards plantations, the 9th-century documents mainly mention the chestnut tree, which produced fruits with a high calorific value in autumn. There were chestnut groves *kurusu-bayashi* or *kuri-bayashi* in the grounds of a number of aristocratic residences,[133] and a few texts make reference to the alternation of trees and cereals. For example, in 845, an owner sold his land in the province of Kii (Wakayama), with a wood of 8 *tan* (2.08 acres), which in fact "had just been cut leaving no trees" *hayashi kiri, ki nashi*. On the other hand, one part of another chestnut grove in Iga (Mie), in 964, was "now wooded" *ima hayashi-chi*, and the other part was cleared of trees, i.e. "empty" *kūchi*. But one also finds, on other land in Kii, a "dry field in a chestnut grove" *kurusu-batake*, suggesting intercropping.[134]

It can easily be imagined that shifting agriculture caused census problems, as is shown by a dispute over land-tax. In 859, the district office of Echi in Ōmi found that several fields previously declared as idle *tsune-arashi* were in fact carrying crops *kenjuku*. The fields in question were on a plot called *Field on a cleared plot* Arakida and near other plots called *Wooded field* Hayashida. This suggests that it was not necessarily a case of deception, but of shifting cultivation, as swidden fields did not come into the administrative categories. A similar kind of change must have occurred on land belonging to the Kōryūji east of Heian-kyō, where there were also plots called *Field on a cleared plot*. One of the blocks on this land had two dry fields each of 1 ha. In 873, one of the two fields was "wooded but not cultivated" *hayashi, arashi*; then in 887, this field was "wild" *no*, and the other field was "idle or abandoned" *tsune-arashi* (as above). In one field there may have been a plantation associated with crops that were no longer grown after a few years, giving it a wild appearance fifteen years later, while the other field, cultivated for a time, was left idle or abandoned.[135]

Other slightly later accounts mention agri-sylviculture practices, such as the register of the property of a certain Fujiwara no Noritaka, in the province of Aki (Hiroshima): it included, in 1154, thirty-one chestnut groves, all with adjoining dry fields that may have alternated periodically. But twenty-two of these fields were "situated in a grove" of chestnut trees *hayashi ni ari*, which implies cereal intercropping between the trees. There may thus have been a system of triple alternating crop rotation.[136] The three villages mentioned earlier called *Wooded field* Hayashida and Haida probably relate to these practices combining the planting of trees and crops. One may also ask whether Haida could mean "field of ashes" *hai-da*. Thus agri-sylviculture gave rise, in our view, to another category of toponyms.

Toponyms possibly relating to agri-sylviculture

Besides *Wooded field* Hayashida, other toponyms seem to relate to agri-sylviculture. The following names are all connected with the district of Echi on the Ōmi Plain, east of Lake Biwa. The plots called Hayashida were next to, or in the village of *Thicket* Yabu, a word meaning 'swidden field' in a modern dialect. Another plot not far from there was called *Cleared field at the entrance to Hirota* Hirotakuchi-kirita, and other plots on this land were named *Field of chestnut trees* Kurusuta, *Field next to the chestnut trees* Kurisakida, *Lowland field of mulberry trees* Kuwaharata, *Lowland field of brushwood* Shibaharata, *Small wilderness field* Onoda and *Field of nettle-tree roots* Enomotoda. On other land in Kii, in the district Nagusa-gun, two plots Hayashida adjoined two other plots called *Field fenced against wild boar* Ikakida, a name associated with the struggle against animals. The two regions, Echi and Nagusa, one in Ōmi, the other in Wakayama, even today have lowland forests *heichi-bayashi*, of which few remain in central Japan.[137] In fact, it might be asked whether the toponym and the shrine name Katano, *Half (cultivated?) waste land* or *Alternating waste land*, could mean alternating cultivation. According to the *Wamyō ruijushō*, *hayashi* means 'lowland forest', but this dictionary does not mention *mori*, 'mountain forest'.

The Kōryūji temple estate, already referred to several times, presents the following: *Field of nettle-tree roots, Lowland field of brushwood, Thicket field* Yabuta, *Field of pine-tree roots* Matsumotota, *Lowland field of oaks* Kashiwaharata, *Willow field* Yanagita/Yakita, *Field of sophora and pear trees* Enjunashita. The map of the rice fields of the district Kadono-gun in

Table 2.7 Toponyms possibly associated with agri-sylviculture: names of districts, villages, blocks and plots

Japanese transcriptions/characters		English
Hayashida/Haida	林田	Wooded field
Hayashibe-ta	林辺田	Field near the wood
Hayashino	林野	Wooded wilderness
Katano	片野	Alternating wilderness
Yabu(ta)	養父・薮 (田)	(Field with) Thicket
Kita	木田	Field of trees
Hirotakuchi-kirita	広田口切田	Cut-off field at the entrance to Hirota
Kurusu	栗・栗栖・栗櫟	Chestnut grove
Kurusuta	栗栖田	Field with a chestnut grove
Kurusushimoda?	栗下田	Field beneath a chestnut grove
Kurusuno	栗野・栗栖野	Wilderness with chestnut trees
Kurita	栗田・栗太	Field with chestnut trees
Kurihara(ta)	栗原 (田)	(Field of) Plain of chestnut trees
Kurisaki(da)	栗前 (田)	(Field) Next to chestnut trees
Kurimoto(da)	栗本 (田)	(Field of/with) chestnut-tree roots
Ekuri	殖栗	Plantation of chestnut trees
Kuwata	桑田	Field with mulberry trees
Kuwahara(ta)	桑原 (田)	(Field of) Plain of mulberry trees
Kuwamotota	桑本田	Field of/with mulberry-tree roots
Asahara(ta)	麻原 (田)	(Field of) Plain of hemp
Shibata	柴田	Field of brushwood
Shibahara(ta)	柴原 (田)	(Field of) Plain of brushwood
Shibakaru?	柴刈	Clearing of undergrowth
Enomoto(da)	榎本	(Field of/with) nettle-tree roots
Enokita, Enata?	榎田	Field with nettle trees
Matsuhara	松原	Plain of pine trees

Japanese transcriptions/characters		English
Matsumoto(ta)	松本田	(Field of/with) pine-tree roots
Yanagita, Yakita?	楊田	Field with willow trees
Kaimotota?	槐本田	Field of/with sophora roots
Enjunashita?	槐梨田	Field with sophora and pear trees
Nashihara	梨原	Plain of pear trees
Nashimotota	梨本田	Field of/with pear-tree roots
Tsukita	槻田	Field with zelkova
Tsukinomoto(ta)	槻本 (田)	(Field of/with) zelkova roots
Haibara(ta)	榛原 (田)	(Field of) Plain of alder trees
Momota	桃田	Field with peach trees
Momomoto(ta)	桃本 (田)	(Field of/with) peach-tree roots
Momohara(ta)	桃原 (田)	(Field of) Plain of peach trees
Kuhara, Kunigihara	柞原	Plain of oak trees
Kunuita	柞田	Field with oak trees
Narahara	楢原	Plain of oak trees
Naramotota	楢本田	Field of/with oak-tree roots
Ichiihara	櫟原	Plain of oak trees
Ichiimotota	櫟本田	Field of/with oak-tree roots
Kashiwahara(ta)	柏原 (田)	(Field of) Plain of oak trees
Kirihara	桐原	Plain of paulownia
Sakurata	桜田	Field with cherry trees
Hagihara	萩原	Plain with lespedeza
Kazurahara	葛原	Plain of pueraria
Kadono(ta)	葛野 (田)	(Field) Wilderness with pueraria
Ehara	荏原	Plain of perilla

Note: ? The reading of this word is uncertain.
Source: The 8th–10th-century documents quoted in this chapter (see notes 137–139).

Yamashiro, also mentioned above, was near the villages *Plain of oak trees* Ichiihara and *Upland wood* Kamutsu-bayashi. The land included plots called *Field near the wood* Hayashibe-ta and *Wooded wilderness* Hayashino. One also finds *Brushwood plain* and *Lowland field of mulberry trees*. Three toponyms relate to resins: *Field of/with chestnut-tree roots* Kurimotota, – *zelkova roots* Tsukinomotota and – *oak-tree roots* Ichiimotota[138] (Table 2.7).

Of the above toponyms, those containing brushwood *shiba* and thicket *yabu* may also relate to the gathering of fuel or foliage intended for green manure. The names referring to resins may allude to agri-sylviculture, or simply to the natural or planted environment. A case of agri-sylviculture is attested by the agricultural treatise *Aizu nōsho* (1684), according to which wheat, perilla and indigo plants were cultivated under tall varnish trees and persimmon trees. It is interesting to note that many toponyms of administrative units of the type *Plain of XY* or *Roots of XY* existed in the small units in the form *Lowland field of XY* or *Field of XY roots*. For example, *Plain of oaks* Kashiwahara becomes *Lowland field of oaks* Kashiwaharata. The same applies to the chestnut *kuri*, the pear *nashi*, the oak *ichii*, the pine *matsu*, the elm *tsukinoki*, the peach *momo*, the nettle tree *enoki*, the mulberry *kuwa*, hemp *asa*, etc. The use of the mulberry tree and hemp is well known, but it is also clear that nearly all the trees and shrubs form part of the supplementary food plants with edible parts.[139] Agri-sylviculture therefore offered a number of advantages: it enabled the recycling of soil nutrients, the formation of a manure heap for the neighbouring fields, and it provided fuel, edible products and industrial raw materials. This form of cultivation was therefore an important source of wealth for the rural populations.

To conclude, we propose a rereading of the myth of the celebrated hero Yamato Takeru in the light of what has been discussed. The versions of the myth may be summarized as follows:[140] In prehistoric times, Yamato Takeru set off to conquer the East, wearing a sword called Kusanagi (*Grass-cutter, Swidden field* in dialect). When he reached Suruga (Shizuoka), the barbarians urged the hero to enter a wilderness *nonaka ni iru/irino* to hunt giant deer with as many legs as the trees in a wood *hayashi*. They then set fire to this land *yaku/hi-tsuku* in order to kill him. However, the hero cleared the vegetation *kusa karu* around him – hence the name *Grass-cutter* of his sword. In his turn he lit a fire using a stone flint, and fled. Then he burned the barbarians and conquered their territory. He called this place *Fire trap* Yakitsu. Later, the sword was placed in the shrine *Burning field* Atsuta jinja in the province of Owari (Nagoya). In this episode of the conquest of Japan, one finds the shrine names Kusanagi, Irino, Yaki, Yakitsu and Atsuta, as well as the semantic elements of the toponyms Nonakada, Hayashida, Yaku and Karunoda. In the early 8th century, the compilers of the Japanese annals seem to allude to the colonization of the country by slash-and-burn: it seems clear that the agrarian conquest was made using fire.

This episode from mythology concludes our catalogue of indicators of the practice of fire techniques. In this chapter, we have surveyed the various types of shifting cultivation as reflected in ethnology, the 17th/18th-century agricultural treatises and the practices of the late medieval period, as well as the administrative texts, lexicography, poetry and toponymy of the 8th–12th centuries. This has enabled us to establish that the early medieval Japanese used several fire techniques. The archaeology must now be considered in order to give an estimate of the beginnings of swidden farming (assartage) and agri-sylviculture. The dating of early agriculture constitutes a problem that is currently being debated, which is why we present the different points of view in the form of a discussion on early agriculture and swidden farming.

Discussion: early agriculture and swidden farming

Japanese specialists ask whether swidden farming was practised from the Neolithic. This question interests us in relation to the link between swidden farming and early agriculture. The ethnologist Sasaki Kōmei and the ethnobotanist Nakao Sasuke place swidden farming as a stage in the evolution of the first forms of cultivation during the Neolithic (Jōmon) and early agricultures of the Neolithic have been termed 'neolithization' by archaeologists and ethnologists in Western countries. However, in Japan, the hypothesis of Nakao and Sasaki has given rise to differing reactions. This question is part of the debate about the beginnings of agriculture in this country. For a long time, the attention of archaeologists was focused on irrigated rice, so much so that the discussion of agriculture revolved around the start of rice growing. Until the 1990s, Sahara Makoto and other archaeologists presented the beginnings of rice cultivation as an "agricultural revolution" and as the origin of Japanese society. In their eyes, there was no agriculture before rice growing.[141] Until 1994, Sahara maintained that agriculture began in the early Yayoi period (then dated 3rd century BC–3rd century AD), linking it with the introduction to Japan of irrigated rice growing from the continent. He also presented a comparative table of the neolithizations of the major cultural areas of the world. He dated them to between 8000 and 2000 BC and showed the striking difference with Japan (where he placed neolithization around 300 BC). However, Sahara later subscribed to the more generally accepted idea according to which the beginning of agriculture is the result, not of a sudden change, but rather of a long process that is today called neolithization.[142] The debate about the neolithization of Japan has followed a long and difficult path in the research carried out in this country. We retrace here some of the stages in this research and present the current state of the debate. It should be noted that in this section we use the dates given for the historical periods by the authors cited, but they are no longer regarded as valid and differ from those in Table A.3 in the Appendix.

Early farming and fire techniques

The first supporters of a "Neolithic agriculture" (Jōmon nōkōron), such as Ono Takeo and Fujimori Eiichi, were not listened to. In the 1950s and 1960s they were accused of lacking evidence.[143] But in 1967, the ethnobotanist

Nakao Sasuke attracted attention with his "Theory of the origin of agriculture" ("Nōkō kigenron"), which placed Japan in the geobotanical context of East Asia. Nakao identified in the region taking in Nepal, southern China, southern Korea and the western part of the Japanese archipelago, a laurel-forest zone with shiny-leaved vegetation, thickly wooded with oaks of the species cyclobalanopsis *kashi* and castanopsis *shii*, etc. This laurel-forest *shōyō jurin* zone is characterized by an evolution of acquisition techniques in five stages: the harvesting of edible plants, nuts, seeds and tubers; the domestication *han saibai* of the same species; the growing of tubers on swidden fields; cereal crops; irrigated rice cultivation. The term *han saibai* and the English equivalent "protoculture" are no longer used. By *han saibai* Nakao means domestication including selection and sowing.[144] In the late 1960s, the ethnologist Sasaki Kōmei completed his research on swidden farming in Japan and other Asiatic countries. Sasaki and Nakao jointly formulated a more general model that distinguishes three stages of evolution: the pre-agricultural stage; swidden farming; agriculture dominated by irrigated rice, and they established the following schema:

Neolithization according to Nakao Sasuke and Sasaki Kōmei:[145]

1   Pre-agriculture (*pre-nōkō*)

- collecting of nuts: (sweet) chestnuts, horse chestnuts, acorns, walnuts;
- gathering of roots: pueraria, bracken, taro;
- domestication: (sweet) chestnuts, Japanese yams/dioscorea? (*jinenjo?*), Lycoris radiata;

2   Swidden farming dominated by cereals

- swidden-field crops: taro, *nagaimo* (dioscorea), *konnyaku* (Amorphophalus Konjac);
- swidden-field crops: barnyard millet, *shikokubie* (an Eleusine), foxtail millet, common millet, dry rice *okabo*;

3   Agriculture dominated by irrigated rice

In this table, Lycoris radiata *higanbana* should no doubt be removed from the tubers, as this plant was shown to be a modern import, following the research of Arizono Shōichirō.[146] Sasaki Kōmei chronologically situated swidden farming "Before Rice Cultivation", the title of his book *Inasaku izen*, published in 1971 (p. 31), and in the late and final phases of the Neolithic (2000–300 BC).

Archaeological discoveries subsequently showed the existence of certain types of Neolithic edible plants. In 1974, in the site of Kōjinyama in Nagano, dating from the mid-Neolithic (3000–2000 BC), were found seeds (perhaps cultivated) resembling foxtail millet, but these seeds were later identified as labiates, perilla (*egoma* or *shiso*), possibly of a self-sown variety. After 1979, excavations carried out on the site of Torihama-kaizuka (discovered in 1972), in Fukui on the Sea of Japan, provided more reliable indicators dating from the early Neolithic (4000–3000 BC). Archaeologists found traces of plant seeds thought to have been cultivated: the gourd Lagenaria *hyōtan*, a leguminous plant Vigna *ryokutō*, Cucumis melo *uri*, perilla *egoma* and *shiso*, a Brassica *aburana*, burdock *gobō* and hemp Cannabis *asa*. (Japanese excavation reports seldom give the Latin botanical names.)[147]

Torihama-kaizuka, one of the most representative sites of the Japanese Neolithic, was the subject of a symposium in 1981 on the topic "Indicators of a Neolithic Agriculture". The theories of Fujimori Eiichi were re-evaluated and the existence "of a form of cultivation or domestication" (*nanraka no nōkō aruiwa saibai*) in the mid-Neolithic was recognized. For the late (2000–1000 BC) and final (1000–500 BC) Neolithic, the paleogeographer Yasuda Yoshinori shared the results of his pollen analyses on radiocarbon-dated sedimentary samples from four sites in the Fukuoka region, analyses from which he inferred forest destruction by fire and therefore forest management by man. At the same symposium, the geographer Katō Shinpei spoke about the cultivation of buckwheat. In conclusion, the model of an evolution in the Asiatic laurel-forest zone proposed by Nakao Sasuke was re-evaluated, and the existence of certain "crop types" including pre-domestic agriculture, notably of tubers, was accepted. This led on to a recognition of the existence of cultural elements for the Neolithic and the characterization of the evolution preceding irrigated rice growing "primitive agriculture" *genshi nōkō*, the conclusion being that one could only speak of a society of farmers from the introduction of irrigated rice cultivation in the Yayoi period (linking up with the early theories of Sahara Makoto, according to which only irrigated rice growing is truly "agriculture").[148] It should be remembered that some researchers of the pre-war generation were accustomed to drawing a sharp conceptual distinction between dry and irrigated crops and between a "primitive" Neolithic and an evolved Yayoi "culture".

Advances in paleobotany have since contributed to changing widely held opinions and clarifying various uncertainties. A symposium organized in 1985 by Sasaki Kōmei made even greater use of the results of analyses of seeds, pollens and phytoliths. The proceedings of this conference were published in 1988 under the title "Birth of a Civilization of Dry Cereal Crops", *Hatasaku bunka no tanjō*. In this work, Sasaki distinguishes two contiguous vegetation zones in the Japanese archipelago: the laurel-forest zone already mentioned, under the influence of the South-East Asiatic continent; and a deciduous forest zone, occupied by fagaceae of the species beech *buna* and oak *konara* and called the Fagus zone *buna-tai*. The latter, previously defined by Ichikawa Takeo (1984), covers the north-eastern half of the archipelago and is under the influence of the north-eastern Asiatic continent including Siberia (Map 2.2). (The Pacific zone taking in Polynesia is omitted from Japanese diagrams.)

With reference to English and American archaeology, Sasaki Kōmei also sees an evolution over time. He distinguishes on the one hand a "small-scale

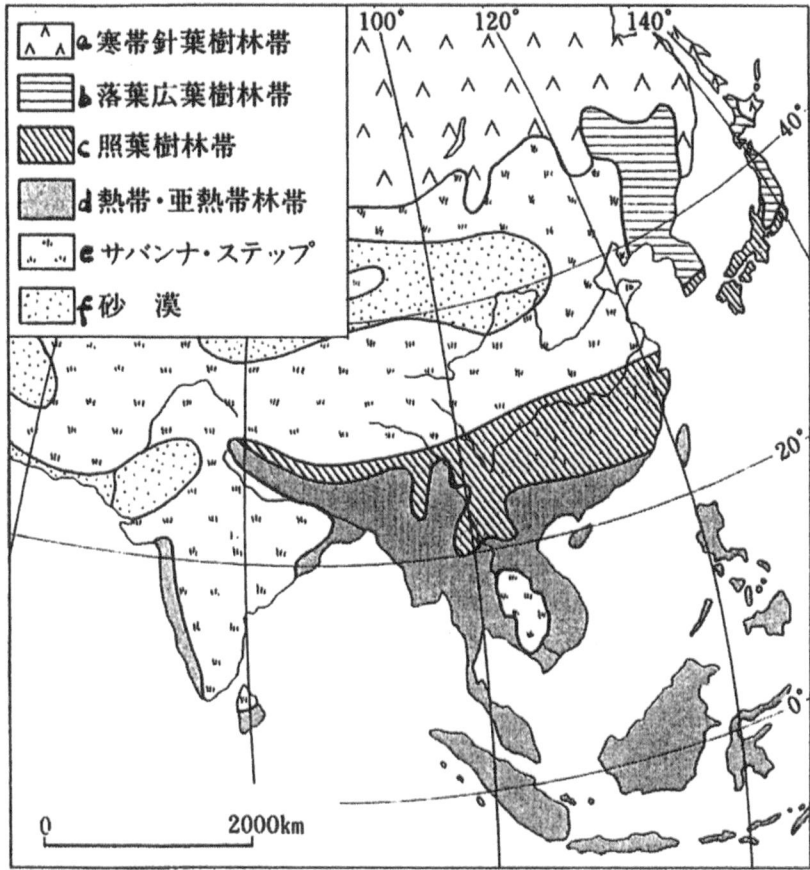

Map 2.2 Vegetation zones of East Asia
Notes: (a) subarctic zone, forest of conifers, (b) cool-temperate zone, forest of broad-leaved deciduous trees (Fagus), (c) warm-temperate zone, forest of shiny-leaved trees (laurel forest), (d) subtropical and tropical zone, (e) savannah, steppe, (f) desert.
Sources: Sasaki Kōmei, *Nihon bunka no kiso o saguru*, NHK Books (1994: 18); map after Kira Tatsuo.

agriculture" *genshoteki nōkō* (Engl. incipient agriculture) during the early and mid-Neolithic, characterized by foraging, hunting, fishing and including "pre-domestic agriculture", and on the other hand a "primary agriculture" *shoki-teki nōkō* (early agriculture) that characterizes above all the laurel-forest zone from the late Neolithic. Sasaki defines this phase as an association of dry crops on swidden fields and wet crops (watered by runoff), referring to certain Western models and to cases known to ethnography. The passage from a "pre-agricultural" society to an "agricultural" society is shown by the introduction of irrigated rice cultivation during the final phase of the Neolithic.

In the Proceedings of this same conference, Yasuda Yoshinori published the results of his pollen analyses from Torihama-kaizuka and other sites. He notes that the destruction of the primary forest is followed by the appearance of cereal and vegetable crops and replantings of alder, chestnut, horse chestnut or hazel trees, with differences in time and space depending on the region. Yasuda infers from this the existence of forest management by fire. He supports the hypothesis of the ethnologist Fukui Katsuyoshi, according to which the periodic cycle of deforestation and reafforestation includes trees with edible fruits and at the same time promotes the regrowth of plants with starchy roots, such as the ferns *warabi* and *zenmai*, lilies *ubayuri*, and pueraria *kuzu*. The archaeologist Koyama Shūzō questioned the material indicators. He noted the insufficient number of finds of plant remains (a few seeds in a small number of sites up to the mid-1980s) and the lack of certainty about the identification of species, except for that of the gourd Lagenaria. He proposed calorimetric estimates, according to which plant food was based on nuts and seeds. In conclusion, scholars agreed that the Neolithic is characterized by a diet derived from self-sown species, but they remained divided on neolithization as set out by Sasaki Kōmei and Nakao Sasuke. Fukui Katsuyoshi put the case for forest management, while Kobayashi Tatsuo and Koyama Shūzō (Sahara Makoto not having attended) opposed the idea of a Neolithic agriculture.[149] Since then, the remains of plants considered by other archaeologists to have been cultivated have come to light in more Neolithic sites in Japan.[150]

The site of Sannai-Maruyama

In 1994, the spectacular discovery of the site of Sannai-Maruyama in Aomori, in far northern Honshū, enriched the body of material evidence. This extensive site (38 ha) was occupied more or less continuously between 3500 and 2000 BC. Here were discovered a total of 500 pit-buildings, some twenty of them very big, 10 m in length, graves, two raised (ritual?) areas, a significant building on pillars each 1 m in diameter, rubbish pits, ritual and household items, silos and organic remains. It was therefore a large-scale settlement as is confirmed by the high number of silos discovered: 350. Twenty-six silos are 5 m in diameter and 1.80 m deep, and 247 silos have a capacity of 1–10 kilolitres. In these silos foodstuffs were probably stored in large earthenware jars. Archaeologists have also found about a hundred pillared-buildings that were probably used as granaries. This significant storage capacity led to demographic estimates that put the population of this settlement at between fifty and five hundred individuals at the same period.

The population of the site lived by hunting and gathering, with, it is thought, chestnuts as the staple food. The pollen diagram drawn up by Yasuda Yoshinori shows a primary forest of alder and fagaceae (we are indeed in the "Fagus zone"), which is systematically replaced by the chestnut from 3500–2800 BC, then followed by a gap in the diagram; later, the fagaceae are again replaced by chestnut trees from 2000 BC. Forest management thus seems to be an established fact. For Takahashi Manabu, a geologist specializing in pedology, the charred remains in the soil layers prove that this management was done by fire. The Sannai-Maruyama site also shows the presence of barnyard millet Echinochloa crus-galli L., which may be self-sown or cultivated. By way of edible plants, archaeologists found on the site some beans, burdock (*gobō*) and the gourd Lagenaria, though in very small quantities.[151]

The discovery of the Sannai-Maruyama site confirms the existence of food planning and forest management. However, as regards agriculture, some remained cautious and others were positive (in 1995). One in fact has the impression that some were speaking of an economic system, others of techniques. Outside Sannai-Maruyama, Neolithic sites have yielded traces of rice from the middle phase (3000–2000 BC). Fujiwara Hiroshi who has analysed the phytoliths and Satō Yōichirō, a biologist who has examined their genetic DNA structure, think that there was rice growing on swidden fields. Fujiwara Hiroshi has also put forward the hypothesis that barnyard millet, phytoliths of which are very numerous at Sannai-Maruyama in some deposits, was intensively harvested.[152]

During a round-table discussion published in 2000, Sahara Makoto, Sasaki Kōmei, Tsude Hiroshi and Nishida Masaki dealt with agriculture in relation to sedentarization.[153] Sahara acknowledged the possibility of swidden farming during the Neolithic, but maintained his distinction in time: according to him, there was *saibai* (cultivation, domestication?) in the Neolithic, then *nōkō* (agriculture?) in the Yayoi period, but he did not clarify the meaning of these two terms. Sasaki Kōmei opposed such a periodization and noted that even the passage from gathering *saishū* to the practice of *saibai* is not clear, as gathering can be systematic, hence the old term *han saibai* (protoculture). For Nishida Masaki, the passage from *saibai* to *nōkō* is a social fact, and concerns an élite that asserts itself by the adoption of agriculture. In the same work, the archaeologist Miyamoto Kazuo highlights this imprecise terminology, by presenting the material data. Taking up the concepts used by European archaeology, Miyamoto establishes a distinction between pre-domestic practices *kanri saibai* (or *han saibai*) and domestication *saibai*. The former include systematic gathering and self-sowing; they imply the familiarity of man with plants, but do not change the plants. Domestication subjects a plant to the pressure of selection and crystallizes the varieties: the plant changes. According to Miyamoto, the passage from a hunting society to an agricultural society in Japan is not a technical but an economic question and it is shown by the increase in food production compared with foraging.[154]

Our summary of the opinions held up to the year 2000 concerning neolithization in Japan (which leaves aside the link with pottery and habitat) shows that Japanese researchers had still not reached a consensus for defining a model. At that time, European archaeologists arrived at the opposite point, namely a questioning of the existing models. For nearly a hundred years, the Near East was considered the cradle of agriculture. Today, it is no longer taken as the epicentre, but as one case among others of a successful experiment with neolithization. This was confirmed by archaeologists from different cultural areas, during a seminar held in 1999 at the Collège de France, chaired by Jean Guilaine and entitled "The first peasants in the world: where? When? How? Why?". The papers questioned the evolutionist models, such as that of Boserup (1970). In fact, archaeologists are no longer sure, as they were previously, that agriculture goes hand in hand with sedentarization, or that demographic pressure brings about domestication. In some cases the domestication of tubers does not precede that of cereals (in Africa), in others it is delayed even in the presence of an edible plant (beans in the Andes), and in still other cases the sowing of cereals does not occur at all (in New Guinea). It is clear that the history of plant conquest is very diverse and does not necessarily obey the rule of progress. This does not prevent a hunting society from having known an economy of plenty, well before moving to the stage of a food-producing society.[155]

With regard to Neolithic swidden farming, evolutionist theories place shifting agriculture chronologically before the intensive and permanent use of fixed plots, but other hypotheses propose a reverse model: demographic pressure led to deforestation and therefore to the gradual change from an intensive agriculture to an extensive shifting agriculture. There are also cases, in tropical environments, of the simultaneous use of permanent and shifting crops (on swidden fields). Swidden-field crops seem to be attested for the Near East and the Americas, but archaeologists seldom mention swidden explicitly, because of the scarcity of material evidence. They admit it implicitly because shifting agriculture is hardly practicable without fire. Discoveries of dated sedimentary layers, including charred remains, are extremely rare. And even in this case, one must be able to distinguish the accidental fire that may have struck a settled area from the fire of cultivation. Fire techniques are therefore difficult to detect by direct indicators. With the current methods, it is pointless to expect other conclusive material evidence of Neolithic fire techniques. Whether in Japan or elsewhere, one way is to evaluate the pollen diagrams that show a sudden change in the vegetation cover at a given time.[156] We saw above that this type of evidence has been provided by several Japanese Neolithic sites in respect of forest management.

Other indicators may support the hypothesis of the existence of some forms of Neolithic cultivation, namely the remains of harvested plants. Miyamoto Kazuo gives a table of the plants he considers to have been cultivated in Neolithic sites up to 1997. Beans *ryokutō* are found in the largest number of sites (twenty), dating from the early phase (4000–3000 BC), and they increase in the western part of the archipelago from the late phase (2000–1000 BC). "Dry" (non-irrigated) rice appears in Kyūshū and southern Honshū in the middle phase (3000–2000 BC) and increases in the late phase. Wheat is found in a single site in Kyūshū in the final phase (1000–500 BC). The dating of barley is problematical at present. Barnyard millet and buckwheat have been identified in the site of Hamanasuno in Hokkaidō, dating from the end of the early Neolithic, and together with common millet and foxtail millet in other sites in Hokkaidō and northern Honshū, from the middle phase. These four plants may therefore have come from North Asia (Siberia), but barnyard millet is thought by some to be an indigenous plant. They spread westwards only in the late phase of the Neolithic. Tubers (taro, yams, etc.) have been found in some sites of the early and middle periods, curiously in central Honshū and not in the south-west of the archipelago, but it is not possible to distinguish between the cultivated and self-sown varieties. Apart from these subsistence food plants, plants harvested as complementary foods have been found in Neolithic sites: perilla, the cucurbitaceae Lagenaria and Cucumis, Cannabis hemp and a Brassica *aburana*. Only four sites have yielded fruits such as the apricot, peach and an anacardiaceae: *chanchinmodoki*, Poupartia Fordii; burdock has been found only at Torihama-kaizuka and Sannai-Maruyama. For Miyamoto Kazuo, the presence of all these plants in Neolithic sites is clear evidence of some types of cultivation. Beyond the technical aspect, Miyamoto proposes taking into account the social organization of the site populations in order to evaluate, in the future, the passage from small-scale pre-domestic practices to an agriculture with domestication and a production economy.[157] For Miyamoto, the presence of "dry" rice rather suggests that some cultivation techniques were known from the middle period (3000–2000 BC).

The early agriculture of the Japanese archipelago is still being debated among Japanese historians, archaeologists and ethnologists. Among the various opinions, certain scholars lean towards "some form(s) of Neolithic cultivation *nanraka no nōkō*" and others opt for the existence of pre-domestic practices *kanri saibai* or *han saibai* (systematic collecting and self-sowing) alone; still others are in favour of active domestication *saibai* (selection and management of varieties), both of these in conjunction with (sweet) chestnuts, or red beans *azuki* or again barnyard millet *hie*, referring to the Sannai-Maruyama site. They do not envisage agriculture *nōkō* or swidden farming *yakihata* during the Neolithic, citing the lack of evidence in support of fire techniques. For Shitara Hiromi, the cereals discovered in Neolithic sites were probably collected or harvested but not cultivated, because the number and volume of these finds are too limited (sometimes only four or five grains in a site) for us to be certain of any agricultural activity. For Nasu Hiroo, there is a need to reconcile different types of evidence, based either on charred grains, pollens and marks on pottery, or on phytolith records. Using the evidence of archaeobotanical weeds, he too disputes the existence of an early agriculture during the Neolithic, before the Yayoi period.[158]

The historian Harada Nobuo brings an ethnological argument to the question. He points out that rice cultivation in the different growing areas began with the gathering of wild plants and passed through the domestication and swidden farming stages before arriving at irrigated rice cultivation. He notes that slash-and-burn is a typical technique of the laurel-forest zone. If the discovery of pollens of edible plants in Neolithic sites, including Sannai-Maruyama, does not, *a priori*, prove that these plants were cultivated, Harada thinks differently concerning the phytoliths of rice found at several sites in

Okayama prefecture. In his view, this rice, which dates to 3000–1000 BC, was grown on swidden fields, arguing that Okayama alone has not just one but several rice archaeological sites and that these are located in the mountains in an environment conducive to swidden fields. As Harada points out, this does not alter the fact that Neolithic society is fundamentally characterized by hunting and gathering and that the use of agriculture was certainly limited, even marginal and localized, in the subsistence economy of this period.[159]

The field of ethnobotany does not restrict itself to rice growing, but studies cereal cultivation in general. In his article on plants and weeds in Asia, Yamaguchi Hirofumi (2011) drew up a table of the edible flora found in the laurel-forest zone comprising West and South-east Asia, and provided a list of the edible plants, some native to Japan and the others imported from the continent from the Neolithic up to the 20th century. Among the two kinds, he distinguishes, in addition, the plants domesticated or cultivated in the Japanese archipelago. This list naturally shows similarities with that of the vegetal remains discovered in the archaeological sites of the Japanese Neolithic, published by Miyamoto Kazuo, given earlier, the dating of which has, however, been called into question by Nakazawa Michihiko.[160]

From 2009 to 2011, the National Museum of Japanese History devoted a research programme to the 'Formation and Development of Agrarian Society: Reconstruction of the Image of the Yayoi Period' and published the results in 2014, edited by Fujio Shin'ichirō. In this volume, Shitara Hiromi discusses "Yayoi Culture as a Complex of Multiple Farming Cultures".[161] He highlights the variety of 'multiple' dry grain and bean cropping that started well before irrigated rice was introduced to Kyūshū in the early 1st millennium. Shitara focuses on the significant regional differences for the beginning of agriculture from the north-east to the south-west of the Japanese archipelago. He also stresses the 'complex' character of Yayoi culture that combines the previous foraging practices with the new agricultural ones. Although giving no dates (because of the regional differences), Shitara situates the first agriculture (before irrigated rice) at the beginning of the Yayoi period.

Let us bring a technological argument to the question of early agriculture. In the case of cereal crops, it may well be asked what the method of fertilization was, if not on swidden fields. Without the ashes from assartage, how could the late Neolithic or early Yayoi peoples have provided seedlings with phosphorous and potassium, without which they cannot grow? It is well known that the harvests of dry crops are roughly halved from the second year in the absence of any fertilization. Permanent dry cultivation without fire would have required a considerable human effort for fertilization that could hardly have been achieved by the early populations. Permanent agriculture seems to us most unlikely for the early Yayoi period. We therefore share the opinion of the ethnologists Nakao Sasuke and Sasaki Kōmei set out at the beginning of this "Discussion" and consider shifting or swidden farming as a common practice of early agriculture.

Another agriculture: swidden farming in premodern Japan

At the end of this chapter, a few general remarks would seem to be in order, firstly with regard to duration. Japanese archaeology has shown the existence of the practice of forest management by fire as well as the presence of edible plants in the Neolithic, from the 3rd millennium BC. Moreover, the land documents of the late medieval period (1200–1600) mention cases of swidden fields; and fire techniques are also attested, as we have seen, by the agricultural treatises of the Edo period (1603–1867). We can therefore speak of a continuity of fire agriculture over the millennia. The evidence that we have presented on the human management of edible plants in prehistoric times, far from consisting of isolated or marginal examples, only confirms a practice common to the subsequent historical periods.

For the historical periods, we have discussed in this chapter several practices involving fire. The study of the administrative and poetic texts, lexicography and toponymy has enabled us to identify several techniques relating to a forest agriculture for the medieval period. Swidden farming (or assartage) involved cleaning by fire a patch of forest on a mountain side and cultivating it for a period limited to three or four years. Agri-sylviculture on a plot that had first undergone selective burning combined the cultivation of useful trees with cereal intercropping. The two systems of forest farming just mentioned were temporary and shifting. There was a third technique that cannot be regarded, strictly speaking, as a cultivation method: each year the Japanese of the early medieval period would set a running fire in wilderness areas intended for the harvesting of edible plants in order to promote their regrowth. As for land clearing aimed at permanent crops, it was done in the same way as for cleared plots intended for swidden fields. Fire was therefore an integral part of the lives of farmers of the medieval period and it is reasonable to speak of "another forest-type agriculture", practised side by side with permanent irrigated and dry crops.

While the existence of systems of forest farming seems undeniable, it is far more difficult to get an idea of its extent. To do so, we must turn to macro-geographical considerations. Forest agriculture is dependent on forests and rainfall. With an average of 1,800 mm of rain a year, Japan receives almost twice as much precipitation as the world average. Moreover, 66.5% of its land area is covered by forests, making it the second most wooded country in the world after Finland (68.7%). Unlike those of Europe, nearly all the Japanese forests are situated at altitude, two-thirds of the country consisting of mountains, some 70% of which have average or steep slopes (8–30°). The Japanese archipelago therefore enjoys ideal conditions for forest agriculture: extensive forests, a rugged relief and abundant rainfall favourable to the growth of trees, i.e. to the recycling of the forest after cropping. If one considers the topography alone, it is evident that the natural environment of Japan is more favourable to forest agriculture than to irrigated rice cultivation. The plains truly suited to rice growing in fact occupy only a tiny part of the land surface,

Japan having no more than eighteen plains and a few intramountainous basins. Everywhere else, irrigated rice cultivation involves major terracing works. In Map I.5 (see the Introduction), the zones suitable for forest agriculture are marked in white and those suitable for irrigated crops are shown in black. In view of the characteristics of the natural environment of Japan, it is unlikely that forest agriculture was a marginal phenomenon in the medieval period.

Generally speaking, for shifting cultivation to take place, there must be a certain balance between the available areas and the population that farms them. For a natural reafforestation period of twenty-five years, the theoretical reckoning is twenty inhabitants per sq. km.[162] No statistical data is available for the early medieval period, but it is reasonable to suppose that the situation was hardly any different from that of more recent times. In the 1950s, a family of Shiiba-mura in Miyazaki prefecture (Kyūshū) provided for all its food needs with 1.3 ha of mountain clearings; as for the families of Kajiwara village at Itsuki, in Kumamoto prefecture, they each had on average an area of 1.8 ha. In Sar Luk, in Vietnam, the family of the village headman cultivated 3.5 ha of swidden fields and the poorest family only 1 ha. Thus a family could live by cultivating plots in an itinerant way on 1–2 ha of forest. And today, thousands of families live by swidden cultivation in Laos. Laos is located, like Japan, in a laurel-forest zone and its topography with an extensive forested mountain area is similar to that of Japan.[163] One may suppose that early medieval Japan had such an environment and that the mountains did not lack suitable areas for shifting agriculture.

Some specialists have drawn up diagrams of agrarian use for the Japan of the medieval and modern periods.[164] These are probably equally valid for the ancient period. These diagrams show that farms would have been divided into two types of arable land: on the one hand, orchards and vegetable gardens situated near dwellings as well as irrigated rice fields (with different land-tax structures) divided into plots and spread around the settled area at a distance that allowed for daily care; on the other hand, mountain plots farmed on a temporary basis that were further away and only needed tending and hence someone staying there when they were brought under cultivation and during the harvest. These mountain crops must have enabled households to ensure their survival far more easily than rice fields subject to taxation. Shifting agriculture on swidden fields was not taxable and had the added advantage of being more likely to go unnoticed by the authorities than permanent cereal cultivation; it also needed less or no fertilizer.

What crops were grown on swidden fields? The cereal most often mentioned by the early medieval texts is foxtail millet. The sources refer as well to the cultivation of beans. The ethnography of contemporary Japan further tells us that buckwheat occupied almost as important a place as foxtail millet. In the 20th century, these two cereals were followed, in order of importance, by soybeans, red beans, barnyard millet, wheat and taro. However, the early medieval administrative texts very seldom mention buckwheat, common millet, wheat and taro. Yet buckwheat has the advantage of being a plant that is easy to cultivate, grows in poor soils and reaches maturity in a relatively short period. To quote the words of an inhabitant of Shiiba-mura, it can "be eaten sixty-five days after sowing".[165] The cultivation of taro is just as easy, as is its preparation for consumption. It occupied the most important place in modern New Year rituals in some mountainous regions of Japan. Up to the 20th century, taro was the staple food for the populations of some Pacific regions including Polynesia, and it may be supposed that it was also appreciated by farmers in Japan, which is, after all, part of the Pacific zone.

Be that as it may, buckwheat is not found any more often in the *Engishiki* than in the *Fudoki*, *Man'yōshū* or Shōsōin documents and taro is only mentioned occasionally. Yet, in the 17th century, buckwheat and taro were strongly recommended as swidden-field crops intended to provide farmers with their basic food needs.[166] Early medieval Japan could therefore have known another forest-type agriculture that fell outside the control of the government and was part of the private economy of the population. It is reasonable to say that swidden farming was a widespread reality in everyday life. It should be considered one of the three poles of the subsistence economy, together with permanent agriculture and the gathering of plant products, throughout premodern Japan.

Notes

1　Sigaut (1975: 18). On swidden-field crops, see Verschuer (1995). Sigaut inspired our thinking on the tradition of swidden fields in Japan.
2　Abe (1995: 15, 16); see also Barrau (1972).
3　Abel (1962). We mention below the chapters by Abel on types of use, "Bodennutzungsformen", pp. 84–93, and special crops, "Sonderkulturen", pp. 97, 98.
4　Bray (1984: 98–101) ('Shifting Cultivation'). Ōsawa Masaaki (1996: 159–193, "Tō Sō jidai no yakihata nōgyō").
5　Yin Shaoting (1994); Yin Shaoting et al. (2000); Daniels and Takeshi (1994). Inoue Makoto (1995). I thank Watabe Takeshi for drawing my attention to the swidden fields of Yunnan, and Yin Shaoting for a question-and-answer session. Ducourtieux (2009) and Harada Nobuo (2011), for Laos. Harada Nobuo (2011) has studied the Tama New Town area near Tōkyō, and also presents the Edo-period policy concerning swidden farming.
6　Steensberg (1979, 1993). I have quoted FAO, ed., "Shifting Cultivation", *Tropical Agriculture* 34 (3), 1957, pp. 159–164, and J. E. Spencer, *Shifting Cultivation in Southeastern Asia*, University of California Press, 1966, pp. 16, 17, after Fukui Katsuyoshi (1983: 235). Ōbayashi Taryō et al. (1990: 39–41, 390, 558–560).
7　For what follows, see Sasaki Kōmei (1971: 88–136).
8　See the figures and general data in Sasaki Kōmei (1972: 3–12, 22, 124–277, 282, 337); and Sasaki (1971: 128–130). See a more complete list of 37 plants grown on swidden fields in 1936, in Fukui Katsuyoshi (1983: 266).
9　Ichikawa Takeo (1961); Ichikawa et al., eds (1984); Inoue Takuya (2011).
10　Fukui Katsuyoshi (1974); Fukui, "Yakihata nōgyō no fuhensei to shinka", pp. 235–274 in Ōbayashi Taryō (1983).
11　Nomoto Kanichi (1984, 1987).
12　Tachibana Reikichi (1995).

13   Sasaki Akira (1998); Saitō Masami and Shiiba Kuniko (1995). See Nomoto Kanichi (1998); Ōga Ikuo (1998); Iida Tasuhiko (2011).

14   Narada at Hayakawa-chō, Minamikoma-gun, in Yamanashi prefecture (40 km south-west of Kōfu city) is a village that also disappeared after the construction of a dam. See Minzoku bunka eizō kenkyūjo ed. (1987); Masuda Shōko (1990). Other fire techniques similar to those identified by Sigaut are found on Sado; see Satō Toshio (1993: 1–11). See an approach to swidden fields through religion by Shiraishi Akiomi (1988).

15   Sasaki Kōmei (1971: 128–130); Sasaki (1972: 3–12, 22, 124–277, 282, 337). See Yamaguchi Sadao (1938: 1–23); Yamaguchi Sadao (1937: 41).

16   Yamaguchi Yaichirō (1943).

17   *Shinpen Musashi fudoki kō*, book 264 "Chichibu", p. 306. See a presentation of this text by Chiba Tokuji, "Hata to hatake, chiikisa to shinkō", in Sasaki Kōmei *et al.* (1988: 313).

18   *Jikata hanreiroku* book 2, p. 100; and *Kojiruien* Sangyōbu 1: 50.

19   *Aizu nōsho*, pp. 101–109.

20   *Nōgyō zue*, pp. 86, 115; *Kōka shunjū*, pp. 165, 166; Furushima Toshio (1975: 329).

21   *Rōnō ruigo*, pp. 170–224. Furushima Toshio (1975: 327–336). See also *Kōeki kokusankō*, 1859, by Ōkura Nagatsune, quoted by Tanaka Kōji, in Watabe Tadayo (1987: 296).

22   Ono Takeo (1942: 261–338); see also Tanaka Toyoharu (1981), and recent studies by Komeie Taisaku, 2002, 2005.

23   *Nōgyō zensho* book 1, p. 56.

24   *Wakansansaizue* book 55, vol. 8, p. 17. I have translated *ta* by "field" in the Chinese sense, since it is a quotation, and not by "rice field" in the Japanese sense. The Japanese readings of *arakihari* and *kochita* are those of the author of 1712.

25   Hatai Hiromu (1981: 1–108). First published in *Ōsaka rekishi gakkai 25 shūnen kinen ronbunshū*, vol. 2, Yoshikawa kōbunkan (1976).

26   Kuroda Hideo (1984: 24–26, 31, 142–146, 268). Kimura Shigemitsu (1992: 23, 107, 203). Farris (1985: 109–113); he quotes Robert Hall, Toshio Noh, "Yakihata, Burned Field Agriculture in Japan...", *Papers, Michigan Academy of Science, Arts and Letters* 38, 1953, pp. 315–322. Nomoto Kanichi (1984: 234–239; 1987: 139–151). Kodama Kōta (1978: 30–33). Morita Tei (1986: 105–109). Abe Takeshi (1995: 67). Yoshida Takashi (1983: 160). Ikeya Kazunobu, Shirōzu Satoshi, eds, *Yama to mori no kankyōshi*, vol. 5 of Yumoto Takakazu, ed. (2011); Harada Nobuo, Kurata Takashi, eds (2011) (vol. 17 of "Library of the Research Institute of Humanities and Nature" *Chikyūken raiburarii*).

27   *Ritsu* (SZKT), p. 162; SY 70: 617; RR Gunbōryō 76; RGi, p. 203; RR Kyūmokuryō 11; RGi, p. 274; RSh, p. 927. See a detailed analysis by Hatai Hiromu (1981). Tanaka Toyoharu (1981). One may ask whether the restrictions are inspired by Chinese texts. A prohibition appears on one of the 1155 lacquered tablets with inscriptions – dating from 217 BC, discovered in 1975 in grave no. 11 at Shuihudi, in Yunmen xian, Hubei – worded thus: "It is forbidden to burn the grass in summer to turn it into ashes" *bu xiayue wu gan shao (ye?) cao wei hui*. See the catalogue *Shuihudi Qin mu zhujian*, 1990, edited by the group Shuihudi Qin zhengli xiaozu.

28   *Nihon shoki* Tenmu 5.5. (677); RSK 1: 9; SNK Jōwa 8.3.1. (841); SJ Jōgan 9.3.25. (867). ES 3: 69. See a compendium of historical sources relating to swidden fields in Harada Nobuo (2011: 533–545).

29   SY 70: 620 (year 814).

30   SY 70: 617, death of four oxen in 1005; Hi 9: 3524, a temple building burned in 995; Hi 4: 1474, a row of trees set on fire in 1104; Hi 5: 1768, fire in a plantation in 1125. See Kimura Shigemitsu (1992: 127, 128, 80).

31 Ono Takeo (1942: 255, 256); Furushima Toshio (1975: 121, 122); Hatai Hiromu (1981). Ikata Sadaaki (1940) found the references in mythology.

32 WR 1: 11. *Yarōden* seems to be a book. There was a *Yelao* in 17 chapters, dating from the Warring States period (4th–3rd centuries BC), according to *Hanshu* 30, Yiwenzhi, p. 1742 (ed. Zhonghua shuju).

33 *Wakansansaizue* book 55, vol. 8, p. 20.

34 See the toponyms in Etsuko Obata-Reiman (1990: 136–138); and the dialects in Chiba Tokuji, "Hata to hatake", pp. 307–323, in Sasaki Kōmei *et al.* (1988). Kuroda Hideo (1995: 142–146); this article first published in *Kamakura ibun geppō* 19, 1980.

35 *Hyōta jiruishō*, p. 204. *Fuboku wakashō* book 22 zatsubu 4, 2: 16-10173 and 10182.

36 *Seiryōki* (ca. 1628), p. 100; *Aizu nōsho* (1684), book 2, p. 217; *Hyakushō denki* (1681–1683), book 3, pp. 73, 75. *Kōka shunjū*, book 3, distinguishes the two terms, but confuses them on p. 106. *Dōbuntsūkō* by Arai Hakuseki, printed in 1760, book 4, pp. 475, 476.

37 *Nōgu zoroe*, p. 21.

38 WR 1: 11. From now on, we will not give references, as words in the 12th-century dictionaries are classified phonetically or according to the character(s). See a practical manual: Nishizaki Tōru (1995); and more details in Veashua (2003: 1–7).

39 See *Niina*, "New Characters" (now lost), in *Nihon shoki* Tenmu 11.3.13. (683); and in *Honchō shojaku mokuroku*, p. 172. Itō Toshikazu (1996: 85) gives the Japanese character *hata* from a document of 1094 quoting *Wakayama kenshi*, Kodai shiryō 1, p. 737; but he in fact thinks that this character occurs earlier in *Nagayoshi shū* (10th century), where *hata* is written in *hiragana* and not in *kanji*; see below.

40 See *haku-den* in Hi 1: 194, 196, etc.; *hatake* in M no. 4122; ES 15: 432.

41 See a poem taken from *Honchō seiki*, in *Kodai kayōshū*, p. 491 (*arata hirakamu*). See the meaning of waste ground invaded by grass in *Ōkagami*, pp. 267, 488; *Kodai kayōshū*, p. 439; and *Makura no sōshi* no. 77, p. 112. The Japanese character *arata* in the *Shinsen jikyō* is missing from the *kokuji* in Obata-Reiman. See the uncultivated/untilled *arata* in poetry, after Toda Yoshimi, "Jū-jūsan seiki no nōgyō rōdō to sonraku", in Toda (1991).

42 *Wakansansaizue* book 55, vol. 8, p. 17. *Man'yōshū* no. 1110 and 3848; *Bungo fudoki*, p. 373. For *araki*, see Sasaki Kōmei (1971: 132); *Nihon kokugo daijiten*; and the toponym Araki in Yanagita Kunio (1956). See Haudricourt (1987: 237–241).

43 Itō Toshikazu (1995).

44 Itō Toshikazu (1996); see also Kuroda Hideo (1984: 143, 144).

45 Itō Toshikazu (1999).

46 Itō Masatoshi, "Chisekizu ni miru Kii no kuni Kada no shō", pp. 227, 250; and Takemoto Toyoshige, "Jitō to chūsei sonraku, Bitchū no kuni Niimi no shō"; and Takemoto and Ishii Susumu, "Niimi no shō o kataru"; all in Ishii Susumu, ed. (1991). Itō Toshikazu (1996: 10) thinks that *yama-batake* include swidden fields from the second year of cultivation when they are no longer burned.

47 Itō Toshikazu (1996). He quotes *Kamakura ibun* vol. 6, no. 4308 and vol. 7, no. 4800, two articles dating from 1232 and 1235 reprinted in the *Shinpen tsuika*.

48 See Itō Toshikazu (1996); Itō Masatoshi (1991); Takemoto Toyoshige (1991) and Amino Yoshihiko (1991: 120, 136) (see note 46). Kuroda Hideo (1984: 268) distinguishes three main kinds of dry fields: permanent fields *sato-batake*, permanent fields with low yields *yama-batake* and swidden fields *yakihata*.

49   *Nagayoshi shū* 3: 69–4, quoted after Heian bungaku rindokukai ed. (1999: 14, 15, 223). The poem reads: katayama ni, hata yaku onoko, hata yakaba, miyama zakura wa, yokite hata yake.

50   *Shūi wakashū* 1: 3–1052. Nomoto Kanichi (1984: 3) and Nomoto (1987: 143) quote the two versions of the poem. Itō Toshikazu (1996: 85) quotes the *Nagayoshi shū* in "Shinpen kokka taikan" 3: 69–4 that transcribes *hata* with the Japanese character, which does not correspond to the earliest versions, see Heian bungaku rindokukai ed. (1999: 223).

51   *Nihon shoki* jō, 5th dan 11, p. 102; but the NKBT editors read *hataketsumono* instead of *hatatsumono*, a reading attested, however, by the *Nihongi shiki* Otsubon (10th century), p. 66.

52   *Man'yōshū* 2: 193; after Sieffert (1997: 189), though he translates "men of the fields".

53   Sano Masami (1963); and Nakanishi Susumu (1993: vol. 1, 137). The NKBT edition says "difficult to explain", and gives the reading *hatake-go* (instead of *hata-go*) from an informant.

54   *Shinsen waka rokujō* 2: 14-479 and *Fuboku wakashō* 2: 16-10186; *Fuboku wakashō* 2: 16-10188. I thank Michel Vieillard-Baron for correcting my French translation of these and the following poems.

55   See the section on "hata" in *Fuboku wakashō* 2: 16-10182 to 10188, which includes three poems from the *Shinsen waka rokujō* 2: 14-17 and 2279, and a poem by Emperor Juntoku (r. 1210–1221) in the *Mandai wakashū* 2: 15-3167; see also *Shinsen waka rokujō* 2: 14-2146, also included in *Fuboku wakashō* 2: 16-901; and other poems in "Shinpen kokka taikan" 4: 15-660, 8: 10-4839, 9: 28-316.

56   Hi 10: 184 (no. ho-139); and on this subject Kuroda Hideo (1984: 144). See Hi 1: 120 (no. 143), and *Jōganji monjo* (which include the *Ninnaji monjo*), facsimile of the Shiryō hensanjo, dated Jōgan 6.1.21. (864), with my thanks to Ishigami Eiichi for having directed me to the *Jōganji monjo*. See Hi 10: 3872 (no. 4999, dating from 1180), quoted by Itō Toshikazu (1995: 4). But the appearance of the character *hata* may date from 1094; see above.

57   Itō Toshikazu (1996: 8, 9), mentions the map "Yamashiro no kuni Kadono-gun handenzu" in Shiryō hensanjo ed. (1995), vol. 2, Kinai 1; the Niijima register in Hi 1: 87 (no. 98); the Jakurakuji certificate in Hi 10: 195 (no. 166); and the Chōfukuji inventory in Hi 8: 2933 (no. 3817). One may well ask whether *nobatake* is not a toponym, see below "The toponyms of small units", pp. 157–158.

58   DNK 4: 252; Fujii Kazutsugu (1986: 288); and an anonymous informant. Virgin lands seem to have been brought under cultivation on other Tōdaiji estates in Echizen; see below, "The toponyms of small units", p. 157. See Kimura Shigemitsu (1996a).

59   *Man'yōshū* 14: 3451, Péronny (1993: 27), M 14: 3364.

60   M 3: 404, 405, translated by Sieffert. See Hatai Hiromu (1981: 20).

61   M 16: 3848. See an interpretation of these poems by Nomoto Kanichi (1984a).

62   M 8: 1110.

63   *Ryōjin hishō* no. 332.

64   M 13: 3270.

65   M 8: 1556.

66   M 7: 1336 (also translated by Sieffert (2001: 119)), and M 14: 3452. The *Man'yōshū* mentions burning in four other poems: M 2: 199 and 230, M 10: 1879, M 12: 3033. Hatai Hiromu, "Man'yōshū ni miru yakihata nōkō bunka", pp. 20–26, in Hatai (1981), inspired us to look in other poetic anthologies.

67   Sakurai Mitsuru, "Omoshiroki no o ba na yaki zo", pp. 418–424, in Sakurai (1995). See above, "Swidden fields in the 8th–10th-century administrative texts", pp. 134–136. *Nobi* are considered a misdeed of farmers who neglect their fields, according to *Owari no kuni gunji hyakushōra no gebumi*, Hi 2: 480.

68 *Kokin wakashū* 1: 1–19. See also burning in *Kokin wakashū* 1: 1–453, and 17 (the latter included in *Ise monogatari* 12th dan); see Nomoto Kanichi (1987: 141).

69 M 10: 1879; see also the translation by Péronny (1993: 182). The girls cook *niru* the asters. See *warabi o taku* in *Kokin waka rokujō* 2: 4–3921.

70 Cobbi (1978: 107); Hirono Takashi (1998: 7–9).

71 *Gosen wakashū* 1: 2–3; see below for *ogi*.

72 *Yoshitada shū*, no. 56. See also burning in *Yoshitada shū*, no. 68, 86 (*yakefu*), 65, and 30, this poem being included in *Izumi Shikibu shū* 3: 73–9.

73 *Kokin wakashū* 1: 1–453.

74 RR Gunbōryō 73. "Musashi no zu shikishi", by Tawaraya Sōtatsu, ca. 1637, illustration of a poem about burning in the *Ise monogatari*, 12th dan; Idemitsu bijutsukan ed., *Kōrin*, exhibition catalogue, 1993, no. 2.

75 *Horikawa-in hyakushu* 4: 26–134.

76 Cobbi (1978: 102), Hirono (1998: 23, 24). A poem by Minamoto no Nakamasa (11th century?), in *Fuboku wakashō* 2: 16-923, mentions digging up *hori-motomeru* bracken. Kagaku gijutsuchō Shigen chōsakai (2000: 94) gives 276 kcal per 100 g of dried stems *hoshi-warabi*.

77 *Horikawa-in hyakushu* 4: 26–137.

78 *Horikawa-in hyakushu* 4: 26–129 to 144 ("sa-warabi"); *Fuboku wakashō* 2: 16-607–630 ("wakakusa") and 890–935 ("sa-warabi"); *Kokin waka rokujō* 2: 4-3544–3550 ("haru-kusa") and 3920–3922 ("warabi"); *Shinsen waka rokujō* 2: 14-1916–1920 ("haru no kusa") and 666–670 ("haru no no").

79 *Goshūi wakashū* 1: 4–33, 80, 822, 823, 824. *Gosen wakashū* 1: 2–3; *Ryōjin hishō* 302; *Wakan rōeishū* p. 197; *Tsurezuregusa*, p. 237 (the toponym Yakeno); *Chōshū eisō* no. 5; *Tadami shū* quoted by Ienaga Saburō (1942: 84). See also *Konjaku monogatari* 17-6; *Kagerō nikki*, p. 297 "Uma no suke". See other references in Hirata Yoshinobu (1998: 380–382, "warabi", and 386–388, "wogi").

80 Fukui Katsuyoshi (1983: 260); Miyamoto Tsuneichi, "Nōkō to Hata-shi no yakuwari" (first published in 1979), pp. 197–208, in Miyamoto (1994: 183, 184).

81 *Shinsen waka rokujō* 2: 14-669, also included in *Fuboku wakashō* 2: 16-950. Some ten poems of the 11th–13th centuries relating to *suguro* can be located on the "Shinpen kokka taikan" CD-ROM edited by Kadokawa shoten, whereas *yakeno, nobi*, etc. are more difficult to locate on this CD-ROM.

82 *Horikawa-in hyakushu* 4: 26–75, also included in *Fuboku wakashō* 2: 16-187.

83 See *shita-warabi* in *Fuboku wakashō* 2: 16-900; see also *wakana* in *Fuboku wakashō* 2: 16-237.

84 *Goshūi wakashū* 1: 4–45.

85 See Chapter 3, pp. 210–215.

86 *Yoshitada shū*, no. 379: *suguro karu* a summer poem. See also *susuki* and *suguro* in *Tsukimōde wakashū* (1182) 2: 12–679, *Fuboku wakashō* 2: 16-1038, *Shūgyokushū* 3: 131-1248, etc. See also *suguro* in *Shinsen waka rokujō* 2: 14-1932, *Fuboku wakashō* 2: 16-186, 13687 and 12988, the latter taken from *Horikawa-in hyakushu* 4: 26-183.

87 *Gosen wakashū* 1: 2–3; *Fuboku wakashō* 2: 16-613, 614 and 618.

88 See Hakone-chō kankōkōsha (1990: 139); Ichikawa Takeo (1985: 83–86).

89 Yanagita Kunio (1936), *Yanagita Kunio zenshū*, vol. 20, pp. 42, 43. For Edo-period philology, see Motoori Norinaga, *Meishō chimei senshiki* (Motoori Norinaga zenshū, vol. 10), *Chimei jion tenyōrei* (vol. 4), vols 8 and 9 of "Zokuzoku gunshoruiju" and the many regional monographs.

90 Yanagita Kunio (1936: 88–90).

91 Yamaguchi Yaichirō, "Yakihata goi to yakihata kankō", "Yakihata to chimei" (1944), in Yamaguchi (1972).

92 Nomoto Kanichi (1984: 301–328, and 304, the swidden field). See the four names for swidden field in other regions, in Nomoto (1987: 147); see a list for the

swidden field in the Japanese dialects in Chiba Tokuji (1983: 202). I thank Wilhem Grootaers for having drawn my attention to the dialects.

93   See the toponymy of land clearings in Nissen Jaubert (1998).

94   See the toponyms in WR books 6–9; ES book 22 (provinces and districts), book 28 (post stations), books 9 and 10 (shrines); and the *Fudoki*. See the "little places" in the Shōsōin documents of DNK 1–25, and in the *Heian ibun*. Reference works: Ikebe Wataru (1981); Naoki Kōjiro (1981); the volume *Sakuin* jō of the *Heian ibun*; Yoshida Shigeki (1991) (including the quotations from the early medieval texts).

95   Ono Takeo (1942: 255). Ide Itaru (1976) and Yoshida Shigeki (1978) discuss the difficulties of toponymy.

96   The references for the names of provinces, districts, villages, shrines and post stations can be easily located in Ikebe (1981), and in the indexes.

97   The village of Hata is written differently in the *Nihon shoki* Suiko 20.5.5., *Nihon ryōiki* ge 32, and *Konjaku monogatari* 11–32; see Ikebe (1981: 233). Haruta is glossed Hata in ES 9: 194 (Kujōkebon) and 10: 252.

98   Yanagita Kunio (1990: 122–130); Yoshida Shigeki (1978); Toda Yoshimi, "San ya (yama no) no kizokuteki ryōyū to chūsei shoki no sonraku", pp. 280–320, in Toda (1967); Seki Kazuhiko (1998: 17–20); Furuhashi Nobutaka (1989); Ashikaga Kenryō (1997: 129–139). The quotation is from Berque (1980: 157).

99   Hi 10: 3868 (12th century).

100 Ikebe Wataru (1981: 748); Ono Takeo (1942: 101).

101 The map of Chimori is found in Tōdaiji monjo, book 4; and the land survey in DNK book 5, see pp. 568, 582, 583. On the map is the entry Atsuhaida on four plots, the adjacent part being damaged. The register notes two plots. Morita Tei (1986: 107) mentions the plots Atsuhaida from the map and omits those from the register. See Ikagabe no Noyaki in DNK 5: 650, 581; and the two other personal names in DNK 5: 567, 580, 590, 647; and the slave Hita in N Tenpyō 15.9.13. (743). Kanasaka Kiyonori, "Echizen no kuni Asuwa-gun Chimori-mura kaiden chizu", pp. 395–428, in Kinda Akihiro *et al.* (1996: 396). See above, "Wilderness fields", p. 144.

102 Hi 1: 52, 53, 68, 69, 357 (Tōji); DNK 4: 84 (Hatasaki-); DNK 3: 330 (Hatayama). I thank Suzuki Yasutami and Ishigami Eiichi for their advice regarding toponyms.

103 DNK 5: 622, 623 (Yakimakida near the village of Sameda); and the Arakida at Sakai in DNK 5: 572 (at Miyata no shō), 604–607, 610, 25: 223, in Kii in Hi 1: 96 (2 plots), in Ōmi in Hi 1: 108, 109, 353 (3 plots), in Yamashiro in Hi 1: 176, 177, 209, 210, in Tanba in the 10th century in Hi 1: 323 (1 plot), and again in Tanba, Higami-gun, Kusakabe, inscribed on a wooden tablet, in Hyōgo-ken kyōiku iinkai, 1990, Plate 21; the Arakida family appears in DNK 5: 596, 599. See Ōarakida in DNK 5: 607, 608; Tera-oarakida in Kokuritsu rekishi minzoku hakubutsukan (1993: 35).

104 DNK 5: 572, 606–609, 612 (Yaefuda); DNK 5: 603, 604, 25: 218, 219, 221 (Korita, Kori); DNK 2: 335 (Korita in Yamashiro). See below, "Agri-sylviculture", pp. 168–170.

105 Shiryō hensanjo ed. (1995), vol. 2., Kinai 1. See *nobatake*, above, under "Swidden fields in the 8th–12th centuries".

106 Hi 1: 197, 176–178, 209, 210. Here, plot no. 32 in the Uesuki block is called Arakida in 873, Arata in 887. See the Hata family in Hi 1: 178. See the other toponyms below, pp. 160–162.

107 Saeki Arikiyo, ed., *Nihon kodai shizoku jiten*. Miyamoto Tsuneichi, "Nōkō to Hata-shi no yakuwari" (1979), pp. 197–208, in Miyamoto (1994).

108 Tōdaiji monjo 2: 78 (Iga); "Ōmi no kuni konden zu" in Tōdaiji monjo, book 4 (Ōmi); "Yamato no kuni Sōnoshimo-gun Keihoku handen zu" in Shiryō

hensanjo 1 (Nara); all from the 8th century. Hi 2: 429 (Kai in 969). For the 8th-century maps, see Kinda Akihiro *et al.* (1996).

109 See the family names in Ōta Akira, *Seishi kakei daijiten.* Noda and Hatano seem to be later than the other family names.

110 ES 9 and 10. See above, "Swidden fields, cleared plots and land clearing", pp. 155–156.

111 ES 9: 189; Shikinaisha kenkyūkai ed. (1977: 2: 322–325). It mentions *Jinja kakuroku*, of which only one copy is known, kept at Niigata University. Present shrine address: Nara-ken, Kitakatsuraki-gun, Ōji-chō, Ōaza, Hatakeda 4–640 (near Ōji station). For the date of the Kujōkebon, see Shikanai (2000: 5).

112 ES 9: 194; Shikinaisha kenkyūkai, 3: 873–876. Present address: Nara-ken, Takaichi-gun, Ōaza, Hōchiji, Azukidani 235 (Ichio station).

113 ES 9: 187; *Yoshitada shū*, no. 56, 68, 86 (*yakefu*, see above, "The poetry of the spring fires", p. 147). Shikinaisha kenkyūkai, 2: 67–71. Present shrine address: Nara-ken, Nara-shi, Daiyagyū-chō, 45 minutes by bus from the centre of Nara. For Ōyamatsumi, see Ikata Sadaaki (1940: 89).

114 ES 9: 194; Shikinaisha kenkyūkai, 3: 844–847. Present address: Nara-ken, Takaichi-gun, Asuka-mura, Ōaza, Fuyuno, Matsuba 152, near Danzan jinja, or in an unspecified place at Takaichi-gun, Takatori-chō. For the family, see Saeki Arikiyo in Miyamoto (1994).

115 ES 9; Shikinaisha kenkyūkai, 2 and 3.

116 ES 9: 187; Shikinaisha kenkyūkai ed. (1977: 2: 48–51). Present address: Nara-ken, Yamabe-gun, Soe-mura, Chūmuzan.

117 *Izumo fudoki*, p. 101, according to the interpretation of the NKBT editors.

118 *Hitachi fudoki*, p. 85, and ES 9: 240. There is another shrine with the same name in Musashi, Kami-gun, in ES 9: 237.

119 *Jidaibetsu kokugo daijiten*, Jōdai hen.

120 *Jidaibetsu kokugo daijiten*; *Nihon kokugo daijiten*; Ueno Makoto, "Araki no miya to yū kenzōbutsu", pp. 226–231, in Sakurai Mitsuru, *Man'yōshū no minzokugakuteki kenkyū.*

121 ES 9: 201; *Heianshi jiten*, art. "Ōtori-gun"; I thank Koyama Yasunori for his advice.

122 See our tables of toponyms (Tables 2.4, 2.5, 2.6 and 2.7). M 10: 2331 and 9: 1809 writes Yata and the word *fuseya* (hut) with the character *hachi*. N Nintoku 1.1. and 30.11. gives the name Yata-hime. WR 1: 11 associates *yakihata* and *yaihata*; RMS, p. 789, associates *yakigome* and *yaigome.*

123 Miyamoto Tsuneichi (1994: 207). ES 3: 64, 72; DNK 7: 464, 24: 316. A late copy of WR records the village of Hachiman (written 'eight' and 'ten thousand') in Awa, according to Ikebe Wataru (1981: 670). It was read Yahata in *Tosa nikki*, pp. 56, 79, and probably in *Ōkagami* 6: 253, 267, 433, 434, 484, 487. In the following texts, the modern editors read Yahata: *Kagerō nikki*, pp. 273, 288; *Sanbō ekotoba* 3: 26; *Makura no sōshi*, p. 287; *Konjaku monogatari* 3: 142. See the shrines of Yahata and Hatsuman under entries *ya* and *ha* in the *Iroha jiruishō.*

124 Poem taken from the *Honchō seiki* (12th century), in *Kodai kayōshū*, p. 491. The editors say that *yahata* is difficult to understand.

125 *Harima fudoki*, p. 263; *Iroha jiruishō*; ES 9: 227, 10: 252.

126 Ichikawa Takeo *et al.* (1985: 50). See other lists of dialectal words for the swidden field in Chiba Tokuji (1983: 201, 202); Nomoto Kanichi (1984: 301–328).

127 Hatai Hiromu, "Hachi kara jusseiki no hayashi nōgyō to yachi keiei", 1976, in Hatai (1981: 237–295). In fact Hatai thinks that Hayashida is not a toponym, but a specific term. See WR 8: 11, 13 (incorrect character) and 9: 6; DNK 2: 621; *Harima fudoki*, pp. 287, 289. Ikebe Wataru (1981: 592, 605, 678).

128   "Yamato no kuni Sōnoshimo-gun Keihoku handen zu" in Shiryō hensanjo (1992); and Hi 1: 21 (Nara). Hi 1: 111 (Kii); Hi 1: 125, 103, 121, 137: adjoining plots no. 20, 21, 27, 28, 33, 34, in Kii, and plot no. 30 in Hi 8: 3318.

129   Hi 1: 61 and 78 (841 and 847, in Yamashiro); Hi 1: 140 (870, in Yamato according to the index Heian ibun sakuin); see also Hi 7: 1605.

130   Hi 1: 191, 192 (883, in Kawachi); DNK 23: 427 (ca. 775, in Ōmi); see also Hi 7: 2616: a chestnut grove and a dry field (1164, in Yamashiro).

131   Abel (1962: 84–86); Menzies (1988: 361–367); François Sigaut provided this reference. I thank Helmut von Verschuer, a student of W. Abel, for his valuable comments.

132   Ichikawa Takeo (1985: 34–54); Ono Takeo (1942: 270, 274, 327–335); *Jikata hanreiroku*, pp. 99–102 (*yamahata, hayashihata*). Sasaki Kōmei (1988: 50, 354, 355) and Fukui Katsuyoshi (1983: 261, 268, 269) think that agri-sylviculture existed from the Neolithic. See *yamahata* above, "Swidden fields in the medieval period", pp. 134–136.

133   Hi 1: 97, 73, 140, 190, 413, see below. Kimura Shigemitsu, "Nihon kodai no hayashi ni tsuite", in Kimura (1992).

134   Hi: 1: 73, 413, 97.

135   Hi 1: 108, 109; and Hi 1: 177, 209.

136   Hi 6: 2327; see Kimura Shigemitsu (1992: 81, 82): the document is incomplete in the *Heian ibun*; it is found in *Hiroshima-kenshi*, Kodai chūsei shiryōhen, vol. 5, p. 444. See also Hi 5: 2029, 2030 (in 1139), and Hi 10: 44 (in 1080); Kimura (1992: 79, 80).

137   Hi 1: 57, Hi 1: 31. See also a block Kurisaki in Hi 8: 3318. See Ichikawa Takeo (1985: 36–42), for *heichi-bayashi*.

138   Hi 1: 176–178, 207–210; "Kadono handen zu" in Shiryō hensanjo (1992).

139   Many references are available. There are twelve toponyms just for the chestnut tree. *Aizu nōsho*, p. 101. See Kimura Yōjiro (1991: 375–401) for the supplementary food plants. Fukui Katsuyoshi (1983) argues for agri-sylviculture in both the modern and Neolithic periods.

140   *Kojiki* chū, p. 213; *Nihon shoki*, 40th year of Keikō Tennō, p. 304. Miyamoto Tsuneichi (1994: 179).

141   We quote Sasaki and Nakao below. Sahara Makoto (1987, 1994: 149, 284) (*nōgyō kakumei* and Yayoi bunka); Sahara, "Nōkō to Yayoijin no tōjō", *Asahi hyakka Nihon no rekishi* 39, Jan. 1987, pp. 162, 163, 192; Sahara, *Nihon bunka o horu*, 1992, pp. 13–15; Kuraku Yoshinori, *Suiden no kōkogaku*, 1991, p. 127 (*nōgyō kakumei*). Tozawa Mitsunori (1983: 261) criticizes this point of view.

142   Sahara (1992: 13, table). He corrects himself in Sahara (1996: 71); and Sahara *et al.*, eds (2000: 30, 36).

143   Ono Takeo (1942); Fujimori Eiichi (1970), a collection of articles from the 1950s and 1960s. See a historical account of the theories on neolithization in Sasaki Kōmei (1971: 34–56); and Yasuda Yoshinori (1987: 170–222).

144   Nakao Sasuke, summary by Ueyama Shunpei (1969: 86–91), of the article: Nakao, "Nōgyō kigenron", published in Morishita *et al.*, eds (1967).

145   Sasaki Kōmei (1982: 31); with details taken from Ueyama Shunpei (1969: 87). See also Sasaki Kōmei (1993: 29).

146   Arizono Shōichirō (1998).

147   Tozawa Mitsunori (1983: 254, 255); Ōtsuka Hatsushige (1995: 378); Sahara Makoto (1994: 150–156).

148   Tozawa (1983: 254–266).

149   Sasaki Kōmei *et al.* (1988).

150   Yoshizaki Shōichi (1992), special issue on carpology.

151   Umehara Takeshi, Yasuda Yoshinori, eds (1995): summary of the excavations and analyses presented by the authors of the analyses. Asahi Graph, ed. (1994).

152 Fujiwara Hiroshi (1998); Satō Yōichirō (2000). Koyama Shōzō summarizes the points of view in "Nihon shoku bunka no kiso", pp. 3–14, in Haga Noboru (1998).

153 Sahara, Tsude, eds (2000); with the round-table discussion *zadankai*, pp. 5–68.

154 On pre-domestic practices *kanri saibai* and domestication *saibai*, see Miyamoto Kazuo, "Jōmon nōkō to Jōmon shakai", pp. 115–138, in Sahara and Tsude (2000).

155 Jean Guilaine, "The first peasants in the world: where? when? how? why?", seminar at the Collège de France, 1998–1999; we have quoted the papers of Danièle Lavallée (Andes), Ginette Aumassip (Sahara) and Michel Orliac (Oceania); Guilaine, ed. (2001); Guilaine (1996: 697–719); Cauvin (1997).

156 Guilaine (1996). For Neolithic swidden farming, see Jacques-Élie Brochier, "Géographie du monde agropastoral", pp. 303–322, in Guilaine, ed. (1991) (in tropical America); G. O. Rollefson, *Le Néolithique de la vallée du Jourdain*, 1994; D. H. Thomas, *Agriculteurs du Nouveau Monde*, 1994; Rollefson and Thomas cited by Mazoyer and Roudart (1997: 75).

157 Miyamoto Kazuo (2000: 118–121); for buckwheat, he records five Neolithic sites, but Yasuda Yoshinori (1987: 214) has identified fourteen (see the Introduction, Map I.4). See other lists of cultivated plants in the Neolithic in Sasaki Kōmei (1993: 223); and Terasawa Kaoru (1986: 25–29).

158 Kimura Shigemitsu (2010: 17, 19, 21) refers to *kanri saibai*; and Tamada Yoshihide (2009: 128–132) mentions *saibai*. Nespoulous (2008: 20) mentions red beans *azuki* and millet *hie*. See Shitara Hiromi (2014: 460); Nasu Hiroo (2014: 100, 101, 110).

159 Harada Nobuo (2006: 40–45). Certain archaeologists, such as Koyama Shūzō (2011: 27), Fujio Shinichirō (2011: 135), Nakazawa Michihiko (2009: 241) and Takase Katsunori (2009: 214–219) mention the "possibility" of rice growing on swidden fields in the Neolithic; and Takase draws attention to fertilizing with ashes, which is an argument in favour of swidden farming. For Niels and Abe (2015: 15), the Jōmon civilization "probably involved swidden (slash-and-burn) methods and took place in dry fields rather than paddies". Niels and Abe (2015: 27) think that human subsistence in East Asian forests "usually depends on slash-and-burn agriculture or other forms of cultivation".

160 Yamaguchi Hirofumi (2011: 155–172), with tables on pp. 160–166; Nakazawa Michihiko (2009: pp. 130–141).

161 Shitara Hiromi, pp. 449–469, in Fujio Shin'ichirō, ed. (2014). In the same volume, Andō Hiromichi, "Suiden chūshin shikan hihan no kōzai", pp. 410, 411, states that the idea of an early agriculture in the form of swidden farming *yaki-hata* was adopted by academic circles in the 1970s, and that since the 1990s this concept has evolved into a wider perception of a 'composite Japanese culture' *Nihon bunka no tagensei* with multiple farming practices including 'non-rice farming' *hiinasaku nōkō*, a 'swidden cropping culture' *yakihata bunka*, a 'laurel-forest culture' *shōyō jurin bunka*, a 'beech-zone culture' *nara-bayashi bunka*, as well as a 'rice-growing culture' *inasaku bunka*. Another collaborative research programme was conducted by the National Museum of Japanese History from 2010 to 2012 and the results were published by Kudō Yuichirō, ed. (2014). This volume highlights a number of articles concerning plant and forest management during the Neolithic, focusing especially on lacquer, chestnuts and hemp.

162 Mazoyer *et al.*, eds (1997: 113). For the geography, see Berque (1980).

163 On Kajiwara, see above "The ethnology of swidden-field crops in Japan", p. 129; on Shiiba-mura, see above "Swidden farming across Japan", pp. 124–125; on Sar Luk, see above "Swidden-field crops in Asia", pp. 121–123. In Laos, in 2010, each family of Koknang village, Luang Prabang province, at 1,100 m altitude, cultivated both 1 ha of irrigated rice and a swidden field with cotton, sesame and dry

rice; see Harada Nobuo (2011: 251–253). In 2014, Verschuer visited about ten villages in Luang Prabang, all situated at high altitude, and noticed that most families lived by means of one swidden plot, a small orchard, pig and poultry farming, as well as plant gathering, hunting and fishing (no permanent crops). On swidden cultivation in Laos, see Ducourtieux (2009).

164 Takemoto Toyoshige (1991); Fukui Katsuyoshi (1974).
165 Sasaki Akira (1998: 75) (quotation).
166 Ichikawa Takeo, "Soba no shokumotsu bunka", pp. 53–71, in Ichikawa Takeo (1978). Tsuboi Hirofumi (1986); Hashimoto Seiji (2001). *Nōgyo zensho* 2: 106 "Soba" (17th century, buckwheat and taro). The cultivation of buckwheat was encouraged by the government in 840, see RSK 8: 328, Jōwa 6.7.21.

3   Biodiversity

Harvesting of wild plants

With an especially rich biodiversity (4,000 plant species today), Japan offers a privileged environment for gathering wild plants. This type of acquisitive activity on the fringe of agriculture in principle characterizes hunter-gatherer societies, but in Japan as in other countries it continued up to the modern period. In our view, foraging constituted one of the three poles of the subsistence economy in the plant kingdom, together with permanent agriculture and shifting crops (or swidden farming). The gathering of wild plants and fruits represented in effect a nutritional supplement alongside crops. However, this area has hardly attracted the attention of historians. Specialists of the early medieval period have taken as little interest in foraging as in swidden-field crops, mainly because of the lack of sources. In fact, the harvesting of wild plants completely escaped the regulatory and administrative control of the imperial court; but it is mentioned in the texts relating to taxation.

An analysis of the annual deliveries from the provinces to the court by way of taxes makes it possible to draw up a descriptive list of the plants obtained by the Japanese from the natural environment. The content of these deliveries is found notably in the "Regulations of the Engi Era", *Engishiki*, completed in 927. Since the information given here is confirmed by the tablet inscriptions *mokkan* used as labels for packs of foodstuffs in the 8th century, it is reasonable to suppose that the *Engishiki* gives a picture of plant gathering in the various regions of Japan corresponding to what actually occurred.[1] (See the catalogue of products mentioned by the *Engishiki* in the Appendix.) The data provided by the *Engishiki* is rounded out by the information in the 8th-century regional monographs *Fudoki* and the Shōsōin documents relating to the Nara court, as well as in the poetry of the 8th–12th centuries. As the available sources (apart from the *Fudoki*) mostly concern the nobility, we will first of all draw up a list relating to the consumption of edible plants at the court. In order to reconstruct the harvesting of wild plants among the rural (peasant) population, we will then present the prehistoric and modern data. We therefore propose a threefold approach: historical, archaeological and ethnological.

As regards the identification of the plants mentioned in the early medieval texts, we have taken as our basic reference the "Illustrated Dictionary of Plant Names", *Zusetsu sōboku meii jiten*, by the botanist Kimura Yōjirō, which identifies each plant found in the Japanese sources and proposes the corresponding modern name(s). This has enabled us to give an English name for each plant encountered in the texts, but archaeobotany will certainly correct some identifications in the future. Moreover, we have not attempted to distinguish between wild and cultivated plants and have noted all the collected plants that occur in the historical sources (see Table 3.1). In fact, it is probable that a number of wild species were also cultivated, notably some vegetables and graminae, which is why we have not excluded these from our list, beginning with tree fruits and nuts.

Fruits

The "Regulations of the Engi Era" mention the nuts that were delivered annually by the provinces to the Heian court. Among these are (sweet) chestnuts *kuri*, walnuts *kurumi*, hazelnuts *hashibami no mi*, castanopsis *shii* acorns and *kaya* (torreya), a taxaceae. These species came from various regions of Japan, where they grew wild. Chestnut trees were also grown in plantations supported by the government. Chestnuts appear very often on the palace tables, in different forms: fresh (*namakuri*), dried (*hoshikuri*), crushed (*kachikuri*) and "flattened" (*hirakuri, hitarikuri*). Fresh and crushed chestnuts, as well as castanopsis acorns, were served to the emperor and high-ranking officials at the seasonal banquets. Like hazelnuts, these foods were supplied to the court as "fresh product specialities" *nie* by some provinces, most notably Tanba (Kyōto, Hyōgo).[2]

Kaya (torreya), delivered by eight provinces, were reserved by the court for medicinal purposes. They grew among other places in the forests of Izumo and were probably eaten in various parts of the country.[3] The provinces also delivered to the court sawtooth oak acorns, which were used as a dye. However, the Chinese character *tsurubami* was also read *tochi*, i.e. the horse chestnut. Horse chestnuts are absent from the tables of the nobility. They were possibly used as food "for the poor", judging by the soldiers mentioned in the *Utsuho monogatari* (10th century), or the pilgrims who appear in the *Taiheiki* (14th/15th century). Horse chestnuts had to be treated before being eaten to get rid of their bitter taste.[4]

Pinenuts *matsunomi* and beechnuts *kashiwa* (of the modern *buna*) do not appear in the sources of the 8th–10th centuries mentioned above. They were, however, an occasional addition to some banquets, at least from the 12th century.[5] Nor did acorns of the *ichii* oak form part of the products supplied by the provinces, but they are listed in the 8th-century texts among the forest products. In fact, in the countryside, the various nuts would probably have been part of the diet during autumn. So it was that the inhabitants of three villages in Hitachi (Ibaraki) found in their surroundings sawtooth oaks, *ichii*

Table 3.1 Wild plants mentioned in Chapter 3 (early medieval Japanese names)

Japanese	English	Japanese	English
Fruits			
kuri*	(sweet) chestnut	kanshi*	mandarin
kurumi	walnut	nashi*	pear
hashibami no mi	hazelnut	moronari	silver berry
shii	castanopsis (shiia)	itabi	fig
kaya	*kaya* (torreya)	natsume*	jujube
ichii no mi	acorn (oak)	biwa*	loquat
hahaso	acorn (oak)	sumomo*	plum
matsunomi	pinenut	momo*	peach
kashiwa no mi	beechnut	kaki*	persimmon
tochi	horse chestnut	yuzu	citron (bitter orange)
tachibana*	orange	karamomo no sane	apricot kernels
yamamomo	red myrica	ume	Japanese apricot
ichigo*	blackberry/ brambleberry	zakuro	pomegranate
akebi	akebia	shitomi	wild quince
mube*	stauntonia		
Vegetables			
kiuri	cucumber	uhagi	(white) aster
uri*	melon	soraji*	nothosmyrnium
hozochi	melon	itadori	polygonum
kamouri	wax gourd	tatarahimebana	?
nasubi*	egg-plant	konasubi	black nightshade
hiyu	amaranth	takana	leaf mustard
azami*	thistle	karashi	leaf mustard
fufuki*	butterbur	yamaararagi	magnolia kobus
aona*	turnip	mega*	ginger
nazuna	shepherd's purse	kurenohajikami*	ginger
chisa*	lettuce	naruhajikami	Japanese pepper
aoi*	mallow	hakobe	chickweed
shibukusa	sorrel	hotokenoza	lapsana
ki*	spring onions	ōdochi	sow-thistle
hiru*	wild chives	ochi*	colza
araragi	a liliaceae?	tara	Japanese angelica tree
konishi*	coriander	tsuchitara	angelica tree
imoshi	taro stems	takamuna	bamboo shoots

Japanese	English	Japanese	English
mame*	soybeans	wasabi	Japanese horseradish
azuki*	red beans	asatsuki	chives
hahako	cudweed	kusabira, take	mushrooms
warabi	bracken		

Aquatic plants

egu	eleocharis	seri*	water-celery
asaza	Nymphoides peltata	mizufufuki	foxnut
tade	water-pepper	nunawa	water shield
nagi*	water-leek	hachisu	lotus
kawahone	water-lily	hishi	water chestnut

Root vegetables

imo*	(satoimo) taro	umafufuki	burdock
imo	(yamanoimo) yam	namai	arrowhead
tokoro	yam	kuzu	pueraria
ōne*	radish	katakago	dog-tooth violet

Seaweeds

mozuku	nemacystus	igisu	campylaephora
miru	codium	funori	gloiopeltis
wakame, me	undaria	nori	porphyra
konbu, hirome	laminaria	aonori	ulva (Ulva pertusa)
ogonori	gracilaria	tsunomata	chondrus (Chondus elatus)
arame	eisenia	tsunomata	chondrus (Chondus ocellatus)
kokorofuto	gelidium		

Oil seeds and nuts

goma	sesame	hemi	Japanese snowball
e	perilla	tsubaki	camellia
asa*	hemp	kurumi	walnut
hosoki	Fagara		

Tinctorial plants

kurenai	safflower	kariyasu	miscanthus
murasaki	gromwell	ai*	Japanese indigo
akane	madder	kihada	Amur cork tree
kuchinashi*	gardenia	tsurubami	sawtooth oak

Japanese	English	Japanese	English
Plants for artisanal use			
urushi*	varnish tree	hakogusa	osier?
tokusa	scouring rush	kuwa*	mulberry tree
koshiabura	acanthopanax resin	take	bamboo
tsuzura	Sinomenium?		
Textile plants/products			
asa*, o	hemp	kaya*	eulalia (miscanthus)
kemushi	ramie, hemp	susuki	eulalia
karamushi	ramie	ogi	sweet miscanthus
taku, kōzō	paper mulberry	ashi	common reed
taku	mulberry	komo*	wild rice
kuzu	pueraria	i*	rush
kuzu	Japanese wisteria	suge*	sedge
shinanoki	Japanese lime	kama	bulrush
yuu	ramie cotton, mulberry fibres, etc.	chi (kaya)	imperata
nio	hemp fibres?	ayame	sweet flag
take	bamboo	yomogi	mugwort
Graminae and plants with elongated stems or leaves			
urushine*	non-glutinous rice	kimi*	common millet
mochi(no)yone	glutinous rice	mino	glyceria
akayone	red rice	hie*	barnyard millet
awa*	foxtail millet	futomugi*	barley
mugi*	(komugi) wheat	sobamugi*	buckwheat

Note: * Wild species also cultivated, according to the texts of the 8th–12th centuries.
Source: © Verschuer, *Rice, Agriculture, and the Food Supply in Premodern Japan*, Routledge 2016.

oaks, "oaks with saw-toothed leaves" *hahaso*, castanopsis and chestnut trees. The gathering of (sweet) chestnuts was the subject of several early poems.[6]

The provinces also supplied the court with many pip and stone fruits. Some came from the five provinces around Heian-kyō: the orange *tachibana*, red myrica *yamamomo*, blackberry *ichigo*, akebia and stauntonia *mube*. The fruit of the modern *tachibana* (a citrus), with a diameter of 2.5–3 cm and six to eight segments, is unfit for consumption, though in the early medieval period this may have been a different species. It grew wild, but was also cultivated in orchards in Hitachi. Three kinds of oranges, *tachibana, hokotachibana* and *hiroitachibana* ('collected orange'), decorated the tables of the emperor and high-ranking officials at the autumn and winter banquets. An orange tree and a cherry tree bloomed, in spring, in the courtyard in front of the Shishinden, one of the buildings of the imperial residence. The orange tree appears in sixty-six poems of the *Man'yōshū* anthology.[7]

By contrast, the blackberry, akebia, red myrica and stauntonia did not inspire the poets, although the last two were eaten at Nara in the 8th century. The akebia, of a pale purple-blue colour, belongs like the stauntonia to the lardizabalaceae family. These four fruit trees and shrubs still grow today in various parts of Japan. In the 10th century, the blackberry and stauntonia were cultivated in the imperial orchards, but only the wild blackberry was served to the emperor and empress, in the 5th month (June).[8]

Other fruits reached the court as regional specialities: mandarins *kanshi* from five provinces; pears *nashi* from Shinano (Nagano), Inaba (Tottori) and Kai (Yamanashi); silver berries *moronari* from Bitchū (Okayama); figs *itabi* from Chikuzen (Fukuoka) and Kawachi (Ōsaka); and dried jujubes *hoshinatsume* from Shinano and Inaba.[9] The mandarin was introduced from China in the 8th century. In 725, two officials were promoted to the 5th rank for having brought back the mandarin from China and successfully cultivating it in Japan. According to the philologist Kariya Ekisai (1775–1835), this was the ancestor of the modern mandarin *mikan*, but the botanist Kimura Yōjirō does not share this view.[10] Be that as it may, the mandarin quickly became widespread in Japan, whereas another plant of Chinese origin, tea, introduced in the 10th century, remained a rare commodity until around the 13th century.[11]

As for the *nashi*, this is the Japanese pear *pyrus serotina*, with round fruits like apples, and not the European pear *pyrus communis*. The silver berry *moronari* corresponds to the *gumi*, a tree that today has some forty species. Jane Cobbi identified two species with edible fruits at Kaida-mura in the Nagano region. The fruits resemble small wild cherries.[12] By contrast, the fruits of the species *itabi* (mod. *itabikazura*) are, it seems, no more eaten today than in the early medieval period. In the documents of the time, they are not found among the food products. The *Engishiki* only mentions fig-tree ashes for artisanal use in the palace Library Office. Fig trees grew in Hizen, as did other industrial plants.[13] Dried jujubes were served on the palace tables and the fresh fruit was used as a medicine. This fruit seems to have been quite common, at least at Heian, for in 1072 the monk Jōjin mentions the jujube by way of comparison with Chinese fruits that he did not know, namely lychees and fruits of the plum of the nightingales.[14]

Apart from the banquets at Heian, the meals of the emperor, empress and crown prince included seasonal fruits. In the 5th and 6th months (June, July), there were loquats *biwa*, plums *sumomo* and blackberries; from the 7th to 9th month (August–October), peaches *momo* and (sweet) chestnuts; from the 9th to 11th month (October–December), *kaki* or persimmons and citrons or bitter oranges *yuzu*. All these fruits, with the exception of citrons, were also eaten at the Nara court in the 8th century.[15]

At Heian, dried persimmons and *tachibana* oranges were included in almost every ritual banquet.[16] Dried persimmons remain part of Japanese customs. Today, persimmons of the astringent *shibugaki* variety are peeled in November–December and fixed with their stems to straw ropes that are hung in front of the house. They are left to dry in the sun in this way for several weeks. Both the wild and cultivated plum *sumomo* grew, in the early medieval period, in the province of Izumo. It is the subject of the following poem in the *Man'yōshū*: "Are these the flowers of the plum tree in my garden that have fallen into the courtyard, or the snowflakes that have stayed there?"[17]

The plum tree, too, had its place in the imperial court orchard, described in the "Regulations of the Engi Era" in these words: "Various fruit trees, four hundred and sixty in total: a hundred pear, a hundred peach, forty mandarin, forty small mandarin, a hundred persimmon, twenty orange, thirty large jujube, thirty stauntonia; and a garden of blackberries *ichigo* of 2 *tan* (0.52 acres)." There were also, in 771, in the orchard attached to a hospice managed by the Tōdaiji temple and located at Tōchi (north of Kashihara) in the province of Yamato: "in total eighty-three trees: seventeen mulberry *kuwa*, nineteen willow, four pear, two Maackia *enisu*, four?, five chestnut, a sawtooth oak [or horse chestnut?], twenty-five paper mulberry, a loquat *bi[wa?]*, nine peach, an apricot *ume*, a persimmon and a [?]".[18]

The Tōchi orchard therefore had a Japanese apricot *ume*, the tree most celebrated by the early medieval poets, with its flowers heralding spring, which the Heian writers admired both for their beauty and their perfume. However, as it is not found in the *Nihon shoki*, the *Kojiki*, or the *Fudoki*, Kariya Ekisai inferred that the fruits were not eaten and that the apricot tree with edible fruits was introduced to Japan by Europeans in the Edo period. Apricots are not in fact listed among the food products in the *Engishiki*; but the presence of an apricot *ume* in the Tōchi orchard strongly suggests that this tree was grown for its fruits. Moreover, we know of three cases where, in 770–771, an office in Nara purchased apricots *umeshi* as food for its employees.[19]

In his description of the Japanese apricot, Minamoto no Shitagō (911–983) writes, quoting a Chinese source, that this tree "resembles the apricot *karamomo*, whose fruit is acidic". The apricot *karamomo* was not eaten as a fruit at Nara or Heian, but is listed in the *Engishiki* among the medicines. It was delivered annually to the court by five provinces, in the form of *karamomo no sane* apricot kernels. Though little known at the court, the apricot *karamomo* appears in six poems of the 12th and 13th centuries. It may not have joined the ranks of the fruits served at banquets until the 13th century.[20] Another fruit, the pomegranate *zakuro*, was ignored by the Heian poets. It is not found in the *Fudoki* or the *Engishiki*. The pomegranate does not appear on the tables of the nobility until the 12th century, but is found no later than the 11th century among the "five fruits" offered to the emperor on certain occasions. The minister Fujiwara no Sanesuke (957–1046) lists pinenuts, acorns, (sweet) chestnuts, jujubes and pomegranates. The name, at least, was known from the 8th century, since it occurs in *zakurokame*, the term used to describe the eleven water bottles that formed part of the tableware of the Daianji temple.[21]

The wild quince *shitomi* is also absent from Heian poetry. The native species has yellow, acidic fruits, 2–3 cm in diameter. The court reserved the quinces delivered by the provinces of Yamato and Ōmi for medicinal use. However, the absence of this fruit in poetry does not mean that it was unknown to the Japanese, since the monk Jōjin mentions it in 1072 as a point of reference for describing a Chinese fruit, a kind of azarole, which he did not know.[22]

The twenty-eight species listed above are the main fruits known to the Heian court. It may be supposed that many fruits were also gathered in the countryside for local consumption, and that diverse species grew in different parts of the country. It is known, for instance, that the inhabitants of Izumo found plum and red myrica trees in their environment.[23]

Vegetables

Besides seasonal fruits, the daily meals of the emperor and empress included vegetables.[24] The *Engishiki* gives useful information about the growing and harvesting periods for vegetables, according to the early medieval Japanese calendar. Approximately one month needs to be added to each lunar month to obtain the modern calendar month. (For example, the 5th month corresponds more or less to June.) Here is the list of vegetables recorded in the "Regulations of the Engi Era":

> Cucumbers *kiuri* (5th–8th months), egg-plant *nasubi* (6th–9th months), amaranth *hiyu* (5th–8th months), thistles *azami* (2nd-9th months), butterbur *fufuki* (5th–8th months), turnips *aona* (mod. *kabu*, all year), crucifer stems *kukutachi* (2nd, 3rd months), shepherd's purse *nazuna* (11th–2nd months), lettuce *chisa* (3rd–5th months), mallow *aoi* (5th, 8th–10th months), sorrel *shibukusa* (mod. *gishigishi*, 4th, 5th, 8th–10th months); the condiment-vegetables are as follows: Chinese chives *komira* (mod. *nira*, 2nd–9th months), spring onions *ki* (mod. *negi*, 4th–5th and 9th–1st months), wild chives *hiru* (mod. *nobiru*, 11th–4th months *hiru* fresh, 5th–9th months *hiru* dried), ginger *kurenohajikami* (mod. *shōga*, 6th–8th months), Japanese pepper *naruhajikami* (mod. *sanshō*, 3rd, 4th months for the leaves, 5th, 6th months for the roots), water-pepper *tade* (4th–9th months), a liliaceae(?) *araragi* (all year), a crucifer of the species *karashi* (leaf mustard) called *konishi* (9th–2nd months); other vegetables listed are: radishes *ōne* (mod. *daikon*, 10th–2nd months), water-celery *seri* (1st–6th months), water-leeks *nagi* (5th–8th months), taro stems *imoshi* (6th–9th months), soybeans *mame* and red beans *azuki* (6th–9th months), cowpeas *sasage* (6th–7th months), cudweed *hahako* (2nd–3rd months), taro *imo* (mod. *satoimo*, 9th–1st months) and a melon *hozochi* (6th–8th months).[25]

Also prepared in the imperial kitchens were conserves, most with salt, some with fermented soy. In spring, the following vegetables were macerated: bracken *warabi*, (white) asters *uhagi* (mod. *yomena*), thistles, water-celery, butterbur, *soraji* that may correspond to *kusamochi* (*Nothosmyrnium japonicum* Miq.), polygonum *itadori, tatarahimebana*(?), black nightshade *konasubi*, melons *uri*, wild chives, Chinese chives and turnips. In autumn, other conserves were prepared: melons, wax gourds *kamouri*, leaf mustard *takana*, turnips, egg-plant, black nightshade, water-leeks, *araragi* (a liliaceae?), soybeans, magnolia kobus *yamaararagi*, water-pepper, foxnuts *mizufufuki* and ginger *mega* (mod. *myōga*).[26] Many plants grew wild in different parts of the country, but the court also had some twenty vegetables cultivated in the imperial vegetable gardens.[27]

Apart from these many vegetables forming part of the emperor's daily diet, it was also customary to offer "twelve vegetables" to the emperor at New Year, on the first day of the rat. These consisted of "young shoots *wakana*, thistles, lettuce, water-celery, bracken, shepherd's purse, mallow, nothosmyrnium *sawasorashi*(?), mugwort *emogi*(?), water-leeks *mizutade*(?), the seaweed nemacystus *mozuku* and water-lilies or foxnuts *kawahone*". (From now on we will call *kawahone* water-lily, to distinguish it from the foxnut *mizufufuki*.) As some plants were not harvested until after the New Year, we may assume that it was offered early for auspicious reasons. On the 15th day, the emperor was served a gruel made with "seven vegetables", this being supposed to ward off the "ten thousand ills". It contained: shepherd's purse, chickweed *hakobe*, water-celery, turnip, cudweed *ogyō*, radish *suzushiro* and lapsana *hotokenoza*. [28]

Certain plants are grown in the Chinese vegetable garden, created in 1997 by Georges Métailié in the garden of the Museum of Food at Vevey in Switzerland. Georges Métailié distinguishes the following categories:

- pulses (beans)
- leaf vegetables (cabbage, lettuce, mustard, water shield, amaranth, shepherd's purse)
- fruit-like vegetables and gourds (egg-plant, gourd, cucumber, calabash)
- root vegetables (radish, yam, pueraria, taro, lotus, arrowhead, water chestnut)
- condiment-vegetables (spring onions, chives, Japanese pepper, ginger)[29]

Besides the emperor, the court nobles also ate vegetables at banquets and in their daily meals. Mention may be made of the banquet of the 4th month held at the residence of the regent Fujiwara no Tadazane (1078–1162). At that time, the regent saw the following foods on his table: dried products: codium seaweed *miru*, ulva seaweed *aonori*, fried cakes *senbei*, water-lilies; raw products: laminaria *konbu*, melon, butterbur, and burdock *umafufuki* (mod. *gobō*), a plant absent from the early medieval regulations.[30]

Most of the seasonal vegetables at the Heian court were also part of the daily menus at Nara in the 8th century, and in the same seasons. But some vegetables do not appear among the provisions of the Nara offices. Missing, for example, are: shepherd's purse, Chinese chives, spring onions, coriander *konishi*, leaf mustard *takana*, foxnuts and cudweed.[31] On the other hand, the Nara court served to the officials some vegetables not included in the Heian regulations, such as sow-thistles *ōdochi* (mod. *nogeshi*). At Nara, bracken sometimes occupied the first place from a quantitative point of view. For example, in 734, the workshops of the Kōfukuji temple purchased for the artisans' meals 5,296 bundles of bracken, together with 170 bundles of mallow and 3,336 bundles of sow-thistles.[32]

Several other vegetables formed part of the diet in the Nara and Heian periods, namely colza *ochi* (mod. *aburana*), angelica tree buds *tara* and bamboo shoots *takamuna* (mod. *takenoko*). As condiment-vegetables, mustard *karashi* (mod. *karashina*), Japanese horseradish *wasabi* and chives *asatsuki* were also eaten.[33]

Mushrooms *kusabira* or *take* seem to have been little appreciated by the nobility. In fact they do not appear either in banquets or in poetry. It is true that the use of the word "mushroom" was imposed on the priestess at Ise to avoid the taboo word "flesh" *shishi*. However, a monograph of the district of Inami (Himeji region) in Harima (Hyōgo) mentions the pleasant taste of the mushrooms that grow at the beginning of the 10th month in the pine forests. In the time of the legendary Emperor Ōjin (r. 270–310), the people of the Kuzu region apparently offered him tribute of mushrooms, (sweet) chestnuts and trout. From the 12th century, the collections of legends also mention the picking of mushrooms *hiratake*. Mushrooms *take* were sometimes served to the low-ranking officials of Nara. By contrast, the juice of the Japanese ivy *amazura* was very much appreciated by the nobility. Together with maltose *ame*, it was one of the two food-sweeteners used by the court. The court received annual deliveries of ivy juice from many provinces, where it was made from the vetch Virginia creeper *tsuta*, a vitaceae.[34]

Aquatic plants form a separate category and played quite an important role in the diet. We have already mentioned the use of water-pepper, water-celery, water-leeks, cudweed and water-lilies, two nymphaeaceae. To these must be added two other nymphaeaceae: water shield *nunawa* (mod. *junsai*) and lotus *hachisu* (mod. *hasu* or *renkon*), as well as water chestnuts *hishi*.[35] It should be remembered that Japan had considerably wetter soil in early medieval times than today. Prior to the drainage campaigns carried out by the Meiji government (1868–1912), there were marshy zones, temporarily flooded, in several parts of the country, including Shizuoka. These regions were characterized by an abundance of aquatic vegetables and aquatic graminae (see below). Even in the capital of Heian, the soil was so wet that the government allowed water-celery, water-leeks and lotus to be grown there.[36]

The nobility of Nara and Heian enjoyed not only fresh-water vegetables, but also seaweeds delivered by the provinces. The seaweeds most often served at the court were: codium *miru*, laminaria *konbu*, eisenia *arame*, undaria *me*, *nigime* or *wakame*, gracilaria *ogonori*, gelidium *kokorofuto*, campylaephora *igisu*, gloiopeltis *funori*, porphyra *nori*, Ulva pertusa *aonori*, Chondrus elatus *tsunomata* and Chondrus ocellatus, as well as *tsunomata*. Many poems allude to the harvesting *karitsumu* and drying *karihosu* of the seaweeds *wakame*, *mirume* and "pretty water grasses" *tama-mo*.[37]

Root vegetables, too, were eaten, beginning with taro *imo* (mod. *satoimo*) and radishes, mentioned above. The provinces also delivered annually to the court a different *imo*, corresponding to the yam of the variety *yamanoimo*, as well as the yam of the variety *tokoro*, two dioscoreaceae. These tubers were also widely used as a medicine. Two other roots were reserved almost exclusively for medicinal use: arrowhead *namai* (mod. *omodaka*) and pueraria *kuzu*.[38]

Pueraria creepers provided resistant textile fibres that were made into a coarse cloth. Pueraria rhizomes containing a nutritious starch were no doubt dried and ground. In the 8th century, the leaves of this plant were also eaten. Pueraria appears in a *Man'yōshū* poem: "On the hill where the deutzia flowers bloom and fall, do you hear the cuckoo's song? Young girl stripping pueraria." It appears, however, that cakes and beverages made with pueraria were unknown in early medieval Japan.[39]

This general survey gives the impression that the Japanese of the early medieval period ate all kinds of vegetables. However, the only ones to benefit were the leading personages of the court, on festive occasions. The daily diet of the lower-ranking nobles was undoubtedly less varied, as can be seen from the list of foods on the menu for the meals of the low-ranking officials of the sūtra-copying office at Nara. They consisted of: rice, salt, sesame oil, fermented soybeans, vinegar, leaf mustard, beans, wheat noodles, seaweeds (codium, eisenia and glioiopeltis), maltose, plums *sumomo*, cucumbers *kiuri*, turnips, thistles, water-leeks, egg-plant and Japanese pepper.[40]

The harvesting of vegetable-plants is confirmed by the poetry, which speaks of the gathering of melons, water chestnuts, water-celery, Chinese chives, cudweed, pueraria, yams of the variety *tokoro*, bracken, (white) asters and three plants not mentioned in the *Engishiki*: erythronium or dog-tooth violet *katakago* (mod. *katakuri*); eleocharis with black tubers *egu* (mod. *kuroguwai*); and Nymphoides peltata *asaza*, a gentianaceae, eaten in the offices at Nara. In fact, the picking of young spring shoots *wakana tsumu* was a favourite subject in poetry and, among the plants gathered, bracken *warabi* was the most often mentioned.[41]

The regional monographs refer to many wild vegetables. For example, in the district of Iishi in Izumo (Shimane) were found yams of the variety *tokoro* and *yamanoimo* and pueraria, tubers that were probably dug up by the local inhabitants, using a digging-stick. In another district of the same province, (white) asters, butterbur, Japanese pepper, taro, bracken, etc. were gathered. Bracken roots were also commonly dug up, according to a poem by Minamoto no Nakamasa (?–1156?). In the province of Harima (Hyōgo) there was Japanese horseradish, in Hizen (Nagasaki) amaranth, and in Hitachi (Ibaraki) poria cocos *matsuhodo* (*bukuryō*). The lotus grew in a number of regions.[42]

Medicinal products

Besides the roots mentioned above, other vegetables such as cudweed, water-pepper and Japanese pepper were used as medicinal plants. The Office of Medicine at the imperial court received each year products from all the provinces in the country, among these sesame seeds and perilla *e* (mod. *egoma*), wild parsley *nozeri*, angelica *tsuchitara* (mod. *udo*), sweet flag or acorus *ayame* (mod. *shōbu*) and atractylis *ukera* (mod. *okera*), in extremely large quantities.[43] A count of the deliveries from the provinces gives a total of more than a hundred medicinal products, not all of which we will list. We limit ourselves to the two regions of Izumo and Shinano.

The *Izumo fudoki*, from the first half of the 8th century, lists the flora of each district in Izumo (Shimane prefecture). Paleobotanists have been able to ascertain that most of the one hundred and ten species mentioned in this text correspond to present-day botanical data.[44] By way of example, in Table 3.2 is the list of the flora of the district Ou-gun in Izumo, in eastern Shimane. There are trees like the Japanese cypress *hinoki* and cryptomeria *sugi*, vegetables like the yam and nothosmyrnium(?), fruits like the plum *sumomo*, castanopsis *shiia* acorns, *kaya* (torreya) and red myrica, as well as medicinal herbs like angelica, wild parsley and atractylis. This list also includes other plants with medicinal uses: the liriope with graminaceous flowers, sophora, phytolacca, fig-wort, phellodendron, etc.[45]

The "Regulations of the Engi Era" mention other plants among the fifty-three products delivered each year by Izumo to the Office of Medicine at Heian. These are the achyranth *inokutsuchi* (mod. *inokozuchi*), peony *kuko* or *numikusuri* (*shakuyaku*), sweet flag *ayame* (mod. *shōbu*), pine resin *matsuyani*, arrowhead, poria cocos, white lotus *yamakagami* (*byakuren*), an iridaceae *karasumafuji* (*hiōgi*), helicia *kamanohana* (*kamanoki*), whitlow grass *hamakarashi* (*inunazuna*), peach kernels *momo no sane* and Japanese pepper. For medical care, the Japanese used, depending on the plant, the roots, shoots, leaves, flowers, seeds, stems or fruits.[46]

The identification of the plants named above is due to Jane Cobbi. Thanks to her, we were previously able to translate a list of fifty-nine medicinal herbs taken with them on their voyage to China by Japanese envoys, which included plants from Izumo. In the 1970s, in the course of her repeated field work at Kaida-mura, Cobbi identified the wild edible plants in Nagano prefecture. She described the use and method of consumption of forty-two species.[47]

The Nagano region corresponds to the old province of Shinano. Its specialities in the 9th–10th centuries were pears, jujubes, leaf mustard, walnuts, safflower and hemp.[48] This province also delivered medicinal products to the Heian court, as shown in Table 3.3.[49] However, the results of the study conducted at Kaida-mura and the lists of medicinal products from Shinano have few species in common: the pear, walnut, safflower, hemp and Japanese pepper. Apart from these plants, Jane Cobbi found some that are not included

Table 3.2 The flora of Ou-gun, Izumo, in the 8th century

	Old name	Modern name	English or Latin name
麥門冬	yamasuge	yaburan	liriope with graminaceous flowers
獨活	tsuchitara	udo	angelica tree
石斛	iwagusuri	sekikoku	dendrobe
前胡	nozeri	nodake	peucedanum
高良姜	kōrahajikami	sanshō or shōga	Japanese pepper or ginger
連翹	itachigusa	rengyō	Forsythia suspensa
黃精	ōemi	narukoyuri	Solomon's seal
百部根	hotozura	byakubu	Stemona japonica
貫衆	oniwarabi?	inuwarabi?	Athyrium niponicum
白朮	ukera	okera	atractylis
薯蕷	yamatsuimo	yamanoimo	yam
苦參	kurara	kurara	sophora
細辛	miranonegusa	saishin	Asarum sieboldii
商陸	iosugi	yamagobō	phytolacca
藁本	sawasorashi	kasamochi?	nothosmyrnium?
玄參	oshigusa	gomanohagusa	fig-wort
五味子	sanekazura	sanekazura	Kadzura
黃芩	hihiragi	koyaneyanagi	Scutellaria baicalensis
葛根	kuzunone	kuzukazura	pueraria rhizome
牡丹	fukamigusa	botan	peony
藍漆	yamaasa	hangonsō	Senecio cannabifolius
薇	warabi	warabi	bracken
藤	fuji	nodafuji	Japanese wisteria
李	sumomo	sumomo	plum tree
檜	hinoki	hinoki	Japanese cypress
杉	sugi	sugi	cryptomeria
赤桐	akagiri	higiri	clerodendron
白桐	aogiri	aogiri	sterculier with plantain leaves
楠	kusunoki	kusunoki	camphor tree
椎	shii	shiinoki	castanopsis
海榴	tsubaki	tsubaki	camellia
楊梅	yamamomo	yamamomo	red myrica
松	matsu	matsu	pinaceae
栢	kae	kaya	*kaya* (torreya)
蘗	kihada	kihada	Amur cork tree
槻	tsuki	keyaki	zelkova

Source: *Izumo fudoki* (p. 117).

Table 3.3 Medicinal products of Shinano, in 927

	Old name	Modern name	English or Latin name
黄連	kakumagusa	ōren	goldthread/coptis
細辛	miranonegusa	saishin	Asarum sieboldii
白朮	ukera	okera	atractylis
藍漆	yamaasa	hengonsō	Senecio cannabifolius
大黄	ohoshi	madaio	Rheum officinalis/ rhubarb
女青	kawanegusa	hekusokazura	Paederia scandens
薗茹	neazami	nourushi	euphorbia adenochlora
干地黄	hoshi-sahohime	(a) jio	dried rhemania
附子		(b) torikabuto	Chinese aconite
蜀椒	naruhajikami	sanshō	Japanese pepper
蕪夷	hikisakura	kobushi	magnolia kobus
石硫黄	iō	iō	sulphur
熊膽		kumanoi	bear's gall
鹿茸		kanowakatsuno	deer's antlers
枸杞	kuko, numikusuri	shakuyaku	peony
杏仁		karamomo no sane	(*anzu*) apricot kernels
大棗		ōnatsume	jujube

Source: *Engishiki* (p. 835).

in the 10th-century list of products from Shinano, suggesting that its flora was probably richer in edible plants than is noted in the early administrative texts.

Moreover, a comparison of the two lists of early medicinal products from Shinano and Izumo shows that the two provinces had just five species in common, namely atractylis, Asarum sieboldii, Senecio cannabifolius, peony and Japanese pepper. The flora of early medieval Japan is in fact characterized by a great geographical diversity.

Two medicinal products played a dual role in court life. The aristolochia *shōmokkō* was highly prized as a perfume. This plant was gathered in the provinces of Owari (Aichi), Sagami (Kanagawa) and Mino (Gifu).[50] The other plant, sweet flag *ayame*, was appreciated for its fragrant leaves shaped like the blade of a sword, with which the palace buildings and officials' caps were decorated during the festival of the 5th month. In the courtyard in front of the imperial residence there was placed a palanquin laden with sweet flag and mugwort *yomogi*, another medicinal plant.[51]

Oils, dyes and textiles

The regulations of the 8th–10th centuries imposed special taxes on the young people of all the provinces, including the supply of the following products:

gromwell *murasaki*, safflower *kurenai* (mod. *benibana*), madder *akane*, coptis (also found in the list for Shinano), so-called Azuma ramie cotton or plant fibre *yuu*, so-called Aki ramie cotton or plant fibre, hemp *asa* or *o*, hemp fibre *nio*, ramie *kemushi*, Amur cork tree *kihada*, Sinomenium *tsuzura*, scouring rush *tokusa*, sesame oil *gomaabura*, hemp oil *mashiabura*, perilla oil *enoabura*, Fagara oil *hosokiabura*, lacquer *urushi*, acanthopanax oil *koshiabura*, Japanese horseradish *wasabi*, sawtooth oak *tsurubami* acorns, and strands for basket-making *hakogusa*.[52]

Oil was obtained by pressing the seeds of sesame, perilla, hemp and a plant called *hosoki* or *itajihajikami*, Fagara mantchuria Honda, that is, *inuzanshō*, a rutaceae like the Japanese pepper *sanshō*. Also pressed, in smaller quantities, were walnuts, the fruits of the Japanese snowball *hemi* and camellia *tsubaki* seeds. Oil was used not only in the diet, but also in handicrafts, with sesame and *hosoki* oil used mainly for lighting.[53]

The tinctorial plants, alone or in combination, gave different colours: safflower, a red, and gromwell, a violet. Madder combined with gromwell produced a bright red. Other plants supplied by the provinces to the court gave an orange colouring: gardenia *kuchinashi* combined with safflower, and miscanthus of the variety *kariyasu* combined with gromwell. The gardenia was cultivated in the gardens of five provinces for the Heian court; and the court itself managed a garden of Japanese indigo *ai*. This plant, used on its own, gave a blue, and combined with miscanthus *kariyasu* or Amur cork tree, a blue-green shade. Miscanthus used on its own produced a yellow.[54] Besides these seven basic plants, the palace workshops used sawtooth oak acorns to obtain a brown dye, which is referred to in six *Man'yōshū* poems, including this one: "The colour of the safflower quickly fades. My familiar garment, dyed with acorns of the sawtooth oak, does not fade."[55]

A number of other products supplied by the provinces as taxes were used in the imperial workshops, where the lacquer tableware and furniture decorated with lacquer intended for use by the leading personages of the court were made. Cultivation of the varnish tree was imposed by the government on all the provinces, as was that of the mulberry tree *kuwa*, used for silkworm rearing. Wooden surfaces were rubbed down with scouring rush. Some objects, such as arrows, were coated with acanthopanax oil, an araliaceae.[56] This oil, called *koshiabura* and probably considered along with camellia oil as a Japanese speciality, was exported to China and Bohai (Jap. Bokkai) by way of the official embassies.[57]

The *tsuzura*, if it corresponds to the modern *tsuzurafuji*, is a plant with twining stems of the menispermaceae family, like Sinomenium diversifolium Diels. Its stems were used for tying packages and sometimes as material for basketry, in the making of boxes. This type of liana was found almost everywhere in Japan.[58] But boxes were more commonly made with silk threads and osier fibres.[59] Osier may well have been included in the "strands for making boxes" *hakogusa*, since this compound is written with the characters for box and osier.

The list of taxes delivered by young people recorded four kinds of textile materials. Hemp *asa* or *o* was the raw material for hempen cloth *mafu/asa no nuno*. Hemp, Cannabis sativa, was the most widespread textile plant in early medieval Japan, in its wild and cultivated form.[60] *Nio*, written like "ripe hemp" *jukuma* and pronounced like "boiled hemp" *nio*, is unclear. However, this material was used for ropes and cords, suggesting that hemp fibres are meant.[61] *Kemushi* had the meaning either of hemp or ramie *karamushi*, i.e. China grass, Boehmeria nivea.[62]

Azuma-yuu and Aki no yuu were textiles delivered to the court by the Eastern provinces and by the province of Aki (Hiroshima), respectively. This material was often found among ritual offerings. Given that it was measured either in units of weight, or in "leaves" *mai*, it is likely to have been compressed ramie cotton. Sometimes *yuu* was used as a tie for fastening envelopes made of paper or tree leaves. It may then have taken the form of raw fibres.[63] These could have been the fibres of a moraceae, namely the paper mulberry Broussonetia papyrifera (L.) Vent. *kōzo*, or a mulberry of the variety Broussonetia kazinoki Sieb. *kajinoki*. In the province of Bungo, *yuu* was produced with a very different fibre, such as commelina *tsuyukusa*.[64]

By analysing the results of the archaeological excavations of the sites from the 2nd century BC to the 7th century AD, Nunome Junrō established correspondences between the principal textile plants and the terms in the early texts: *asa, o* correspond to hemp (mod. *asa*); *karamushi, kemushi* to ramie (mod. *karamushi*); *kōzo* or *taku* to the paper mulberry (mod. *kōzo*); *kuzu* to pueraria (mod. *kuzu*); and *kuzu* or *fujikazura* to the Japanese wisteria (mod. *nodafuji*). A few rare textile fragments from the Japanese lime *shinanoki* have also been discovered.[65] A tree from Izumo, glossed as *kaji* by the *Fudoki* editors, no doubt corresponds rather to the *kōzo*, as the term *kaji(no ki)* does not appear in texts before the 13th century.[66] All the plants mentioned in the *Fudoki* probably grew wild in various parts of Japan.[67] It remains for us to touch on the question of cotton. In 799, a foreigner brought some seeds to Japan and the government ordered its cultivation in several provinces. However, this crop did not become widespread and Japan reimported cotton from Korea in the 15th century.[68]

Cereals, straws and stems

The provinces delivered annually to the imperial court six species of edible graminae: rice, foxtail millet, wheat, common millet, glyceria and, in lesser quantities, barley and barnyard millet.[69]

Rice *ine*, Oryza sativa japonica, was the main cereal from the point of view of the volume delivered. The court imposed its cultivation on the entire population of the country, production being intended first and foremost to feed members of the provincial governments and the aristocracy of the capital. A distinction was made between non-glutinous rice *urushine* or *uruchi(no)yone* and glutinous rice *mochi(no)yone*, mod. *mochiine*, Oryza sativa L. var. glutinosa Matsum. The starch of the latter contained more amylopectin.

According to the encyclopedist Terashima Ryōan (active ca. 1700), *ine* denoted non-glutinous rice in his day, whereas it meant glutinous rice in the early medieval period. *Sekimai* or *akayone* (mod. *akamai*) was an indica glutinous rice, with a red caryopsis, a dry wild rice. It gave rise to the tradition of "red rice" *sekihan*, i.e. white rice cooked with red beans served on festive occasions, from the late medieval period. Besides these two kinds of irrigated rice and red rice, it is almost certain that dry rice was grown on swidden fields, judging from the discovery made by Satō Yōichirō of tropical Oryza sativa japonica rice in the archaeological sites of early Japan.[70] The texts also record many names for rice, relating to the early and late varieties, the stages of preparation for consumption and the methods of storage.[71]

Foxtail millet *awa*, Setaria italica Beauv., was the second cereal from the point of view of the quantities supplied by the provinces to the court. The Chinese noted this crop in Japan as early as the 3rd century AD. A work on Japan states that the inhabitants "grow foxtail millet, rice, ramie and hemp". The Heian court encouraged the cultivation of *awa* and occasionally levied small amounts by way of tax. There was a glutinous variety of this graminae, called *mochiawa*, as for rice.[72]

Wheat *komugi*, Triticum aestivum L., and common millet *kimi* (mod. *kibi*), Panicum miliaceum L., formed part of the food supply at the imperial court, like rice and foxtail millet. Wheat was served to the nobility as crushed(?) grains and flour noodles *muginawa*.[73]

According to the "Regulations of the Engi Era", another cereal, *mino*, had its place on the palace tables, but its name disappears in the 10th century from the Japanese flora and from the literature in general. The botanist Makino Tomitarō identified *mino* with glyceria *mutsuoregusa*, Glyceria acutiflora Torrey. For their part, the botanist Kimura Yōjirō and the historian Torao Toshiya think that it is *kazunokogusa*, Beckmannia erucaeformis Host, which has edible seeds. In the 10th century, Minamoto no Shitagō asks "where this plant, mentioned in the regulations, comes from (*izuru tokoro*)". Glyceria grows today in rice fields and irrigation channels. An Edo-period manuscript of the *Engishiki* notes that this cereal "grows in abandoned rice fields". It was therefore a wild cereal harvested in certain provinces for the court.[74]

Barley and barnyard millet were two very rare foods on the palace menus. Barley *futomugi* (mod. *ōmugi*), Hordeum vulgare L., was supplied annually to the court by six provinces, but the quantity delivered amounted to just 12 *koku* (1,020 litres), as against 207 *koku* (17,595 litres) of wheat. As in China, the texts often use the same term, *mugi*, for barley and wheat.[75]

Barnyard millet *hie*, Panicum crus-galli L., may have been regarded by the nobility as a medicinal product, as it is found among the court's supply of medicines. It was also part of a gruel consisting of seven cereals, served to the emperor on his enthronement and each year, on the 15th day of the New Year. Two *Man'yōshū* poems tell us that barnyard millet grew wild in the rice fields. The population may have harvested both cultivated and wild millet *inubie*.[76]

Buckwheat does not appear either among the provincial taxes destined for the court, or in the imperial kitchens. Buckwheat *sobamugi* (mod. *soba*), Fagopyrum esculentum Moench, is not a graminae, but a polygonaceae. The court encouraged its cultivation from the 8th century, as a suppletive cereal for famines. This autumn crop, requiring little fertilizer and labour, was perhaps more widespread in the provinces than the texts might suggest. It is also possible that the rural peasant population harvested wild buckwheat as well, as implied by a legend of the 13th century.[77]

Besides the edible cereals mentioned above, the court needed a certain number of graminae and stems for handicrafts. It received bamboo *take* and rice straw *wara* from the provinces. The palace workshops used the ashes from straw, among other things, as a fixative in dyeing.[78] The working of bamboo was a speciality of Hayato artisans from southern Kyūshū. From it they made baskets, lattice frames for paper-making, as well as bamboo chopsticks for eating and saké containers *hasō* with a bamboo spout.[79]

The Heian court also used eulalia or imperata *kaya* and reeds *ashi* (mod. *ashi* or *yoshi*) to form the roofing, partition-walls and flooring of the temporary pavilions for the enthronement ritual.[80]

To meet its other requirements of stems for artisanal use, the court maintained a wet field of rushes *i* (*igusa*) of 1 ha in Yamashiro and a pond of 190 ha in Kawachi, where wild rice *komo* (mod. *makomo*), sedge *suge* and bulrushes *kama* (mod. *gama*) were harvested.[81] Rushes were made into straw mats, hats, cloaks and small rice containers.[82] Sedge was used to make palanquin blinds, fans, flabellums, portable daises and round mats.[83] Bulrushes were used especially to make arrows for the *tsuina* ritual at the end of the year, the case for a sacred arrow at Ise, screens for the residence of the Ise priestess and mat-seats for the Sekiten ceremony in honour of Confucius.[84]

Wild rice, Zizania latifolia, the only graminae among the four aquatic plants, was far more widely used. It was the raw material for many kinds of straw mats serving as partition-walls, dais coverings, packaging, and especially as seats in all the court offices and at all official ceremonies.[85]

The court therefore made very specialized use of the various plant stems. But these played a more general role, outside of aristocratic circles, in the Japanese countryside. Aquatic stems were abundant in early medieval Japan. The *Fudoki* often mention these plants: in one spot, reeds, wild rice and sedge grew thickly all around a pond; elsewhere, sweet miscanthus *ogi*, reeds and imperata *chi* grew in profusion in a marsh.[86] The *Fudoki* most often mention bamboo and reeds. They give less attention to rushes, bulrushes and eulalia. Perhaps these plants were too commonplace.[87]

Reeds have given their name to dozens of toponyms of the 8th–9th centuries, like that of "Reed plain" Ashihara. They were considered the very symbol of the Japanese landscape: the mythology relates that the gods descended into the "central country of the reed plains" *ashihara nakatsukuni* to found Japan.[88]

Wild rice *komo* and eulalia or imperata (*kaya* variously denoting one or other of these two plants) seem to have been especially prized by the Heian nobility. The poetry makes reference to the harvesting of these wild graminae. Here is a poem by Ōe no Masafusa (1041–1111): "Though the water of the Yodo marsh, where they reap the wild rice, is deep, the moon's reflection retains its purity down into the depths." In the 7th century, Prince Kusakabe (662–689) wrote: "When you, Ōnago, harvest the eulalia *kaya* (written like grass *kusa*), down on the moor, at the moment when you bind the sheaves, is there an instant when you forget me?"[89] In fact, this kind of scene was so familiar to poets that the "harvesting of wild rice" *karikomo* and the "harvesting of eulalia or imperata" *karikaya* were typical subjects in the poetic anthologies. In poetry, the harvesting of wild graminae characterized autumn and the picking of young shoots the activity of spring.[90]

Why such zeal in harvesting wild graminae? What did the Japanese do with these stems in the early medieval period? Let us first take some examples from modern Japan. An exhibition organized by the School of Fine Arts at Musashino in 1994 displayed around a thousand objects made from plants forming part of the School's collections. The catalogue of this exhibition distinguishes straw *wara*, lianas *tsuru*, stems *kusa* and the leaves of trees *ki no ha*. It lists thirty-eight plants, including those found in the *Engishiki*. Here are the objects made with these plants according to the catalogue: objects made from rice straw: sandals, capes, shoes for horses and oxen (as Japan did not know of the horseshoe until the 18th century), back-bands (matting used to protect the backs of men who carried heavy loads), ropes, sacks, baskets, envelopes, rice sacks *tawara*, mats, roofing material, ritual objects (symbolic décors, figurines); objects made from wheat straw: hats, caps, open baskets, envelopes, figurines; objects made from reeds: baskets, nets for drying sea-weeds and fish, fishing nets, blinds, screens; objects made from rushes: sandals, capes, hats, back-packs, back-bands, pillows; objects made from bulrushes: sandals, capes, hats, back-bands, open baskets for carrying on the hip; objects made from imperata: capes, sacks for ashes, saucepan lids, cereal containers, mats, roofing material, mats placed under hand-mills, covered baskets containing rice boxes; objects made from sedge: sandals, capes, hats, baskets for rice boxes; objects made from eulalia: brooms, ritual décors; objects made from wild rice: back-bands, figurines and ritual décors.[91]

The encyclopedist Terashima Ryōan mentions some ways of using graminae in Edo Japan. According to him, roofs were covered in order of importance with eulalia, reeds, rice straw and wheat straw. Rice straw was also used to make mats, ropes, sandals, mud walls and fertilizers, with the ashes used to wash bitter plants before consumption. Eulalia was made into ropes, blinds, sandals and dusters. Reeds were used for hats, blinds, torches and brooms. Wild rice *komo* was used for mats *komo* and sedge was the material for hats.[92]

We have given examples of modern uses to fill in the gaps in the early medieval texts. These texts do, nonetheless, provide some information. According to one *Fudoki*, sedge hats were a speciality of the village of Shikigusa in Harima.[93] The early poetry refers to hats and mats made from sedge, as well as pillows, bedding, mats and blinds made from wild rice. The poets see roofs covered with reeds and eulalia. Eulalia was woven into sacks for ashes and sandals, and made into brooms.[94] The poetry mentions eulalia (Miscanthus sinensis), called variously *kaya* or *susuki,* and sweet miscanthus *ogi* (Miscanthus sacchariflorus Benth. et Hook.). In fact, the name *kaya* did not apply to only one plant and denoted other plants like imperata *chigaya.* Imperata played a ritual role when it was plaited into a rope *shimenawa* delimiting a ritual place.[95]

The poets show less interest in the use of rushes and bulrushes and in fact ignore many objects in everyday use. However, the rural population could not do without utensils made from straw or stems. In farming activities, straw made its appearance after the harvest. Without straw mats, the grains could not be collected while cereals were being threshed. To separate the grains from the husks, winnowing-baskets and sieves made from plant stems were used. To transport the tax rice, it was put into straw sacks. But even apart from taxes, all transporting of grains from the fields and products gathered from virgin lands could only be done using plant containers.[96] In their daily activities, farmers protected themselves from the sun and rain with hats and capes. Lastly, plant stems allowed them to have a roof over their heads, since rural dwellings consisted of pillared- and pit-buildings, all thatched.[97]

The usefulness of stems was not limited to everyday objects. Wild graminae also played a role in the diet, notably hemp and pueraria that provided at the same time textile fibres, oil seeds and a nutritious starch; bamboo that gave artisanal stems and edible shoots; and some aquatic plants that were also edible. In the 18th–19th centuries, young reed shoots, the roots and leaves of imperata shoots, the roots and stems of young bulrushes, eulalia and wild rice seeds formed part of the substitute foods during famines.[98] Terashima Ryōan, mentioned above, writes that children are fond of imperata seeds and that eulalia seeds could be steamed like foxtail millet. He stresses that wild rice, a suitable food in times of scarcity, was eaten in a gruel mixed with foxtail millet, or as steamed cakes *mochi.*[99]

For the early medieval period, the *Man'yōshū* makes reference to the consumption of two stemmed plants. Sedge roots and seeds were dug up and collected and the young ears of imperata were eaten.[100] Wild rice was served in the offices at Nara. Wild rice seeds are in fact found among the vegetables, such as ginger and sow-thistles, in the purchase lists of provisions for the sūtra-copying office and the construction office for the Kondō Hall of a temple, in the 8th century. These seeds were called *komo no mi* or *komo no ko* (with two ways of writing, *komo* wild rice, or *komo* mat, followed by the character for child). Depending on the context, a third way of writing, that of *ko* (written like basket *kago* with the grass radical instead of the bamboo radical), followed by the character for child, also denoted wild rice seeds.[101] In the late medieval period, these seeds were considered a supplementary vegetable.[102]

After having surveyed the graminae and the other wild stems, it is clear that these plants played a significant role alongside the cultivated graminae. These wild plants were very important for the houses, clothing and diet of the population of early medieval Japan.

We have focused our attention on the fruits, nuts, vegetables, medicinal herbs, oil plants, industrial plants and graminae harvested, essentially from the point of view of deliveries from the provinces to the imperial court. The useful and edible plants in the provincial flora must, however, have been far more numerous than the species that were sent to the capital. We can get some idea of this, just from the number of edible substitute plants (more than seven hundred) mentioned in the works of the 18th–19th centuries, which gave some protection against poor harvests.[103] Our analysis therefore gives only a general idea of what the forests and virgin lands must have offered the Japanese of the early medieval period.

The archaeological context

The catalogue of plant products discussed above gives an idea of the flora that played a role in the annual cycle of provisioning in early medieval Japan, within the limits of the species gathered in the provinces for the imperial court. The court's requirements were based on a number of factors: the diet of the nobility, the desire to obtain for its own use certain rare specialities, the transportation conditions, the distances that separated the provinces from the capital, the possibility of storing foods, etc. Our catalogue of plants sheds light on the court customs and, in a more limited way, on how the rural population ensured its own basic food needs. This explains why it does not include all the edible and useful species. In view of the lack of sources on this subject, we give a summary of the current archaeological knowledge concerning foraging. The Neolithic provides information about the practices of hunter-gatherers, and the following periods up to the 7th century tell us about the practices of a society of farmers.

Foraging and diet in the Neolithic

During the Jōmon period, i.e. the Japanese Neolithic (10000–300 BC, from this point on we follow the dates given by the authors cited), hunting, fishing and the harvesting of plants were the principal means of obtaining food for the inhabitants of the archipelago. To these practices were added the first forms of agriculture that appeared at a still unclear point in time. The situation with regard to certain practices relating to the plant kingdom during the Neolithic is discussed below.

Up to 1975, the archaeologist Watanabe Makoto identified the remains of 39 edible plants in 104 sites of the Neolithic (see Table 3.4). He considered that horse chestnuts and the other nuts played a crucial role in the diet (*shushoku*, staple food) of Neolithic people. His theories were supported by other archaeologists.[104] According to Watanabe, the fruits of the following trees were

Table 3.4 Edible plants discovered in Neolithic sites (Jōmon); after Watanabe Makoto (1975, 1999)

Japanese	English	Latin name
mizunara #	oak konara	Quercus crispula Blume
konara #	oak konara	Quercus serrata Thunb.
kunugi #	oak konara	Quercus acutissima Carruth.
kashiwa #	oak konara	Quercus dentata Thunb.
buna #	beech buna	Fagus crenata Blume
kuri*#	chestnut	Castanea crenata Sieb. et Zucc.
tochi #	horse chestnut	Aesculus turbinata Blume
(oni)gurumi*#	walnut onigurumi	Juglans mandshurica Maxim.
himegurumi*#	walnut onigurumi	= var. cordiformis Kitam.
akagashi #	Cyclobalanopsis akagashi	Quercus acuta Thunb.
ichiigashi #	Cyclobalanopsis	Quercus gilva Blume
arakashi #	Cyclobalanopsis	Quercus glauca Thunb.
sudajii*#	castanopside	Quercus sieboldii Makino
tsuburajii*#	castanopside	Quercus cuspidata Makino
matebashii*#	matebashii	Lithocarpus edulis Nakai
kaya*	*kaya* (torreya)	Torreya nucifera Sieb. et Zucc.
inugaya	Cephalotaxus	Cephalotaxus drupacea Sieb. et Zucc.
haiinugaya	idem	idem
hashibami*#	hazel tree	Corylus heterophylla Fisch.
kajinoki*	mulberry	Broussonetia papyrifera Vent.
tsubaki*#	camellia	Camellia japonica L.
yamagobō	phytolacca	Phytolacca esculenta van Houtt.
yamamomo*	red myrica	Myrica rubra Sieb. et Zucc.
hasu*	lotus	Nelumbo nucifera Gaertn.
hishi*#	water chestnut	Trapa natans L.
azumabishi?		
uri no shu	cucurbitaceae	
sharinbai	Rhaphiolepis	Rhaphiolepis umbellata Makino
sanshō*	Japanese pepper	Zanthoxylum piperitum DC.
inuzanshō	Fagara	Fagara mantchuria Honda
akamegashiwa	Mallotus	Mallotus japonicus Muell.-Arg.
chanchinmodoki	Poupartia	Poupartia fordii Hemsl.
nobudō #	Ampelopsis	Ampelopsis brevipedunculata Trautv.
matatabi	Actinidia	Actinidia polygama Miq.
nobiru*	wild chives	Allium grayi Regel
makomo*	wild rice	Zizania latifolia Turcz.

Japanese	English	Latin name
sasarui*	(dwarf bamboo)	
kuroguwai*	eleocharis	Eleocharis kamtschatica Komarov
inubie**	wild millet	Panicum crus-galli L.
ine**	rice	Oryza sativa L.

Notes: * Plants mentioned in the *Engishiki* and/or Heian poetry.
** Discovered in 1994 at Sannai-Maruyama, Aomori prefecture.
# Belong to the species most often found in Yayoi sites; after Terasawa Kaoru (1999: 36–37).
Source: © Verschuer, *Rice, Agriculture, and the Food Supply in Premodern Japan*, Routledge, 2016.

eaten: various kinds of fagaceae *buna*: oaks *konara*, Castanopsides Shiia *shii*, cyclobalanopsis *akagashi*, the beech Fagus *buna*, Lithocarpus *matebashii* and of course the (sweet) chestnut *kuri*; the horse chestnut *tochi*, a hippocastanaceae; the walnut *kurumi*; the hazelnut *hashibami*; and *kaya* (torreya). Geographically, the oak of the genus *konara*, species *konara*, the horse chestnut, (sweet) chestnut and walnut occupied the north-east of the archipelago characterized by decid-uous forests *rakuyō jurin* and also called the Fagus zone *buna-tai*. The oak of the species *akagashi*, Castanopsides Shiia and Lithocarpus occupied the south-western half of the country, i.e. the evergreen laurel-forest zone *shōyō jurin* (see the Introduction, Map I.1).

Some nuts, however, needed treating to make them edible. Acorns of the oak species *konara*, *mizunara*, *kunugi* and *kashiwa* contain tannin, and horse chestnuts saponin and aloin, bitter substances that had to be removed by a complicated process of washings, soakings, rincings, drying and repeated decoction, combined with potash in the form of wood ash. To get rid of the bitterness of cyclobalanopsis *akagashi*, they simply needed to be washed several times. Watanabe Makoto thinks that Neolithic people mastered these practices in the middle period (3000–2000 BC). It is easy to understand the choice of the Heian court, which had only edible, untreated nuts delivered and passed over horse chestnuts and most acorns.

From a nutritional point of view, walnuts, *kaya* and beechnuts were oleagi-nous plants rich in lipids. *Kaya*, acorns and other nuts were a source of starch; and Watanabe drew attention to two more starchy plants discovered in the archaeological sites: the lotus *hasu* and water chestnut *hishi*. In his opinion, the Jōmon people extracted starch from a number of roots: three liliaceae, namely a lily *yuri*, another lily *ubayuri* and the erythronium or dog-tooth violet *katakuri*, as well as yams *yamanoimo*, taro *tennanshō* (*satoimo*), pueraria *kuzu* and bracken *warabi*. Most of these starches were also eaten at the court, but instead of extracting the starch from pueraria and bracken, the young shoots were eaten as a vegetable.[105]

Since the 1970s, Sahara Makoto, Nishimoto Toyohiro and others have car-ried out calorimetric estimates on some archaeological sites. The results differ according to the environment and whether it is in the mountains or near the sea.

In certain cases hunting accounts for more than a third of the calorie intake; in other cases it is fish. At the site of Toriyama-kaizuka in Fukui, dating from 4000–3000 BC, walnuts and the other nuts together provide more than a third of this intake, and fish another third. For the site of Kosaku in Chiba, occupied from 2000–1000 BC, it is estimated that 80% of the calories came from plants. However, roots are excluded from this type of analysis as they are difficult to identify in the sites.[106]

The site of Sannai-Maruyama

Since the studies of Watanabe Makoto, cultivated plants have also been found in Neolithic sites. But the spectacular discovery, in 1994, of the site of Sannai-Maruyama in Aomori covering 38 ha, once again confirmed the importance of nuts in the diet. This settlement was occupied over a long period (3500–2000 BC) and yielded up to the archaeologists building structures, two raised (ritual?) areas, graves, refuse pits, objects and traces of organic materials. Located near the sea, this site contains many remains of fish and whales as well as of small game (hare and flying squirrel *musasabi*) and a few remains of large game (deer and wild boar). The analyses of the sedimentary layers have shown a proportion of 60% of chestnut pollen for certain periods. The late period is characterized by the appearance of phytoliths of wild barnyard millet *inubie*. The site also contains remnants of acorns of the oak *konara*, walnuts and horse chestnuts, as well as seeds of edible plants: Sambucus sieboldiana Blume *niwatoko*, probably used to make a fermented beverage, a vitaceae *yamabudō*, a bramble *kiichigo*, Cornus kousa Buerg. *yamaguwa*, Actinidia arguta Planch. *sarunashi*, beans *mame*, two graminae, Setaria viridis Beauv. *enokorogusa* and polygonaceae of the genus *tade-ka*, as well as a Lagenaria of the species *hyōtan*. It is thought that chestnuts were a major source of calories during the early and late periods of the Sannai-Maruyama site.[107]

We know that the nutritional value of some nuts is quite high, when compared with that of whole rice which is 351 kcal (see Table A.5 in the Appendix). This is especially so for horse chestnuts (281 kcal per 100 g), *kaya* (427 kcal), walnuts (337 kcal) and (sweet) chestnuts (109 kcal). For fish, shellfish and game, these figures are from 10–90 kcal, but these foods are higher in protein than nuts. In this sense, only walnuts, with 24 g of protein, and *kaya*, with 12 g of protein, could compete with foods of animal origin.[108] (See Table A.5 in the Appendix.) According to the archaeologist Koyama Shūzō, Neolithic people could live perfectly well just by foraging and fishing. He gives the following examples: a daily ration consisting of 1,282 g of chestnuts and 123 g of salmon was equivalent to 2,205 kcal and 60.1 g of protein; and 587 g of oak acorns and 185 g of salmon corresponded to 2,311 kcal and 59.9 g of protein.[109] Thus it was certainly possible for these people to subsist in large part on nuts and seeds. As for the following periods, opinions remain divided on the role of gathered wild plants in the society. The debate revolves around finding out to what extent an agricultural society could have continued to harvest wild plants.

Discussion: foraging in an agricultural society

We have been discussing the Neolithic. For subsequent periods, we are less well informed and opinions are divided. In fact, as Terasawa Kaoru noted, archaeological research devoted to the post-Neolithic focuses on rice growing and neglects the acquisition of the other cultivated or wild plants. Terasawa studied the plant situation in Japan after the introduction of irrigated rice growing from the continent. This was transmitted to Japan during the final Neolithic, between 1000 and 500 BC. Yet rice was far from being the main source of nutrition at that time. In 1981, Terasawa identified 298 plant species in 224 archaeological sites dating from the Yayoi period (then dated 3rd century BC–3rd century AD). Among these plants, found in the form of seeds and pollens, 173 belong to edible species and 37 were probably cultivated, notably foxtail millet, common millet, barnyard millet, buckwheat and soybeans (all with a calorific value comparable to that of rice). Another 136 wild plants were no doubt also gathered for food by prehistoric people.[110]

We still lack calorimetric studies for the sites of the Yayoi period, but Terasawa drew up a list of the 14 plants most often found in 224 sites (the figure in brackets corresponds to the number of sites where traces of the plant in question have been discovered): acorns and beechnuts (168), rice (128), peaches (95), beans (51), gourd-calabash (42), walnuts (39), (sweet) chestnuts (35), barley and wheat (34), water-pepper (33), melons (30), horse chestnuts (28), vitaceae (grapes, etc. 26), camellia (21) and water chestnuts (20). This table shows the presence of nut trees, both in the mountainous regions and on the lowlands where most of the sites are located. It also shows the association, in the same places, of tree fruits and rice. This means that many acorns, nuts, etc. must have been eaten, even in places where rice was cultivated. However, the picture is incomplete, as plants break down in different ways. Thus, the hard casings of nuts are better preserved in the soil than beans and the grains of cereals without phytoliths. Starchy roots leave hardly any trace. Even with these reservations, it is reasonable to suppose that nuts and seeds continued to play a crucial role in the diet after the Neolithic. This is also confirmed by the presence of acorn silos in the Yayoi sites.

The archaeologist Terasawa Kaoru and the ethnologists Sasaki Kōmei and Matsuyama Toshio have drawn attention to the presence of acorn silos in over twenty sites associated with rice growing. In their view, this is proof of the coexistence of acorn gathering and rice cultivation in the same places. For example at Uryūdō, on the Kawachi Plain, near Ōsaka, a hundred silos have been discovered containing remnants of acorns of cyclobalanopsis and castanopsis *shii*, trees of the laurel-forest zone (in the south-west of the archipelago), as well as remains of horse chestnuts and walnuts, characteristic of the forests of the Fagus zone (in the north-east of the country). This diversity is also found in other sites, such as that of Karako, in the centre of the Nara basin. Matsuyama Toshio infers from this that Yayoi people went far afield to collect tree fruits and nuts. In fact, it may be supposed that the primary forest must have been dotted with areas planted with different species, judging by the results of analyses of the timber found in some sites of the same zone.[111]

Terasawa assesses the degree of dependence of the ancient inhabitants of Japan on various foods by means of an estimate of rice yields. According to him, Yayoi people were obliged to turn to starches other than rice: for a good part of their food in the 3rd century BC; for half from the 2nd century BC to the 2nd century AD; for a third from the 3rd century (but in our view, even this relatively small proportion still overestimates the role of rice). Terasawa rightly thinks that Yayoi people must have eaten mainly tubers, various cereals and acorns. In any case, it is probable that rice cultivation did not suddenly take hold in the archipelago over a wide area, but spread gradually and unevenly depending on the region. In other words, certain privileged areas were able to benefit from rice growing, while most regions had to continue to depend mainly on the gathering of acorns and tubers.

Foraging and rice cultivation in the Yayoi period

Terasawa Kaoru produced evidence proving that the population of the country continued, even after the introduction of irrigated rice growing, to feed itself largely with collected fruits and nuts, as well as cereal and vegetable crops. But his opinion, which must be considered pioneering, is still not shared by most other archaeologists. In particular, Sahara Makoto and Koyama Shūzō pay almost no attention to the diversity of the wild plants harvested. They think that Yayoi society was an agricultural as distinct from a hunting society, and that cultivated rice, generally speaking, was the "staple food" *shushoku* of the Japanese throughout the Yayoi period.[112]

In 1995, an exhibition organized by the Kawasaki Municipal Museum titled "The Food of Yayoi" and subtitled "What people ate in the time of Himiko", presented a reconstruction of the "daily meal" of the Yayoi period.[113] It was a veritable feast comprising several serving dishes containing a soup made with 70% red rice and 30% chestnuts, dried fish, quenelles of acorn starch, a piece of fresh fruit (a peach), berries (blackberries), a piece of cucumber and cooked bracken stems. According to Terasawa Kaoru, this type of reconstruction, which was very common in Japanese archaeological museums, overestimated the importance of rice.

In the text of the exhibition catalogue, Watanabe Makoto (who was not responsible for the reconstruction) presented a comparative diagram of the diet in the Jōmon and Yayoi periods: in the Neolithic (Jōmon), acorns, horse chestnuts and the other nuts form the "staple food", and fish and game a "subsidiary food" *fukushoku*; during Yayoi, it is "rice with other things?" (*kome nado*) that forms the "staple food", with a proportion of a third to half of other foods "in case of scarcity"; and fish becomes the only subsidiary food. Watanabe Makoto explains that Yayoi people ate rice mixed with "all kinds of things ... and even nuts and seeds, which by then had lost their predominant place in the diet".[114] Watanabe's text reflects the widely held opinion among archaeologists, according to which rice suddenly became the main staple food in the Yayoi period, relegating the other plant foods to a secondary role.

In 1999, the Museum of Yayoi Culture in Ōsaka published a book called "The Food of Himiko".[115] It quotes the table and definitions regarding "staple food" of Watanabe Makoto. Like the Kawasaki Museum catalogue, it offers no calorimetric analysis. This shows that Terasawa Kaoru, who had identified all the edible plants in the Yayoi sites, was still not being listened to and that opinions persisted concerning the predominance of rice, whereas this grain seems to us to be negligible in the diet of Yayoi Japan. Terasawa repeated his call for a rethinking of the role of rice cultivation in another collective work published in 1999.[116]

Terasawa's call has been heeded since 2000. Thanks to the progress of research into irrigated rice cultivation, it has become clear that this grain did not feed the population of the Yayoi period. The yields from rice growing were in reality very much lower than the estimates given by archaeology up to the 1990s. This was revealed following the research of biologists such as Satō Yōichirō, which showed that irrigated rice, rainfed rice, dry rice and swidden-field rice existed in parallel. Satō Yōichirō (2002) and Kimura Shigemitsu (2010) republished the tables of edible plants drawn up by Terasawa (1981), in order to show that the Yayoi people were essentially hunter-gatherers, who only occasionally supplemented their diet with rice.[117] Thus they overturned the general opinion that Yayoi marked the arrival of a society of farmers in the Japanese archipelago. Scholars therefore became aware of the continuity of foraging after the Neolithic. These practices continued over the centuries and up to modern times.

Wild plants in the modern period

We have a number of accounts by writers of the 17th–19th centuries on the customs relating to foraging. Between 1681 and 1683, an anonymous author from the region of the present-day prefectures of Aichi and Shizuoka wrote the following, in a work called *Hyakushō denki*:

> The peasants *domin* store various foods for many years to provide for their daily needs, but these foods end up losing their taste and going rotten, causing wasteful losses. It is important to know how to conserve the products obtained from the mountains, the virgin lands *no*, the sea and rivers, the village surrounds and lowlands, so as to avoid losses. Among the plants gathered in the mountains, bracken keeps for a long time when it is cut and dried. They also dig up its roots to make into flour, which allows it to be kept for a long time. They also dig up the root of pueraria *kuzu* and grind it into flour to keep it. In spring, they pick young leaves that are dried in such a way that they will keep. They also pick the young leaves of Clethra barbinervis *ryōbu*, wisteria *fuji*, Rosa multiflora *nobara*, polygonum

itadori, and royal fern *zenmai*, and dry them, after cutting them into pieces, which allows them to be kept for several months. The most suitable time for drying plants is *doyō* (18 days before the end of spring); but most of the grasses gathered in virgin areas and lowlands are not suitable for prolonged storage. Among the seaweeds, eisenia *arame* and ecklonia *kajime* are easy to keep. They are a good food for the farmers, but seaweeds can become stuck in the throat if they are not eaten with the five grains *gokoku* or various cereals *zakkoku*. Lotus and water chestnuts, which grow in the lakes and rivers, will keep for several years, provided they have been dried after being boiled in water. Yams *nigatokoro* in virgin areas and lowlands are dug up in the 12th, 1st or 2nd month, and will also keep for a long time, if they are dried, after having removed their bitterness using a decoction of wheat glumes or the ashes of rice straw. They also dig up poria cocos *bukuryō*; they cut it into shavings, and dry it in such a way that it will keep for several months.[118]

Given that seaweeds are mentioned in this text, it quite possibly concerns the coastal region of Aichi or Shizuoka. Also worth noting is the importance of roots compared with stems (called leaves) where length of storage is concerned. According to this text, stems keep for several months and roots for several years. Most green plants were totally unsuitable for prolonged storage. Tree fruits and nuts, which are not mentioned here, were probably plentiful in all parts of the country. Another document, relating to the Hida region, gives some interesting information on the use of wild plants.

The ethnologist Matsuyama Toshio has examined the Hida monograph, *Hida gofudoki*, completed in 1873 at the request of the mayor of Takayama. The old province of Hida corresponds to the mountainous region around present-day Takayama city in Gifu prefecture. The work describes in detail the agricultural and artisanal products of 415 villages of this region, or a population of 92,600 people (ca. 1873). According to the calorimetric estimates of Matsuyama Toshio, rice supplied 35% of the calorific needs of the population, barnyard millet 17%, and the other cereals, tubers and soybeans together 12%. The province imported from other regions cereals with a calorific value of 15%. There thus remained a deficit of 21% which, in Matsuyama's view, was made up by foraging. In fact, the quantity of rice eaten was probably less, after deducting taxes.[119]

The *Hida gofudoki* mentions some forty wild plant species, among which are distinguished green plants, fruits, mushrooms, roots and nuts; but only the last two were suitable for storage. At that time, bracken flour was sold outside the region to buy rice. Bracken, which cost 5 *ryō* per *koku* (180 kg), as against 5.3 *ryō* for a *koku* of rice, was very sought after. On the other hand, pueraria roots were eaten locally. In the Hida region, people mainly gathered horse chestnuts, (sweet) chestnuts and oak acorns, and, in far smaller quantities, walnuts *kurumi*, hazelnuts *hashibami* and *kaya* (torreya).

Precise figures are available for the harvests of two villages: at Kanaki-mura (today Kamitakara-mura), the annual harvest amounted in total to 61 *koku* of rice, barnyard millet, foxtail millet, buckwheat and soybeans, 80 *koku* of acorns and 3 *koku* of (sweet) chestnuts. In the neighbouring village, Akaoke-mura, they harvested 40 *koku* of horse chestnuts, 50 *koku* of oak acorns and 30 *koku* of cereals and soybeans. Thus the quantity of nuts collected far exceeded that of the cereals. Overall, Matsuyama estimates the average annual production for 109 villages in the district Yoshiki-gun, where the two villages were located, at the following quantities for each household: 1.1 *koku* consisting of (sweet) chestnuts, horse chestnuts and acorns, and 4.2 *koku* consisting of rice and barnyard millet. In other words, in this district, one-fifth of the food was made up of nuts and seeds (a very small proportion if one takes account of the taxes deducted for rice).

This proportion of nuts and cereals is the same as that found in a text from another region. This is a diary kept by Yamashita Tadajirō, an inhabitant of Shiramine, edited by his son Yamashita Kōjirō (1895–1964); one hundred photocopies were made in 1965. The ethnologist Chiba Tokuji has examined this document. The village of Shiramine, located on Mount Hakusan in Ishi-kawa, is famous among ethnologists for its swidden-field crops, grown into the 1970s.[120]

Every year, Yamashita Tadajirō and his family gathered the following plants, in the different seasons: bracken, royal fern, another bracken *kugumi* (*kogomi*?), angelica *udo*, butterbur, thistles of the variety *azami*, yams *yamanoimo*, lilies *yuri*, pueraria roots, a vitaceae *yamabudō*, akebia, lotus, foxnuts *tanifutagi* (a dialectal name for *onibasu*), potatoes *fudo* (a dialectal name for *jagaimo* or *hodoimo*) and an unidentified plant, called *geboshi*.[121]

Chestnuts played a particuliarly important role in the life of the Yamashita family. Yamashita Tadajirō knew the exact number of trees and the location of the wild chestnut groves *kurihara* near the village. He had planted others, called *kuribayashi*. He could tell the moment of ripening, the size of the nuts and their volume, in each place. Some chestnut trees even had names, such as Nanagikuri, Kayabakuri and Hyohyokuri. In early September, Yamashita Tadajirō cut the grass in the planted chestnut groves to facilitate gathering. The chestnuts began to fall around 10th September, and the last nuts were collected around the second half of October. Every morning, two or three members of the family set out on a half-day round that took them from one chestnut grove to another. These trees fruited only every second year. There were no outside buyers for the chestnuts, or if so at a very low price. But in fact the Yamashita needed their harvest for their own subsistence. Chestnuts were their primary source of calories for two months every autumn. The Yamashita did not grow rice. They lived by swidden-field crops and commercial products, such as silk-worm rearing, cryptomeria wood and charcoal. This is the information that can be obtained from the diary of Yamashita Tadajirō. We have limited ourselves to these three examples concerning the regions of Aichi, Hida and the village of Shiramine on Mount Hakusan, even though there are other reports, like that of

Jane Cobbi, mentioned above, on the role of foraging in the lives of modern Japanese.

In the light of the information available to us concerning the Neolithic on the one hand and modern Japan on the other, we can interpret the data given above for the period from the 8th–12th century. It is necessary, for this period, to distinguish between the nobility and the rural population. At the imperial court, wild fruits were simply an addition to the diet, rice being the staple food. The choice of wild fruits was probably made based on the possibilities for transportation and storage. The products eaten by members of the court enable us to get an idea of the diversity of edible and useful plants in early medieval Japan, since all the provinces had to supply their share.

By contrast, the situation in the countryside seems to have been very different from that at the court. A look at the archaeological and modern data suggests that foraging played a very important role in the rural areas of early medieval Japan. Of course, in the nutritional intake, variable proportions between cultivated cereals and collected fruits must be taken into account, depending on the region. For example, nuts and seeds were eaten in larger quantities in the non-rice-growing regions and even in rice-growing regions located in the mountains. This has led some archaeologists to say that the cases of Hida and Shiramine are exceptions. Yet we should not forget that mountains occupy more than two-thirds of Japan's land surface. The role of roots as starches must also be taken into account. In our view, the importance of nuts, seeds and roots in the 8th–12th centuries was somewhere in between Yayoi and the modern period. To this we must add the many other edible plants known thanks to the texts. We have stressed the importance of certain wild graminae, such as wild rice, which played a part not only in the diet, but also in dwellings and everyday material life. For this reason we think that, in the plant kingdom, the harvesting of wild products must be considered one of the three poles of the subsistence economy of early medieval Japan.

Notes

1   Sakamoto Shintarō (1979, 1980); Watanabe Akihiro (1996). Makino Tomitarō (1974) lists 3,996 plants.
2   *Engishiki* (ES) book 24 gives a list of the provincial taxes. For an overview of the Shōsōin documents, see Farris (2007). For fruits and nuts, see Sekine Shinryū (1969), which provides an analysis of the Shōsōin documents, published in DNK vols 1–25; we quote this work by Sekine, unless otherwise stated; Nihon gakushiin ed. (1980); Hirata Yoshinobu, Misaki Hisashi (1998). For chestnuts, see Sekine, pp. 132–134; ES 31: 753, 761, 33: 775, 39: 865–870, and the many other references that can be found in the index of the *Engishiki* in the Rinsen shoten edition. According to ES 33: 773, 1 *koku* of *namakuri* gives 1.25 *to* of *hirakuri* without boiling or the addition of other ingredients. See ES 33: 779 for hazelnuts.
3   ES 37: 828–842; *Izumo fudoki*, in *Fudoki, Fudoki itsubun*, NKBT vol. 2, pp. 117–245; Sekine, p. 135.
4   See below, note 55; ES pp. 599, 600, for the deliveries of *tsurubami*. The character *tsurubami* is read *tochi* in ShJ and in RSh, p. 385. The RMS gives both

readings. *Tochi* are not mentioned in DNK 1–25. According to *Utsuho mono-gatari* 1: 85, *kuri* and *tochi* are gathered; according to *Sangōshiiki* ge p. 118, *tochi no ii* rice is prepared; according to *Taiheiki* 1: 169, *tochi no kayu* and *awa no ii* are served to pilgrims. In these three cases, however, the gloss reading *tochi* is not contemporaneous with the early text.

5 Pinenuts and beechnuts are absent from DNK 1–25 and in ES, but are found in *Ruiju zatsuyōshō* 1: 512, *Shisseishoshō*, pp. 432, 449 and *Chūjiruiki*, p. 748. *Kashiwa* in the early texts has two meanings: Fagus *buna* and Mallotus *mitsunagashiwa*.

6 For *ichii*, see Sekine, pp. 134, 135; *Hitachi fudoki*, p. 87; the *mokkan*, quoted by Sekine; and DNK 16: 299. ES p. 455 mentions *ichii* as a timber for making palanquins. See oaks in *Hitachi fudoki*, pp. 55, 63, 71. See *kuri* in Hirata, Misaki (1998: 105), who quote *Fuboku wakashō* 2: 16-14012, 14013, 14015.

7 See *tachibana* in Kimura Yōjirō (1991: 245); Péronny (1993: 156); Sekine, p. 124; *Hitachi fudoki*, pp. 53, 71; ES pp. 761, 779, 865–870.

8 There is no poem of the 8th–12th centuries about these trees; see Nakanishi Susumu, *Man'yōshū jiten*; and Hirata Yoshinobu (1998); Sekine, pp. 130, 131. ES pp. 753, 779, 873, 878.

9 ES pp. 753, 779; SJ Ninna 3.2.9. 887.

10 SN Jingi 2.11.10. 725; Verschuer (1985: 265). Sekine, pp. 123, 126, 127. Kimura Yōjirō (1991: 156, 171).

11 ES pp. 719 and 590, mentions *chakago* and *chamari (chawan)*. Tea also appears in poems written in *kanbun*. The monk Eichū served tea to Emperor Saga, after his return from China. Later, a tea garden *chaen* was established at the Heian palace. Verschuer (1985: 205–206); here, however, it is necessary to correct the "2 *koku* of tea" quoted in DNK 6: 136, 225, 277, to "2 *koku* of sow-thistles" *ōdochi*.

12 ES pp. 753, 779, are the only mentions of *moronari* in the *Engishiki*. WR 17: 9 assimilates this to *gumi*, but with a different character from the modern *gumi*. Cobbi (1978: 83).

13 ES p. 387. *Itabi* is absent from the index of DNK 1–25 by Sekine (2001). *Hizen fudoki*, p. 401.

14 Sekine, p. 130. See the many references in the *Engishiki* index. Verschuer (1997: 360).

15 ES p. 873 and ES books 32, 33, 39. Sekine, pp. 119–122, 128–130.

16 ES pp. 866–870, 758–768.

17 *Izumo fudoki*, in *Fudoki, Fudoki itsubun*, NKBT vol. 2, pp. 117–245. M 4140 (read: *Man'yōshū* no. 4140), after Péronny (1993: 153).

18 ES p. 878 at Heian-kyō; DNK 6: 121 at Tōchi. There are two unidentified trees.

19 Hirata Yoshinobu (1998: 46–53). Sekine, p. 122, quotes DNK 17: 303, 6: 176, 188.

20 WR 17: 10, definition copied from Chinese sources, cf. *Daikanwa jiten* 6: 358. ES pp. 125, 586, 816–835. Six poems in *Shinsen waka rokujō*, cf. Hirata Yoshinobu, pp. 87, 88. *Chūjiruiki*, p. 748, banquets.

21 *Shōyūki* Chōgen 5.11.26. 1032, cf. Hérail (1991: 3: 259). *Ruiju zatsuyōshō*, p. 512; *Shisseishoshō*, pp. 432, 449; *Chōjiruiki*, p. 748. DNK 2: 633.

22 ES pp. 829, 834. *San Tendai Godaisanki* Enkyū 4.9.22. 1072 writes *moke* (quince). A number of fruits are absent from the 8th-century Shōsōin documents: *akebi, itabi, gumi, moke, yuzu*. Branches of *moke* were used for the festival of the day of the hare at New Year; cf. ES p. 969.

23 *Izumo fudoki*, in *Fudoki, Fudoki itsubun*, NKBT vol. 2, pp. 117, 133, 221, 245.

24 Bibliography relating to vegetables: Aoba Takashi (1991); Hirose Tadahiko (1988); Asai Keitarō, "Sosai engei", in Nihon gakushiin ed. (1980); Sekine; Hirata Yoshinobu (1998).

25 ES p. 873. See the index of the *Engishiki*, Rinsen shoten edition, for the many other references concerning each plant.

26   ES pp. 873, 874. *Tatarahime* is also mentioned in DNK 11: 352.
27   See the imperial vegetable gardens in ES pp. 880–881; and the translation in Chapter 1, pp. 17–21. According to Sekine, pp. 53, 86, most plants were wild. Nihon gakushiin ed. (1980: 354–356) lists the plants in the *Fudoki*. Aoba (1991: 112) lists the wild and cultivated vegetables in Iyo in the 17th century. For the geographical distribution, it is necessary to consult the agricultural treatises of the Edo period.
28   *Shisseishoshō*, p. 432; *Moromitsu nenjūgyōji*, p. 331; *Nenjūgyōji*, pp. 155, 159; *Morotō nenjūgyōji*, pp. 206, 207; *Moromoto nenjūgyōji*, pp. 238, 239.
29   Métailié, Stäuble Tercier (1997).
30   *Shisseishoshō*, p. 449. The *Ruiju zatsuyōshō*, p. 512, presents some differences.
31   Sekine, pp. 40–87, 214–223; Aoba (1991: 23–38); and a calendar of vegetables in Sekine, p. 84, and Aoba (1991: 34).
32   DNK 1: 555.
33   See DNK quoted in Sekine, pp. 45–83, 214, 219, 222; and the ES index, Rinsen shoten edition. *Asatsuki* appears only in ES pp. 599, 614.
34   Sekine, p. 81, quotes DNK 13: 281 and 15: 492, to which may be added DNK 13: 349, 351. On the subject of mushrooms, see ES pp. 100, 131, 144; *Harima fudoki*, p. 267; N Ōjin 19.10.; *Konjaku monogatari* 28: 17; *Uji shūi monogatari* 1: 2 (p. 54); Sahara Makoto (1996: 35–47) thinks that *matsutake* appears only in the late medieval period, but there is a reference in the *Ryōjin hishō* (12th century), p. 425. On the subject of *amazura* and *ame*, see ES pp. 753, 779, 773.
35   ES pp. 768, 771, 779; Sekine, pp. 57, 60, 61, 77.
36   On this point, see Chapter 1, p. 36. SNK Jōwa 5.7.1. 838, and ES p. 921, allow the cultivation of aquatic vegetables.
37   ES pp. 591–594, 599, and the ES index. Sekine, pp. 87–117 has determined the species. M 12: 3177, *Shoku gosen wakashū* 1: 10-1055, *Fuboku wakashō* 2: 16-11732, 11443, 11682, 15571, 16673, *Shinsen waka rokujō* 2: 14-1095, *Shinkokin wakashū* 1: 8-1725.
38   Sekine, pp. 71–78. ES pp. 763, 768, 771, and ES 37: book dealing with medicines.
39   Péronny (1993: 101, 102); M 10: 1942; the DNK 15: 458 (*kuzu* leaves) has only one reference; *Izumo fudoki*, in *Fudoki, Fudoki itsubun*, NKBT vol. 2, p. 117 (*kuzu* rhizome); Sekine, p. 62.
40   DNK 6: 223 et seq., quoted by Iyanaga Teizō (1980: 476–479). On occasion, more fruits were served, cf. DNK 14: 349–358.
41   See the names in Péronny (1993: 70, 90, 102, 107, 142, 187, 197); the poems in Hirata Yoshinobu, pp. 11, 46, 55, 68, 97, 178, 227, 273, 287, 381, 384. See *asaza* in DNK 14: 303. *Horikawa-in hyakushu* 4: 26–65 to 80: poems about *wakana*, 4: 26-129–144: poems about *warabi*, etc. See *wakana* and *warabi* in Hirata Yoshinobu (1998: 273–375, 380–382).
42   *Izumo fudoki*, in *Fudoki, Fudoki itsubun*, NKBT vol. 2, pp. 117, 133, 139, 159–161, 177, 191, 221, 245; *Fuboku wakashō* 2: 16-923 (*horimotomeru*); *Harima fudoki*, p. 323; *Hizen fudoki*, p. 401; *Hitachi fudoki*, p. 71. See Asai Keitarō, "Sosai engei", in Nihon gakushiin ed. (1980: 354, 355). See the lotus in the *Fudoki*, pp. 71, 161, 401, 411, 457.
43   ES book 37.
44   Yamamoto Kiyoshi, ed. (1995: 3: 44).
45   *Izumo fudoki*, in *Fudoki, Fudoki itsubun*, NKBT vol. 2, p. 117.
46   ES p. 838.
47   Verschuer (1985: 42, 83–85); Cobbi (1978).
48   ES pp. 609, 753; SJ Ninna 3.2.9. 887.
49   ES p. 835.

50   ES pp. 431, 586, 587, 830–839, and the treatise on perfumes *Kōjishō*, p. 478.
     Verschuer (2014: 64–66). DNK 25: 48–50; *Shinsarugakuki* "Hachirō Mahito".
51   ES pp. 423, 752, 958.
52   RR Buyakuryō 1 *sowatsumono*; see the same products in ES pp. 599, 600, *chūnan
     sakumotsu*. SN Yōrō 1.11.22. 717.
53   Sekine, pp. 223–227; DNK 12: 180 *e o shiboru*. RR Shikiinryō 54 Abura no
     tsukasa. ES 23: 591, 24, 36: 809, 810, 813. Oils from nuts and the Japanese
     snowball are absent from the RR. For Torao Toshiya, trans., annot., *Engishiki*, 2
     vols, Shūeisha, vol. 1, 2000, *Engishiki*, vol. 1, p. 343, *hosoki* is *inuzanshō*.
54   ES pp. 400–405, 418–427, 429; and the gardens of Japanese indigo and gardenia
     in ES pp. 431, 432.
55   ES pp. 403, 419–428. M 4109; Péronny (1993: 175). See above, note 4.
56   ES pp. 448–463, for lacquers; see RR Buyakuryō 9, and Chapter 1 in this
     volume, for the crops. ES pp. 111, 428, 453, 455, 794, 988, for the scouring rush;
     Shinano scouring rush is mentioned in poems, cf. Hirata Yoshinobu, pp. 225,
     226. For resin, see ES pp. 660, 988, and Kobayashi Yukio (1982: 169–174).
57   ES p. 738; SN Hōki 8.5.23; Verschuer (1985: 63).
58   ES pp. 394, 409, 429, 825, 826, 871, 889, 890, 951. RSK 6: 279. See the *Fudoki*,
     pp. 101, 303, 321, 323, 325, 401, 501.
59   ES p. 451.
60   See the many references to wild hemp in the *Fudoki*.
61   *Nio* was used for *o, tsuna* and *nawa*, according to ES pp. 423, 451, 820; but the
     Nihon kokugo daijiten explains *nio* as being sesame boiled for a long time, and
     Endō Motoo, *Orimono no Nihonshi*, NHK, 1971, p. 61, asks whether this is boiled
     hemp. See also ES p. 592; DNK 14: 310, etc. For Torao Toshiya (2000: 1: 343), *nio*
     is boiled hemp fibres.
62   WR 14: 13; and *Ritsuryō*, p. 586.
63   ES pp. 409–416 and ES books 1 and 2 (*Aki no yuu*). ES pp. 25, 424, 759, *yuu* is
     used for tying *musubu*.
64   *Bungo fudoki*, p. 371; ES p. 420, mentions *tsuyukusa no yuu*.
65   Nunome Junrō, *Kinu to nuno no kōkogaku*, pp. 176–179. He quotes DNK 1: 574,
     634, for *kuzu no nuno*, and the *Kojiki* chū for *fujikazura*.
66   *Izumo fudoki*, in *Fudoki, Fudoki itsubun*, NKBT vol. 2, pp. 211, 221, 231, 245;
     Hirata Yoshinobu, Misaki Hisashi (1998: 68–70).
67   *Fudoki*, pp. 37, 57, 117, 121, 133, 139, 159–163, 173, 191, 193, 221, 231, 243,
     245, 371, 397, 529.
68   *Ruiju kokushi* book 199: 377; Enryaku 19.4.12. 800; NK Enryaku 18.7.28. 799.
     Verschuer (1985: 357, 450, 2006: 126 et seq., 2006: p. 126 et seq.).
69   See the deliveries in ES pp. 591–594, 127, RR Shikiinryō 42; see the consump-
     tion at the court, in ES pp. 109, 119, 869, 941, 761–765, 867–869. See Chapter 1,
     on the subject of cereals, pp. 25–27. See Ikata Sadaaki (1985); Sekine, pp. 14–40;
     Tsukuba Hisaharu, "Hatasaku", in Nihon gakushiin ed. (1980).
70   ES p. 761 and other occurrences, etc.; DNK 1: 465, 610, 6: 74, 14: 356. *Sekimai*
     is found in DNK 1: 398–411, 608, 4: 367, and on the *mokkan*, cf. Tatsumi
     Kazuhiro, 1986. *Chūjiruiki*, p. 742, seems to be the first occurrence of *sekihan*.
     On rice varieties, see Watabe Tadayo, *Ine no daichi*, 1993; *Wakansansaizue* book
     103, art. "urunokome"; Smolarz (1997: 23, 130). For tropical rice, see Satō
     (2001: 126–134); see also the Introduction p. 5.
71   See Chapter 1, pp. 39–43 and pp. 47–56.
72   ES pp. 799–806, etc., RR Shikiinryō 42; SN Reiki 1.10.7.; DNK 1: 396–460, 2:
     12–19, 44–129, 3: 469, 14: 358. Ikata (1977: 19–42); Sekine, pp. 27, 28.
     Sanguozhi Weizhi book 30, in Wada Sei *et al.* (1976: 44). See *mochiawa* in
     ES pp. 866, 868, 869, 941.

73   Wheat is frequently mentioned in the DNK, see Sekine, pp. 24–29, and the DNK index by Sekine (2001). See Ikata (1977: 220–290); ES p. 773 (noodles), p. 871 (a kitchen for grinding wheat *mugi-dokoro*), p. 878 (a plot of wheat in the imperial vegetable gardens). Wheat was pounded (DNK 13: 277, 305), but also boiled (DNK 25: 233). See *kimi* in DNK 15: 313 and 16: 482; M 3834. The glosses reading *kibi* in ES pp. 109, 802, are later than the 13th century.

74   See note 69. Sekine, pp. 36, 37, quotes DNK 1: 465, 609, 616, 14: 358. Ikata, pp. 194–219, mentions (p. 207) a note in the "Nihon koten zenshū" edition of the *Engishiki* Minbu ge, "kōeki zatsumotsu", which is not found in ES p. 591. WR 18: 7 (Minamoto no Shitagō). Torao Toshiya (2000: 1: 343).

75   The only references to *futomugi* in ES and DNK are: ES pp. 591, 593, 878, 802; and DNK 2: 76, 90. Sekine, p. 23. ES p. 591 (wheat).

76   Ikata (1977: 112–130); Sekine, pp. 35, 36. M 2476, 2999. *Hie* is seldom referred to: DNK 1: 609, 616; ES pp. 591, 830, 842, 898, 899.

77   Thereis no mention of *sobamugi* in ES, DNK vols 1–25, RR, F, and M; but there is in SN Yōrō 6.7.19. 722; SNK Jōwa 6.7.21. 839; RSK p. 328; *Shinsarugakuki* "San no kimi no otoko" (see a translation at the beginning of Chapter 1, p. 14). The *Kokon chomonjū*, a collection of legends from the 13th century, 12: 444, p. 358, mentions *hatake*, and 18: 616, p. 476, mentions wild *sobamugi*. Ikata (1977: 292–311).

78   There are no studies on artisanal stems in the early medieval period. Here are our references: ES p. 796; RR Buyakuryō 29; RSK 6: 280, for deliveries of *wara* from the provinces. For handicrafts, see: ES pp. 109, 118, 136, 139, 388, 394, 395, 401, 405, 418, 420, 423, 426, 457, 791, 793, 796, 803. Straw is often mentioned with fuels and charcoal; see also DNK 1: 560, 2: 179. Mayuzumi Hiromichi (1985: 85–89) explains the use of ashes.

79   ES pp. 719, 720, 886–899, and ES book 32.

80   ES pp. 86, 144–149, 848. See Ichikawa Takeo *et al.* (1985: 83, 84), on the subject of eulalia *susuki* and imperata *chigaya*.

81   ES pp. 857, 663, 575.

82   ES pp. 369, 429, 431, 594, 602, 659, 812, 575, 857, 663, 855, 859, 861.

83   ES pp. 454, 461, 813, 861, 600.

84   ES pp. 88, 122, 444, 857, 1002.

85   ES pp. 580, 600, 611, 619, 775, 812, 820, 857–861, 868, 848, and ES books 1 and 2.

86   *Izumo fudoki*, in *Fudoki, Fudoki itsubun*, NKBT vol. 2, pp. 159, 161, 321.

87   There are many references to *take* and *ashi*. See the *Fudoki*, pp. 135, 137, 161, 275, for *komo*; pp. 289, 321, for *suge*; p. 159, for *ogi*; pp. 55, 141, 145–151, 159, 489, for *chi*.

88   See the toponyms in Ikebe Wataru (1981: 788); and in DNK vols 2, 4 and 5. See the myths about *ashihara no nakatsu no kuni* in the *Kojiki* and *Nihon shoki* books "Jindai-ki". See the harvest *karitsumu ashi* in *Gosen wakashū* 1: 2-625.

89   *Shinkokin wakashū* no. 229, French translation by Pigeot (1998: 78); and M 2: 110.

90   The subject of *karikaya* appears in *Horikawa-in hyakushu* 4: 26-641 to 656; *Kokin waka rokujō* 2: 4-3785 to 3787; *Shinsen waka rokujō* 2: 14-1996 to 2000; *Fuboku wakashō* 2: 16-4444 to 4456. There are poems about *karu komo* or *makomo karu* in *Kokin wakashū* 1: 1-485, 587, 759; *Gosen wakashū* 1: 2-483; *Goshūi wakashū* 1: 4-1144 and 206 (*makomogusa karihosu*); *Kin'yō wakashū* 1: 5-135; *Kokin waka rokujō* 2: 4-3810, 3811. See Chapter 1, note 127.

91   Musashino bijutsu daigaku Minzoku shiryōkan ed. (1994: 13).

92   *Wakansansaizue* book 103 art. *wara, mugiwara*; book 92 art. *kaya, susuki* and *chigaya*; book 94 art. *okaashi, ashi, sudareyoshi*; book 94b art. *suge*; book 97 art. *komo*. Matsuyama Toshio, "Inawara to shokubutsu sen'i", in Sasaki Kōmei (1988: 289–305), has examined several Meiji-period monographs on this subject.

93   *Harima fudoki*, p. 321.

94 Hirata, Misaki (1998: 12–16, 83, 84, 115–119, 163–171, 385–388).
95 M 3050 *chi-shime*. Hirata, Misaki (1998: 94–198), poems about *chigaya*. See
    Hagiwara Hidesaburō (1996: 170–180); and Ōgata Tōru (1998), on *chigaya* in
    rituals. See the references to *shimenawa* in poetry, in Chapter 1, p. 38.
96 See a "mat for transportation" *tawara komo* in DNK 13: 279, 281, 349.
97 Verschuer (1993).
98 Kimura Yōjirō (1991: 379, 381, 390, 397).
99 *Wakansansaizue* book 92, art. *tsubana*; book 92b, art. *karukaya*; book 103, art.
    makomo no mi; book 97, art. *komo*.
100 M 414, 1250, 948, on *suge*; M 1449, 1460, 1462, on *tsubana*.
101 Sekine, p. 80, quotes DNK 6: 176, 188, 16: 299. See also DNK 13: 277, 310, 314,
    317, 349, 351, 25: 232.
102 *Kokon chomonjū* 18: 621, p. 478; *Teikin ōrai* "Jūgatsu no jō kaeshi", p. 266;
    Honchō shokkan 1: 252. See Aoba Takashi (1991: 222–226).
103 Kimura Yōjirō (1991: 374–412).
104 For what follows, see: Watanabe Makoto, "Kenkarai" (1986: 32–35), "Jōmon
    jidai no kaizai kiban" in Ōtsuka Hatsushige *et al.* (1996: 2, 27–38), and "Jōmon
    jidai no tabemono" (1999: 15–32). In these three articles, Watanabe summarizes
    his book *Jōmon jidai no shokubutsu shoku* (1975).
105 Shiomi Hiroshi (1986) discusses the history and problems arising from this
    research; see p. 441, regarding *warabi* and *kuzu*.
106 Sakurai Junya, "Kaizuka to karorii keisan", in Ōtsuka *et al.* (1996); Umehara
    Takeshi, Yasuda Yoshinori, eds (1995: 134, 135).
107 Umehara Takeshi's (1995) book, subtitled "Kyōi no Sannai-Maruyama iseki", is
    a summary of the excavation reports of the archaeologists who worked on this
    site. For the cultivated plants, see Chapter 2 in this volume.
108 Sakurai Junya, in Ōtsuka *et al.* (1996: 174–185); Sahara Makoto (1994:
    143–149, 174).
109 Koyama Shūzō, Gotō Yoshiko (1999: 80–93), with the quotation, p. 86.
110 Terasawa Kaoru, "Inasaku gijutsu to Yayoi no nōkō", in Mori Kōichi, ed. (1986:
    291–350), esp. pp. 336–350; and Terasawa, "Hatasakumotsu", *Kikan kōkogaku*
    14, 1986, pp. 23–31. In these two articles, Terasawa summarizes his study:
    "Yayoi jidai shokubutsushitsu shokuryō no kisoteki kenkyū", *Kōkogaku ronshū*
    5, March 1981, pp. 1–130. See the species cultivated in Chapter 2 in this volume.
111 Matsuyama Toshio (1990: 90–92); Sasaki Kōmei (1997: 282); Kajiwara iseki
    chōsakai ed. (1988: 28–39): the objects from this site in the Ōsaka region are
    made of wood from the laurel-forest and Fagus zones.
112 Sahara Makoto (1994: 309–314, 1995, 1996: 227–233). Koyama Shūzō, Gotō
    Yoshiko (1985: 473–500); Koyama Shūzō, "Kaisetsu: Nihon shoku bunka no
    kiso", in Haga Noboru, Ishikawa Hiroko, eds (1998: 3–14); Koyama Shūzō,
    Gotō Yoshiko (1999: 80–91). See Chapter 4 in this volume.
113 Kawasaki-shi shimin myūjiamu ed. (1995).
114 See notes 112–113; and Watanabe Makoto (1999: 15–32), with our quotation,
    p. 32.
115 Ōsaka furitsu Yayoi bunka hakubutsukan ed. (1999).
116 Terasawa Kaoru (1999).
117 Harada Nobuo (2006: 61, 75–77); Satō Yōichirō (2001, 2002), for the DNA
    analyses; Satō Yōichirō (2002: 112), and Kimura Shigemitsu (2010: 28–30) (table
    of Terasawa Kaoru). Nespoulous (2008: 40–42), for Yayoi, argues in favour of
    parallel practices, irrigated and dry, and the continuation of the Neolithic
    foraging practices.
118 *Hyakushō denki* 14: 320, 321.
119 Matsuyama Toshio (1990: 176–196). See Koyama Shūzō, in Sasaki Kōmei (1988:
    281), for the reference to Akaoke-mura. However, Koyama Shūzō, "Hida

gofudoki ni miru Edo jidai no shoku seikatsu", in Haga and Ishikawa (1999: 212, 213), estimates the calorific proportion of cereals throughout the province at 90%. These publications follow a collective project: Koyama, Matsuyama, *et al.* (1981: 363–596). Ishii Susumu in Ishii and Amino (2000: 40–45) disputes these figures and points out that the authors have not deducted the rice taxes of 40–50% paid to the State.

120   Chiba Tokuji (1986). See a presentation of Shiramine in Matsuyama Toshio (1990: 132–175); and in Tachibana Reikichi (1995).

121   See the dialectal names of the plants in *Nihon shokubutsu hōgen shūsei*, 2001. I thank Mechtild Mertz for this reference.

4 Food security

How much rice did they eat?

After having studied agricultural production from the point of view of its various technical aspects, we now examine the relationship between production and consumption in terms of food security. We are interested in knowing what constituted the nutrition and diet of the Japanese of the early medieval period. One question arises: how much rice did they eat? Though Fernand Braudel has stated, rightly so, that "rice does not become king, in the Japanese diet, before the 17th century",[1] the general opinion is far from being unanimous on this point. On the contrary, until the 1980s, it was thought in Japan that rice had been the staple food of the country for two millennia. This conviction was so widespread among the public and in academic circles that the term "rizicentrism" was used. The notion of rice as the central element of the Japanese diet had gradually been taking shape since the Edo period, but it is to the ethnologist Yanagita Kunio (1875–1962) that we owe the image of rice growing as the driving force of Japanese culture. This notion influenced post-war historical and archaeological research. In fact, agrarian archaeology focused on rice fields and studies on economic history mainly dealt with the management of rice, while the management of the other cereals was largely neglected. The debate about the importance of rice in Japan continues and opinions remain divided. Before proposing our conclusions, we present the way ideas have progressed and the current state of the debate on this question.

Discussion: rice in the dietary history

The theories of Yanagita Kunio were called into question by the ethnologist Tsuboi Hirofumi, in his book "The Japanese and Taro" (*Imo to Nihonjin*, 1979). In his fieldwork across Japan, Tsuboi analysed the New Year menus because they reflect dietary customs. He observed that in many regions people ate not the "traditional" rice cakes *mochi*, but taro *imo*. According to Tsuboi, Japan maintained two parallel traditions in the course of its history, one based on rice cultivation and the other based on swidden-field crops, with the latter preceding the former in time.[2]

In the 1970s, two historians sensitized academic circles to dry cereal cultivation. Amino Yoshihiko (1973) drew attention to the existence of dry fields
alongside rice fields on the medieval estates; this fact, though evident, had
been neglected by historians. Then Kimura Shigemitsu (1977) published an
article on dry cereal crops in early medieval Japan, this being followed by
other studies that were brought together in the first Japanese book devoted to
non-rice-growing agriculture published in 1992.[3] The archaeologists lent their
support, with the first discoveries, in the 1980s, of the remains of dry fields
in the early medieval Japanese sites. In 1981, Terasawa Kaoru published a
report on 224 archaeological sites dating from the 3rd century BC to the 3rd
century AD; he put forward the hypothesis that rice provided only half of the
food needs up to the 2nd century, and only two-thirds in the 3rd century.[4]

After this initial stage, indicating a growing awareness of the existence of
dry cereals (i.e. other than irrigated rice), Amino Yoshihiko began to conduct
a veritable offensive and he was soon joined by Ishii Susumu, another
medievalist. In 1980, Amino criticized in very explicit terms the "rizicentric"
view *suiden chūshin shikan* that was current among scholars. He thus rallied to
Tsuboi's view and opened up the debate on the role of rice cultivation in Japanese history. According to Amino, rice predominated only in the tax system. In
the reserves of the rural population in the early and late medieval periods, dry
cereals exceeded those of rice. Amino based his argument on some ten documents of the 12th–15th centuries, previously published by Kuroda Hideo, and
he emphasized geographical differences. The arable areas show a preponderance of
dry fields in the east of the country, i.e. in the Fagus zone, and a higher proportion
of rice fields in the west belonging to the laurel-forest zone. But, he says, in the
countryside rice appeared only on festive tables. Amino accepts the analyses of the
historian Harada Nobuo, according to whom, in the medieval and modern periods, the Japanese lived not only from the products of the fields, but also by hunting,
fishing and gathering, and he places special emphasis on fishing. These different
arguments form the basis of Amino's criticisms regarding the "rizicentric theory"
inasaku ichigenron.[5]

Harada, like Amino, distinguishes the role of rice in taxation on the one
hand and in everyday life on the other. Besides the documents mentioned by
Amino, he quotes a dozen examples of polyculture and low consumption of rice
dating from the 13th–16th centuries. For Harada, rice is above all an imperial
attribute that is revealed in mythology and ritual.[6] By contrast, the argument of
Kimura Shigemitsu (1992) is based primarily on the agrarian management of
the 10th–15th centuries, with the emphasis on land areas. Even in 1885, he says,
dry fields still occupied 41% of the arable areas: a relatively large proportion
compared with 59% of rice fields. In his second book, "The Japanese and Dry
Fields" (*Hatake to Nihonjin*, 1996), he also discusses the role of the "five
grains", i.e. graminae and beans in the mythology, and he notes the diversity of
agricultural products in the taxes and in the ceremonies of the Heian court.
According to Kimura, all the cereals were important, not only for the rural
population but also at the imperial court.

At a conference organized in 1993, historians, ethnologists and archaeologists began to "re-examine the conventional image of Japanese culture" *Nihonzō o toinaosu*, as indicated by the title of the Proceedings. However, concerning the primacy of rice (one of the topics discussed), the participants on the contrary confirmed certain received ideas. While deploring the biased rizicentric arguments *henjūron* and recognizing the low proportion of rice in the diet up to the mid-20th century, the ethnologists Ōbayashi Taryō and Miyata Noboru agreed with Amino Yoshihiko in admitting the existence of a "structural hunger" for rice *kome o hossuru shikō* specific to Japanese history. The shortage of rice would thus have had a psychological effect, meaning that rice would not have dominated in the diet, but in the minds of the early medieval Japanese. Both scholars justify their point of view by a "cultural predominance" of rice, which they perceive through the rituals and the calendar of festivals. They appear to attach special importance to the rituals and festivals of the modern period and to give insufficient weight to the studies on the ritual role of dry cereals published by the ethnologists Nomoto Kanichi (1984) and Shiraishi Akiomi (1988).[7]

In 1996, the archaeologist Sahara Makoto summarized the various arguments. He defends a viewpoint different from that of Amino Yoshihiko and observes that anti-rizicentric opinions are also distorted by the study of exceptional or extreme cases. Sahara places the question of food in the context of demographic growth, which saw four phases (according to the estimates of Koyama Shūzō): during the Neolithic, acorns were the staple food *shushoku*; from the Yayoi period (starting in the 5th century BC) to the 12th–13th centuries, the staple food was rice; from the 14th–18th centuries, it consisted of dry cereals, owing to the commercialization of rice and the increase in the yields of dry cereals; – in the 19th century, the sweet potato *satsumaimo* was added to rice and the other cereals to sustain the demographic growth brought about by industrialization. According to Sahara, the arguments advanced by Tsuboi Hirofumi and Amino Yoshihiko are limited to the late medieval period, i.e. the time of the third wave of population growth. To those who point out that rice did not provide even half the needs of certain regions in the modern period, as for example in Hida in 1873, Sahara replies that these are marginal cases of mountainous areas unsuitable for rice growing, and that at the same period the urban population ate rice every day. (It should be remembered that the mountainous regions occupy more than two-thirds of the entire country; moreover, the demographic data quoted by Sahara is far from being accepted by historians.) For the Yayoi period, the high consumption of rice is documented, in his view, by the discovery of tens of thousands of rice grains and barely a few hundred grains of other cereals in the archaeological site of Hashibara in Nagano dating from the 2nd–3rd centuries and located, as it happens, in a mountainous region. Although many Yayoi sites have acorn silos, as shown by the analyses of Terasawa Kaoru, these silos, according to Sahara, were not reserves of everyday foodstuffs, but reserves intended to guard against rice shortages.[8]

Sahara also puts forward some general considerations. He thinks that no food other than rice could have sustained the extraordinary demographic and cultural growth that occurred with the introduction of rice growing in Japan. Iron and bronze working spread rapidly, and only "six or seven hundred years after the beginnings of irrigated rice cultivation there appear tumuli of a considerable size, even on a world scale; these tombs show the very rapid emergence of powerful centres of authority, an evolution that took several millennia in Central Asia, China and Europe".[9]

In 2006, Harada Nobuo published his book "Japan Chose Rice: A History" *Kome o eranda Nihon no rekishi*. Harada presents a historical picture from the introduction of rice in the 1st millennium BC up to the end of the 19th century. He bases his work on the historical data in the written sources, which are mainly concerned with the public economy and the fiscal history of Japan. It is in fact an unusual feature of Japanese history that this country did not mint coins (except for a limited period). From the founding of a State in the 7th century, rice was established as the basis of public accounting. It served as a value standard for loans and commercial exchanges until the 19th century. Rice remained a method of payment even after the spread of Chinese money in Japan from the 14th century and simultaneously with Japanese coin minting from the end of the 16th century. In modern Japan, public taxes were levied in rice and this even applied to non-rice cereal production. Thus, for Harada Nobuo, the Japanese "chose rice" because they adopted it as the value standard in the country's economy.[10]

Rice has indeed played a crucial role throughout the economic history of Japan. However, in our view, it has not occupied a predominant place in the dietary or ethnological history of the country. In this book, we therefore propose a different approach: a cultural and material history that concerns the daily life of the Japanese population, namely the place of rice in the overall food supply.

From 2009 to 2012, the National Museum of Japanese History devoted a research programme to the 'Formation and Development of Agrarian Society: Reconstruction of the Image of the Yayoi Period'. In the report published in 2014, Andō Hiromichi states that academic circles adopted from the 1970s the idea of an earliest agriculture in the form of swidden farming *yakihata*, and that since the 1990s this concept has evolved into a wider perception of a 'composite Japanese culture' *Nihon bunka no tagensei* with multiple farming practices including 'non-rice farming' *hiinasaku nōkō*, a 'swidden cropping culture' *yakihata bunka*, a 'laurel-forest culture' *shōyō jurin bunka*, a 'beech-zone culture' *nara-bayashi bunka*, as well as a 'rice-growing culture' *inasaku bunka*.[11] Research results of this kind may soon gain wider acceptance and the rizicentric viewpoint might then belong to the past.

We now present our estimates concerning the production and consumption of rice in early medieval Japan, starting with yields and production in the 8th century.

Yields and production in the 8th century

Yields have been estimated by different criteria depending on the period. In the early medieval period, the Japanese government was mainly concerned with the relationship between the harvest and the cultivated area. It did not pay much attention to the quantity of seed or the investment in labour. The notions of productivity and returns were still foreign to a mentality based on zeal. Since the public economy was based on rice, the State imposed rice growing on the population. In the 8th–9th centuries, it distributed lands to the entire population by way of public holdings, called "household fields". The State made a distinction between soils of good *jō*, average *chū* and mediocre *ge* quality, but it largely reserved the first two categories for the use of the administrations and the officials of the provincial governments themselves.[12] An estimate of yields was issued at the latest in the Kōnin era (810–824), with a view to calculating the taxes (one-fifth of the harvest) on the public rice fields leased by the provincial governments. Here is the scale of yields per *chō* (1.13 ha):

Forecast rice yields per *chō* for the Kōnin era (810–824)

- good rice fields *jōden*:   500 sheaves *soku*
- average rice fields *chūden*:   400 sheaves
- mediocre rice fields *geden*:   300 sheaves
- poor rice fields *gegeden*:   50 sheaves

To guard against fraud, quotas were determined for each province by a proportion expressed in seven parts: one part of good lands, for two parts respectively of average, mediocre and poor lands. It was therefore assumed that good soils were half the number of each of the other qualities.[13]

The grades of *jō*, *chū* and *ge* were used in the land surveys and other administrative documents, even outside the context of public field leases.[14] Thanks to several accounts, we know that these figures reflected a certain reality and were not simply made up by the authorities. For example, in 735, several plots on the property of the Gufukuji temple in Sanuki produced respectively 400, 450, 490 and 500 sheaves per *chō*. In 810, a temple estate located near the old capital of Fujiwara (Nara prefecture) harvested on one plot 478 sheaves per *chō* and on another 420 sheaves per *chō*. In Kyūshū, the provincial governments, in 823, leased rice fields estimated at 460 sheaves in Higo and at 400 sheaves in the other provinces. Around that time, the Ise Shrine was expecting a harvest of 420 sheaves in the rice fields that provided the offerings. A few good-quality lands are also attested for the 11th–12th centuries. While all these cases occurred in a privileged context, lower yields are also known in the same milieu. Some rice fields on the Tōdaiji temple estate at Kuwabara, Echizen (Fukui prefecture), produced 400 sheaves per *chō* and others only 300 sheaves. In 876, the rice fields leased by Satsuma had yields of only 300 sheaves.[15]

However, on the public holdings of tenant farmers, "good rice fields" were probably rather the exception. For example, the province of Hitachi points out that it has few good lands and many soils of average quality. In Izumi, there was also a shortage of good lands. Yamato and Yamashiro were so lacking in soils of good and average quality that they had only soils of mediocre and poor quality to lease, and these lands did not even find a lessee. Certain regions of Kawachi requested allocations of additional lands, because many rice fields produced only 200–300 sheaves. The province of Aki, for its part, requested a tax exemption on poor-quality rice fields.[16] According to this information, there were apparently few good rice fields; but poor rice fields, which benefited from special advantages, were also considered exceptional. It therefore seems that mediocre-quality rice fields made up most of the peasant holdings, and we adopt for the moment the figure of 300 sheaves per *chō* as the average yield.

Theoretical production of an average peasant household

Takigawa Masajirō gave an estimate for peasant households of the 8th century. For convenience and in the absence of other estimates, we adopt Takigawa's model of the average household as the basis for what follows. According to him, the average farming household consisted of ten people, namely three young or adult men, five young or adult women and two children under six years of age.[17] The yearly production of this same household is calculated in the following manner. The regulations allocate to the peasant population, by way of public tenure, "household fields" amounting to 2 *tan* (0.6 acres) for men and 1.33 *tan* (0.4 acres) for women.[18] An average household of ten people thus has a holding of 12.66 *tan*, i.e. 1.266 *chō* (1.43 ha). The production of this average holding, in the case of mediocre-quality rice fields with a yield of 300 sheaves per *chō*, theoretically amounts to 379.8 sheaves *soku* of rice. Here, therefore, is what our estimates are based on:

- an average household of ten people;
- an average holding of 1.266 *chō* (1.43 ha);
- an average yield of 300 sheaves per *chō*;
- a theoretical annual production of 379.8 sheaves of rice.

However, the actual harvest was less than these estimates owing to technical and climatic factors and, in addition, our average household could not use its entire harvest for its own reserves since its production was subject to a number of deductions, including taxes.

Technical and climatic factors

We know that all holdings included uncultivated areas and that they were often hit by bad weather. Harashima Reiji gave estimates on this point. According to him, these lands, whether idle (short-term) or abandoned (long-term), accounted for between 12% and 44% of the total surface of holdings.[19] Allowing for an average of one-third of uncultivated areas, our average holding of 1.266 *chō* is reduced by 0.422 *chō*. This uncultivated one-third includes soils left lying idle because of management difficulties. Moreover, the lands allocated to farmers did not always come up to the standard defined by the regulations. For example, in certain households in the provinces of Buzen, Bungo and Chikuzen, all in Kyūshū, the share of lands allocated to the men was reduced by one-third (240 *bu*) and that of the women by one-eighth (60 *bu*). Furthermore, the regulations permitted the distribution of fields to vary according to the lands locally available.[20] Still other arable lands could therefore be deducted. Even the household fields actually received had surface losses due to paths, ridges and channels between the rice fields; but for our purposes we will keep to one-third of lost areas. There then remains a sown area of 0.844 *chō* (0.954 ha) which, with a yield of 300 sheaves, should theoretically produce 253.2 sheaves.

However, harvests that reached the ideal forecast of 300 sheaves per *chō* were not the norm. Bad weather often hit the rice fields and destroyed part of the harvest. Natural disasters included floods, droughts, typhoons, frosts and insect invasions. The government made provision for tax exemptions or distributions of food, if losses affected more than half the harvest, at a regional level. It did not take into account losses of less than half the harvest, or individual cases. Since losses gave rise to administrative measures, the documents are full of reports on droughts, floods, etc.[21] For example, in the first half of the 8th century, there were twenty-four bad years out of fifty in the central Kinai region; and for twenty years, from 784 to 805, natural disasters struck one or more provinces of Japan sixty-six times.[22]

Harashima Reiji studied the variations in harvests due to bad weather, taking into account different factors relating to administrative management, and gave the following average estimate: over a period of ten years, there are three good harvests (according to the yields provided for by the regulations), two very good harvests (increased by 20%), three poor harvests (reduced by 30%) and two very poor harvests (reduced by 60%). Based on this average, he established a coefficient of 17% to be deducted from the annual yields per *chō*.[23] The theoretical harvest (300 *soku* per *chō*) of 253.2 sheaves on a sown area of 0.844 *chō* is then reduced by 43.04 sheaves (17%) and reduced to an actual average harvest of 210.16 sheaves.

Deductions for taxes and seed

To ascertain the volume of household reserves for the average peasant family, taxes and seed must be deducted from the actual average harvest. The land-tax *so* amounted to fifteen sheaves of rice per *chō*.[24] Our holding of 1.266 *chō* therefore paid 19 sheaves a year to the provincial government.

The peasants were also subject to a compulsory loan *suiko*, which was a kind of disguised tax. The district administrations distributed each year, in spring, rice from the public granaries to every tenant household. The interest rate was fixed at 50% during the 8th century. The volume of the loans was fixed at 10–100 sheaves for each male taxpayer or a minimum of 30 sheaves per *chō* of land with a minimum of 15 sheaves as the interest payment. Taking the figure of 30 sheaves per *chō*, the compulsory loan of our average household with its holding of 1.266 *chō* will be 38 sheaves and an interest rate of 19 sheaves a year, assuming that the debts has not increased.[25]

For seed, we have one example: in 730, the Sōnokami district office kept 20 sheaves (2 *koku*) of rice per *chō* for the rice fields of a shrine under its control and for those cultivated by the guards. This gives a yield of 7.5 to 25 times the seed depending on the soil quality, which seems a reasonable figure judging by the agricultural treatises of the modern period. In another case, in 723, a directive was sent to the provinces with a view to distributing seed to tenants by way of special farming aid. The quantity was then 2 *koku* (20 sheaves) of rice per household, which perhaps corresponded to the seed for 1 *chō* of rice fields. On a holding of 1.266 *chō* with a sown area of 0.844 *chō*, the quantity of seed therefore amounts to 16.88 sheaves of rice.[26]

Actual harvest and household reserve: an estimate

The actual harvest was subject to deductions that we have estimated as follows: 15 sheaves per *chō* for the land-tax; 15 sheaves per *chō* as interest on the compulsory loan; and 20 sheaves per *chō* for seed. Our average holding of 1.266 *chō* therefore paid a tax of 19 sheaves, interest of 19 sheaves and stocked 16.88 sheaves in order to be able to sow, the next year, 0.844 *chō*, in all 54.88 sheaves of rice. Consequently, the average household of ten people, which benefited from a plot of 1.266 *chō* of arable land and cultivated 0.844 *chō* (slightly less than 1 ha), had 155.28 sheaves a year as its own reserves for consumption, after making the various deductions to which its actual average harvest of 210.16 sheaves was subject (Table 4.1).

Consumption of rice in the 8th century

What were the food requirements of an average household? Were the reserves of 155.28 sheaves of rice sufficient to feed a household of ten people for a year? To estimate consumption, the daily rations adopted by the imperial court and public administrations may be used as a guide.

The case of rice as the staple food

We give some examples of daily rations in order to estimate the average consumption, even though our data is fragmentary and from different milieus. The provincial governments allowed for 2 *shō* (1 *shō* = 0.85 litres) of hulled rice a day for each labourer working in the public rice fields. In the capital, the court supplied the men employed in the workshops with 1.5–2 *shō* of rice, and the women who worked hemp with 0.8 *shō*. The sūtra-copying office at

Table 4.1 Production and consumption of rice in the 8th century – estimates of the volume of hulled rice

	Household A	Household B	Household C
Total area	1.266 *chō*	2.15 *chō*	0.6 *chō*
	= 1.43 ha	= 2.43 ha	= 0.678 ha
Theoretical production	379.8 *soku*	1,020 *soku*	120 *soku*
Uncultivated 1/3 of area	−0.422 *chō*	−0.72 *chō*	−0.2 *chō*
Losses 17% on 2/3	−43.04 *soku*	−97.24 *soku*	−13.6 *soku*
Actual harvest	210.16 *soku*	474.76 *soku*	66.4 *soku*
Tax (15 *soku/chō*)	19 *soku*	32.25 *soku*	9 *soku*
Interest (15 *soku/chō*)	19 *soku*	32.25 *soku*	9 *soku*
Seed (20 *soku/chō*)	16.88 *soku*	28.6 *soku*	8 *soku*
Total deductions	−54.88 *soku*	−93.1 *soku*	−26 *soku*
Household reserves	155.28 *soku*	381.65 *soku*	40.4 *soku*
(hulled rice *hakumai*)	= 7.764 *koku*	= 19 *koku*	= 2.02 *koku*
	= 660 litres	= 1,615 litres	= 171.7 litres
Annual needs	26.28 *koku*	38 *koku*	16.1 *koku*
	= 2,234 litres	= 3,230 litres	= 1,368.51 litres
Reserves/needs*	122 days	183 days	47 days
	= 30%	= 50%	= 13%

Notes: Household A: 10 people, yield of 300 *soku* per *chō* (Takigawa, 1988).
Household B: 15 people, yield of 400 *soku* per *chō* (DNK 1: 163–164).
Household C: 7 people, yield of 200 *soku* per *chō* (DNK 1: 262–263).
* Annual household reserves divided by annual needs.
Units: 1 *chō* = 1.13 hectares; 1 *soku* = 1 sheaf; 1 *koku* = 85 litres.

Nara, in 772, gave to supervising monks 2.4 *shō*, to copyists 1.2–2 *shō*, to female workers 1–1.2 *shō*, and to assistants 0.8 *shō* of rice. The temples followed these guidelines. In 941, the Tōdaiji allocated 2 *shō* of rice to monks, 1.6 *shō* to novices, 1.1 *shō* to young acolytes, and 2 *shō* of less pounded rice (*kokumai*) to the monastery corvée labourers.[27] In the workshops attached to the Ishiyamadera, artisans received 2 *shō*, office workers 1.2 or 0.8 or 0.6 *shō*, and women 1.2 or 0.8 *shō* of hulled rice. The rice was accompanied by small quantities of salt, fish and/or fermented soybean paste *miso*. In the residence of Prince Nagaya (684–729), daily rations of 2 or 1.5 or 1 *shō* of rice were given to artisans and employees in various posts, and quantities of 0.5–0.75 *shō* to messengers, domestic servants and female workers.[28]

Daily rations therefore varied significantly, depending on the grade of the recipient. Quantities higher than 0.8 *shō* probably included surpluses granted by way of allowances. From the point of view of nutrition in daily calories, 0.8 *shō* of rice supplemented by small quantities of side dishes was supposed to feed an adult for one day. For our purposes, we have provisionally used this volume, which corresponds to the case where rice was the staple food, and we have fixed at 0.8 *shō* of hulled rice per day the food ration per person, for the next part of our estimates.[29]

If an average peasant household of ten people ate rice almost exclusively, the daily quantity was 0.8 *shō* respectively for three adult men and five women. For the two children under six, we have divided the quantity by two, i.e. 0.4 *shō*, taking into account that the quantities distributed to adults in the residence of Prince Nagaya are up to 0.5 *shō*. (The quantity of 0.4 *shō* may be insufficient for adolescents and that of 0.8 *shō* excessive for old people: the first and second will strike the balance in our estimates.) In the case of 0.8 *shō* for adults and 0.4 *shō* for children, the daily needs of hulled rice for an average household of ten people amount, in total, to 7.2 *shō* (6.12 litres), or 0.72 *shō* per day per person. In this case, such a household required 26.28 *koku* (2,234 litres) of hulled rice a year. We therefore examine whether the yearly production of this average household supplied its dietary needs.

Reserves and dietary needs of an average household

The average yearly harvest of an average household amounts to 155.28 sheaves of rice, after the various deductions. This volume represents the annual rice reserve for this household. To be able to eat the rice, it had to be pounded. According to the standard of the time, one sheaf gave 1 *to* (8.5 litres) of grains with their husks and 0.5 *to* (4.25 litres) of hulled rice, pounded at 50%, i.e. 'white' rice *hakumai*.[30] A yearly reserve of 155.28 sheaves therefore gave 7.764 *koku* (660 litres) of hulled rice. If our average household had eaten rice every day, its annual consumption would have amounted to 26.28 *koku* (2,234 litres a year, with an average of 0.72 *shō* per day per person). The yearly reserve therefore supplied only 30% of its needs, corresponding to 122 days a year. Thus an average household of ten people (called Household A below), who cultivated its household fields received from the State as a public holding, could eat hulled rice for a third of the year, in terms of volume and with moderately good harvests. However, there were also households better off than those in the category of the average peasant family (Table 4.1).

The case of a well-off household

Let us take the example of an actual case recorded in the household register for Buzen for 702. This concerns a household of fifteen people headed by a certain Yoboro no suguri Umate, aged forty-four. This household lives in the district of Nakatsu (Yukuhashi-shi, Fukuoka prefecture) in Kyūshū. It consists of seven men, four women and four children. It is normally entitled to 1 *chō* 8 *tan* of rice fields, but the land survey has given it 2 *chō* 1 *tan* 171 *bu*, i.e. 2.15 *chō*

(2.43 ha).[31] We will call it Household B. Its holding theoretically produces 1,020 sheaves, if it has soils of "average" quality (400 sheaves *soku* per *chō*). It pays in principle a land-tax of 32.25 sheaves, interest on its loan of 32.25 sheaves and uses 28.6 sheaves as seed (for two-thirds of the area). The deductions thus amount to 93.1 sheaves. With 0.72 *chō* of lost areas, there remain 1.43 *chō* of sown lands. With average losses of 17% (97.24 sheaves) due to bad weather, the actual average harvest on 1.43 *chō* is 474.76 sheaves. After deducting 93.1 sheaves, the Umate household has a yearly reserve of 381.65 sheaves for its own use. This reserve gives 19 *koku* (1,615 litres) of hulled rice. With consumption of 0.8 *shō* per day per adult, the needs of this household of fifteen people amount to 10.4 *shō* (8.84 litres) a day, or 38 *koku* (3,230 litres) a year. In terms of volume of rice, the yearly reserve therefore supplies 50% of its needs or 183 days in the year. This is an example of a rather privileged household, which benefits from a plot of land larger than normal, with soils of "average" quality that can be considered quite good in the context of peasant holdings. Now let us look at an example of a disadvantaged household.

The case of a disadvantaged household

As an example of a disadvantaged household, we have chosen that of Anahobe no Okuni (called Household C below), recorded in 721 in the household register for Shimōsa, district of Katsushika (north-west of Chiba, east of Tōkyō). Okuni, a handicapped man, aged twenty-seven, is exempt from taxes in kind and from corvée labour, but not from the land-tax. He has in his household an old woman aged sixty-seven, two other adult women and three children.[32] This household of seven people is entitled to a plot of 0.6 *chō* (1.76 acres) of rice fields. Assuming that this household only has lands of rather poor quality (200 sheaves per *chō*), its plot should theoretically produce 120 sheaves. Okuni pays a tax of 9 sheaves and uses 8 sheaves for seed. He takes a loan of 18 sheaves and pays interest of 9 sheaves. The deductions therefore amount to 26 sheaves. On a sown area of 0.4 *chō*, he harvests, on average, 66.4 sheaves (after average losses estimated at 17%). After deductions, he has 40.4 sheaves left for his own consumption. The 40.4 sheaves give 2.02 *koku* (171.7 litres) of hulled rice. If Anahobe's household, with seven people, eats only rice, it needs 4.4 *shō* (3.74 litres) a day, or 16.1 *koku* (1,368.5 litres) a year. However, his yearly reserves of 2.02 *koku* can supply just 13% of his needs in hulled rice. In other words, Anahobe's family could eat rice on only 47 days a year (Table 4.1).

Good and bad years (Household A)

The estimates given above concern an average household, a privileged household and a disadvantaged household, each time in the case of an average harvest, i.e. less than the expected ideal yield, but higher than in cases of significant losses. We now deal with the question of good and bad years. In very bad years, the regulations provided for tax exemptions, if an entire region was hit by bad weather. The areas concerned were exempt from the land-tax if losses were more than half of the expected harvest; they were also exempt from the tax in kind if at least 70% of the harvest was lost; and they were exempt from corvée labour if losses amounted to 80% of the harvest.[33] We do estimates for an average household of ten people (Household A) in three cases: a loss of 60% of the harvest, a loss of 35%, and no loss. It may be recalled that we earlier deducted a coefficient of 17% to average out losses due to bad weather.

With a loss of 60% of the harvest, an average household produces 120 sheaves instead of 300 sheaves per *chō*. On a sown area of 0.844 *chō*, that gives 101.28 sheaves. This household pays no land-tax, but 19 sheaves interest on the loan, and it uses 16.88 sheaves for seed (totalling 35.88 sheaves to be deducted from production). It therefore has 65.4 sheaves left for its yearly consumption, giving 3.27 *koku* (278 litres) of hulled rice. With annual needs of 26.28 *koku* (2,234 litres), its rice reserve supplies 12.5% of them, i.e. 46 days of eating rice.

In the years when an average household loses 35% of the harvest, it does not benefit from a tax exemption. It harvests 195 sheaves per *chō* instead of 300 sheaves, or 164.6 sheaves on 0.844 *chō* sown. After deducting 54.88 sheaves, this average household has a yearly reserve of 109.72 sheaves, or 5.48 *koku* (465.8 litres) of hulled rice. With this reserve, it can meet 20.8% of its needs, or eat rice on 76 days a year.

By contrast, in good years, an average household harvests 300 sheaves of rice per *chō*, i.e. 253.2 sheaves on 0.844 *chō*. After deducting taxes and seed (totalling 54.88 sheaves), this leaves a yearly reserve of 198.3 sheaves. This volume corresponds to 9.9 *koku* (841.5 litres) of hulled rice and supplies 37.7% of the annual needs. An average household can therefore eat rice on 138 days in good years, during which it has not suffered any loss (Table 4.2). Finally there is the case of a very good year when the harvest is increased by 20%; the harvest then supplies 48% of rice needs and gives 175 daily rations.

A factor of one to four in terms of volume

It is clear that there are significant variations within a single household (number of people, areas worked, soil quality) and also significant variations depending on the external conditions (climate, natural disasters). According to Harashima Reiji, the very good and good years occur respectively twice and three times in ten years, the poor and very poor years respectively three times and twice in ten years. An average household of ten people that works lands of "average" quality (which, however, make up the majority of public holdings) can eat rice on 175 days in very good years, 138 days in good years, 76 days in poor years and 46 days in very poor years. The annual reserves of this household may therefore vary by a factor of one to four from year to year.

Moreover, the situation differed from one household to another for the same year. For a disadvantaged household with few people, a small holding

Table 4.2 Good and bad years (household A) – estimates of the volume of hulled rice

	Loss of 60% of harvest	Loss of 35% of harvest	No loss	Loss of 17% of harvest
Total area	1.266 *chō*	1.266 *chō*	1.266 *chō*	1.266 *chō*
Sown area	0.844 *chō*	0.844 *chō*	0.844 *chō*	0.844 *chō*
Harvest/*chō*	120 *soku*	195 *soku*	300 *soku*	249 *soku*
Harvest	101.28 *soku*	164.6 *soku*	253.2 *soku*	210.16 *soku*
Tax		19 *soku*	19 *soku*	19 *soku*
Interest	–19 *soku*	19 *soku*	19 *soku*	19 *soku*
Seed	16.88 *soku*	16.88 *soku*	16.88 *soku*	16.88 *soku*
Deductions	–35.88 *soku*	–54.88 *soku*	–54.88 *soku*	–54.88 *soku*
Reserves	65.4 *soku*	109.72 *soku*	198.3 *soku*	155.28 *soku*
Hulled *hakumai*	3.27 *koku* = 278 litres	5.48 *koku* = 465.8 l.	9.9 *koku* = 841.5 l.	7.764 *koku* = 660 litres
Needs	26.28 *koku* = 2,234 l.	26.28 *koku* = 2,234 l.	26.28 *koku* = 2,234 l.	26.28 *koku* = 2,234 l.
Reserves/needs*	46 days = 12.5%	76 days = 20.8%	138 days = 37.7%	122 days = 30%

Notes: Household A = an average household of ten people in the 8th century.
* Annual household reserves in hulled rice, divided by the annual needs.
Units: 1 *chō* = 1.13 hectares; 1 *soku* = 1 sheaf; 1 *koku* = 85 litres; l. = litre.

and soil of fairly poor quality, the yearly rice reserves supply its needs, on average, for only 47 days, i.e. one-eighth of the year. By contrast, a well-off household with many members and working quite a large area of fairly good-quality soils could, on average, eat rice on 183 days, i.e. half the year. Thus rice reserves could vary, in the same year, by a factor of one to four.

A factor of one to four in terms of calories

After having discussed the production and consumption of rice in terms of volume, we present below our estimates of calorific values for the early medieval period. Japanese historians have provided estimates of this type for the prehistoric and modern periods and European historians for the medieval period in Europe,[34] but there have been no systematic calculations for early Japan since Sawada Gōichi (1927) and Takigawa Masajirō (1943).[35]

We adopt, for our estimates concerning nutrition, an average amount of 2,000 kilocalories per day per person. This is the same as that of Koyama and Gotō,[36] which takes account of the practices in thirteen Asian countries in the 1970s, where the average is 2,137 kcal per day per person. We also use a conversion rate from capacity to weight of 150 kg for 180 litres (where 1 litre = 834 g) of whole rice *genmai*, a rate generally adopted for the Edo period.[37]

For the calorific values, we follow, as does Koyama Shūzō, the scale established by the Japanese government Scientific Agency.[38] Based on this data, 0.8 *shō* (0.68 litres) of hulled rice, distributed in the offices at Nara, corresponds to 567 g and 1,990 kcal per day per adult. Supplemented by small quantities of other foods, these daily rations were therefore more or less sufficient to feed the employees of the imperial court in the 8th century. However, the rural population did not have access to these quantities of rice.

Above, we estimated the production of rural households and noted the inadequacy of their yearly reserves in terms of the volume of rice. We continue with our estimates in terms of nutrition, based on the following: 1 *shō* = 0.85 litres = 709 g = 2,489 kcal. An average peasant family of ten people (Household A) who worked an average plot of rice fields, after a moderately good harvest had a yearly reserve of hulled rice (7,764 *koku*) that gave a daily ration of 0.2127 *shō* per person, amounting to 530 kcal. In calorific value, hulled rice therefore supplied 26% of the needs (2,000 kcal per person) of this household after an average harvest. In a good year, the reserve (9.9 *koku*) provided 0.2712 *shō* or 675 kcal per day per person, i.e. 34% of the calorific needs. In a very good year (with a harvest increased by 20%), the yearly reserve supplied 42% of the needs of Household A. However, in a bad year, this household had only 0.1763 *shō* or 439 kcal per day per person, or 22% of its nutritional needs. In a very bad year, there remained just 0.0896 *shō* or 223 kcal, i.e. 11% of its needs. Thus the rice ration for an average peasant family varied by a factor of one to four, depending on the type of year. Its plot of rice fields provided it with between one-tenth minimum and over one-third maximum of its needs.

In the case of a fairly well-off household (Household B), an average harvest gave daily rations of 0.347 *shō* or 864 kcal per person and supplied 43% of its needs. The quantities were higher or lower, depending on the harvest. By contrast, a not so well-off household (Household C), after an average harvest, had only 0.079 *shō*, or 197 kcal, per day per person, which supplied just 10% of its needs, with variations depending on the year. Thus, a fairly poor family ate four times less rice than a well-off family and three times less than an average family, after an average harvest (Table 4.3). As for rice production, one can estimate that in terms of calories it supplied the needs of peasant households by a factor of one to four, depending on the household situation and the type of year.

Taking the evaluations of Harashima Reiji regarding harvests,[39] the consumption of rice may be estimated as follows. Over a period of ten years, an average household could supply 34% of its rice needs over three years, 42% over two years, 22% over three years and 11% over two years. With the coefficient for these years, this gives an average of 26% or one quarter. The difference between an average year for well-off households (with 43% of needs supplied) and an average year for poor households (with 10% of needs supplied) likewise gives an average of 26%. It therefore seems reasonable to say that in the 8th century, rice provided, as a general average, one quarter of the calorific

Table 4.3 Production and consumption in calories in the 8th century – estimates in calories of hulled rice

Harvest	Reserve/year per person*	Reserve/day per person	Reserve/day = in %**
Household A: 100%	0.99 *koku*	0.2712 *shō* = 0.217 litres = 192.3 g	675 kcal = 34%
Household A: +20%	1.245 *koku*	0.341 *shō* = 0.29 litres = 241.8 g	848.7 kcal = 42%
Household A: –60%	0.327 *koku*	0.0896 *shō* = 0.076 litres = 63.5 g	223 kcal = 11%
Household A: –35%	0.548 *koku*	0.15 *shō* = 0.13 litres = 106 g	373 kcal = 19%
Household A: –17%***	0.7764 *koku*	0.2127 *shō* = 0.18 litres = 150.1 g	530 kcal = 26%
Household B: –17%	1.27 *koku*	0.347 *shō* = 0.296 litres = 246.7 g	864 kcal = 43%
Household C: –17%	0.2886 *koku*	0.079 *shō* = 0.067 litres = 56 g	197 kcal = 10%

Notes: Household A: 10 people, yield of 300 *soku* per *chō*.
Household B: 15 people, yield of 400 *soku* per *chō*.
Household C: 7 people, yield of 200 *soku* per *chō*.
* Annual reserves divided by the number of people in the household.
** Production of unhulled rice in calories per day per person, divided by 2,000 kcal: gives the percentage of calories produced as compared with daily needs.
*** –17% corresponds to an average harvest.
Units: 1 *koku* = 100 *shō*; 1 *shō* = 0.85 litres = 709 g = 2,489 kcal.

needs of the Japanese population. This figure also agrees with the estimates for the Edo period (see below, pp. 256–257).

About our estimates

Are our estimates exaggerated? Are they limited to the 8th century? Our figures in fact mainly concern the 8th century, because of the availability of sources. We now review the elements on which our estimates are based, in order to re-examine their relevance and validity in time. First of all, regarding yields, the official forecasts of the 8th–9th centuries are hardly more than half the yields of the modern period. This seems justified to us from the point of view of farming techniques including fertilization. Farming techniques did not evolve significantly between the 9th and 16th centuries.[40]

The volume of rice available for consumption varied according to the method of hulling. The 8th/9th century court defined rice pounded at 50% as hulled "white" rice *hakumai*. One sheaf gave 1 *to* (8.5 litres) of unhulled rice and 0.5 *to* (4.25 litres) of hulled "white" rice. During the same period, the court also distributed less pounded rice, called "black rice" *kokumai*. With pounding at 40%, one sheaf gave 5 litres of rice, instead of 4.25 litres. In this chapter, our estimates have been based on *hakumai* rice, but if the population ate less pounded rice, it had quantities higher than those of our estimates. Not knowing the varieties of early medieval rice, it is difficult for us to determine at what pounding ratio the rice was edible. The degree of pounding also differed according to the quality of the rice harvested. It may be noted by way of comparison that at Aizu (in 1685), good-quality rice gave 5.4 litres of hulled rice, rice of average quality 4 litres, and rice of poor quality 2.7 litres of hulled rice.[41] There were certainly variations in quality and quantity, but one can accept the official figure of around 4.25 litres of hulled rice per sheaf.

As regards land areas, we have assumed that one-third of each holding was uncultivated. This seems to us to be conservative, even for the large monastic holdings of the 8th century. However, the unused areas subsequently decreased. Kimura Shigemitsu estimates them at 20% in the 12th century.[42] The uncultivated areas included unusable lands, as well as losses due to management problems. The plots of 2 *tan* (0.6 acres) per man provided for by the imperial court in the 8th century no doubt represented the maximum of arable land per adult, and also included the plots of young people from the age of seven and those of men absent from their household. For the modern period one accepts an average of 3 *tan* (0.78 acres) per person,[43] but in the early medieval period it is probable that households could barely manage 2 *tan* per person in terms of available labour. It is therefore hardly conceivable that tenant-farmer households would have increased their rice-growing areas by clearing land or taking on leases, as in so doing they increased their tax obligations. Yet one can imagine that farmers would have tried to increase their grain reserves by growing dry cereals that were not subject to tax. Up to the 12th century, any increase in rice-growing areas penalized the tenant-farmer. It may be assumed that the uncultivated lands gradually decreased, but that the proportion of rice-growing areas per tenant-farmer (not in absolute terms) remained more or less stable up to the 12th century.

There was, nevertheless, a gradual change in agrarian organization after the 8th century. The figures for land areas that we have given are based on the system of public holdings and the distribution of plots of land by the government to the population. However, these distributions ceased in the 9th century and gave rise to a more complex structure of aristocratic holdings on the one hand, and holdings of public lands on the other, both managed by agents who held cultivation rights and/or taxation rights. At the same time, there emerged a class of intermediaries at different levels, consisting of wealthy cultivators and local notables. Thus there came into being a class of well-off cultivators. These wealthy peasants, who the texts tell us owned 10–20 *chō* of

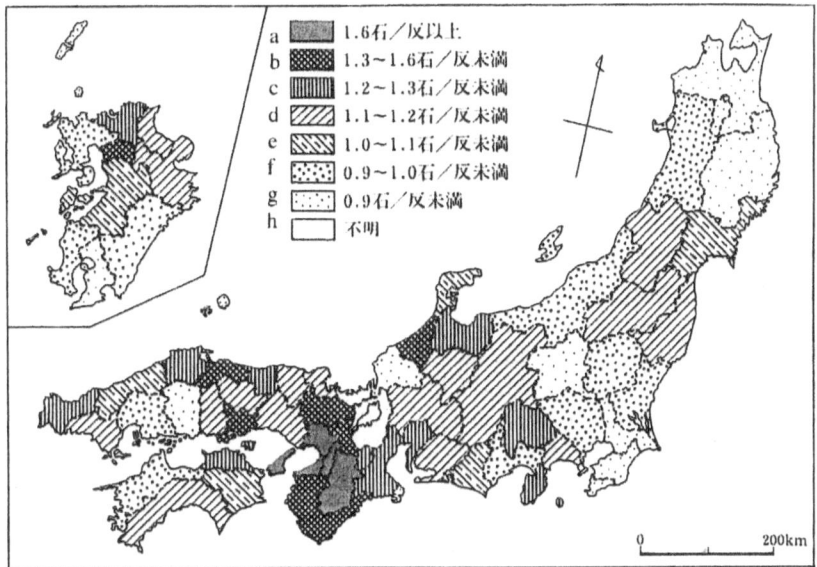

Map 4.1 Yields in each region in 1877
Notes: These are the yields in hulled rice *genmai* per *tan* (0.26 acres) in *koku* (180 litres): (a) +1.6 *koku*; (b) 1.3–1.6 *koku*; (c) 1.2–1.3 *koku*; (d) 1.1–1.2 *koku*; (e) 1–1.1 *koku*; (f) 0.9–1 *koku*; (g) less than 0.9 *koku*; (h) unknown.
Sources: *Meiji jūnen zenkoku nōsan hyō*, 1877; after Watabe Tadayo *et al.*, *Ine no Ajiashi*, Shōgakukan, 1987, p. 300.

rice fields, employed very large numbers of individual tenant-farmers. With the transfer of rights over plots of land, under different guises, to intermediaries and various agents, the number of large holdings gradually increased. At the same time, the distribution of the harvest, it seems, moved in the direction of a growing gap between the rights holders and individual cultivators.

The tax system evolved with the change in agrarian organization. In the 8th century, the roughly equal proportion of the land-tax and interest on the compulsory loan is discernible, at the provincial level, in the yearly account registers for the provinces.[44] This rice tax (to which other taxes were added), amounted to around 30 sheaves per *chō* (land-tax and interest included). It was in fact fixed in relation to the land area, and therefore variable in relation to the harvest. For an "average" rice field, with a yield of 400 sheaves, the rice taxes of 30 sheaves represented 7.5% of the harvest; for a "mediocre" rice field (with a yield of 300 sheaves), these taxes represented 10%; for a "poor" rice field (producing 150 sheaves), the rice taxes were equivalent to 20% of the harvest. This tax rate is very low, compared with the practices of medieval societies that often taxed at least one-third of production. In Edo Japan, taxes were as high as four-tenths to half of the harvest, depending on the region. In the 8th century, the rice taxes of 30 sheaves per *chō* exclude the accumulated debt and other tax loans. After the 8th century, the tax system changed. In the 10th century, taxes of 3 *to* of rice per *tan*, i.e. 6 sheaves per *tan* (0.3 acres),[45] twice as much as in the 8th century (3 sheaves per *tan*), were considered normal. In some regions, the taxes were even higher. For instance, in 988, the local notables of Owari complained about the tax of 13 sheaves per *tan* and put the normal rate at 8, 9 or 10 sheaves per *tan*.[46] Viewed in this way, our estimates of the taxes deducted do not appear to be exaggerated, but rather conservative. It is possible that household reserves were in fact subject to more taxes. Generally speaking, in the medieval agrarian economies, production was siphoned off from the farmers to the élites (nobility, wealthy cultivators) through the tax system and the rate was between one-third and one-half of production.

One must also take account of regional differences. According to the national register of 1877 (see Map 4.1 and below, pp. 251–253), yields depended on the region. The best yields at that time were achieved only in the central region, that of the old capitals of Nara and Heian (Kyōto), and in four other provinces. It may have been the same in the early medieval period, since the imperial court carried out large-scale irrigation works in the central region from the 7th century. If so, the fairly high yield of 400 sheaves per *chō* that we have adopted in our estimates for the well-off household (Household B) was no doubt relatively rare, outside the central provinces. The yield of 300 sheaves that we have adopted for the average household (Household A) may actually reflect the average for the provinces of early medieval Japan. We should also keep in mind that there are significant variations within each region. For instance, in Aizu, yields varied by a factor of one to five in 1685 (see below, pp. 254–255).

In view of these considerations, it seems to us that the figures we have proposed do not underestimate the importance of rice in the early medieval Japanese diet. According to our estimates, rice supplied on average a quarter of the nutritional needs. This average is not, however, representative of the entire population of early medieval Japan; it is only an intermediate figure between the different scenarios, as there was a significant disparity between the well-off classes and those of lower economic status, and between the regions suited to rice growing and those that were not. It is possible that the élites had rice reserves equivalent to half their needs and that rice made up only a quarter on average of the food of the tenant-farmer population.

In order to evaluate the data discussed above, we present, by way of comparison, the estimates for the prehistoric period, then the modern period.

Yields and consumption of rice in the prehistoric period

Rice has been a focal point of interest for Japanese archaeologists since the mid-20th century. We give below a brief overview of past research. Terasawa Kaoru, a specialist in the Yayoi period (in the 1980s, dated from the 3rd century BC to the 3rd century AD), whom we have mentioned previously, bases his arguments, as we do, on the yields provided for by the regulations of the 8th–9th centuries.[47] He points out that yields vary by a factor of one to three in the early medieval period, and by a factor of one to two in the modern period (17th–19th centuries), depending on the soil quality. Based on the calculations done by Sawada Gōichi in 1927, Terasawa converts the yields expressed in sheaves to kilogrammes of unhusked rice. After deducting seed, he obtains the following weights per *tan* (1/10 *chō*):

- good rice field *jōden*:   50 sheaves = 105.75 kg
- average rice field *chūden*:   40 sheaves = 84.625 kg
- mediocre rice field *geden*:   30 sheaves = 63.5 kg
- poor rice field *gegeden*:   15 sheaves = 31.75 kg

Terasawa compares these quantities of unhusked rice with the production in the 1970s for a number of Asian countries, which is, per *tan* (here 1 *tan* = 0.26 acres): in Assam, India: 75.4–150 kg; in the Philippines, on Luzon, region of Kauayan: 130 kg; still on Luzon, but in the region of San Valez: 19–62 kg; in Indonesia, on Java in the region of Pamkasa: 109.2 kg; in northern Thailand: 40–70 kg; on Taiwan, in the village of Qulucun: 125 kg; on Okinawa, in the north-west of Iriomote island: 87–100 kg. Terasawa thinks that the yields in early medieval Japan were not so very different from these countries, with 40–100 kg per *tan*. (It may be noted that the average yield in present-day Japan is 500 kg per *tan* or 0.26 acres.)

Terasawa also draws on the results of experimental archaeology. Based on the rice found on stalks and in ears in twelve sites, he thinks that the length of stalks, size of ears and average number of grains per ear correspond to those of rice grown today in a paddy left lying idle for one or two years. Experimental crops have yielded 93 kg (7.5 *to*) of unhusked rice per *tan* (here 1 *tan* = 0.26 acres) in a paddy abandoned for one year and 50 kg per *tan* in a paddy abandoned for two years.

For the Yayoi period, Terasawa has given an estimate for three archaeological sites: Tsushima in Okayama (3rd century BC–2nd century AD), Dainakanoko-minami in Shiga (1st century BC–1st century AD) and Toro in Shizuoka (2nd–3rd centuries). He bases this on five people in each pit-building, giving a total population of ten people in the first two sites and sixty people in the Toro site. He estimates the daily consumption of hulled rice at 3 modern *gō* (1 *gō* = 18 cl), i.e. 54 cl per person, and takes the example of yields from poor, mediocre and average soils (according to the 8th/9th-century norms). By comparing the data for the areas of rice fields uncovered with the number of pit-buildings and the putative yields, namely: (a) 31.75 kg, (b) 63.5 kg and (c) 84.62 kg per *tan*, Terasawa obtains the number of daily rice rations for the inhabitants of the three sites. After deducting seed, Table 4.4 can be drawn up.

Terasawa infers from these calculations that rice fields did not provide all the starch needs of people in the early phase of Yayoi (5th–2nd century BC), that they supplied roughly half of them in the middle phase (first centuries around our era) and at least 70% in the 2nd–3rd centuries. (Terasawa's estimates take account of the deduction for seed, but leave out possible taxes and climatic variations.)

Table 4.4 Estimated production of daily rice rations in three sites

Site	*Area* tan	*Population people*	*Yield per* tan	*Days of food*
Tsushima	1 *tan* = 0.26 acres	10	(a) 31.75 kg (b) 63.5 kg	6 15
Dainakanoko	9.27 *tan* = 2.4 acres	10	(a) 31.75 kg (b) 63.5 kg	58 136
Toro	71.154 *tan* = 7.06 ha	60	(a) 31.75 kg (b) 63.5 kg (c) 84.62 kg	74 174 241

© Verschuer, *Rice, Agriculture, and the Food Supply in Premodern Japan*, Routledge, 2016.

Terasawa also points out that quantities of still immature rice have been discovered. These make up 10% of the volume found at Itazuke (800–500 BC) and 27% in the site of Ayaragi-gō (3rd century AD). At the site of Ikegami the proportion of immature rice is 22% for the 1st century BC, 42% for the 1st century AD, then 29% for the end of the 1st and the 2nd century. Terasawa infers from this that harvests often amounted to no more than half the expected yields. In his view, Yayoi people still ate mainly roots, acorns and dry cereals. He asks whether it was in fact only from the 2nd century, and only in certain regions, that the inhabitants of the Japanese archipelago began to eat rice. This also explains why, in half of the Yayoi sites, the remains of rice are associated with those of millets, wheat and barley.

In Terasawa's view, rice, introduced to Japan via the south-west of the archipelago, gradually spread to a certain extent until the 3rd century AD, in the west and centre of the country, but the population no doubt remained dependent on other cereal crops. It is for this reason that Terasawa rejects the widely held idea of a Yayoi culture characterized by rice growing.

The archaeologist Kuraku Yoshiyuki defends the opposite point of view by referring to the site of Hashibara in Nagano, in a region that is not, *a priori*, suitable for rice growing. In pit-building no. 59 dating from the 3rd century AD were discovered 350,000 grains of rice (estimated number), 2,100 grains of millets and perilla, and 130 soybeans. With two other buildings dating from the medieval period (a thousand years later), one yielded 1,685 grains of rice and 838 other grains, the second 333 grains of rice, 123 other grains, 13 beans and 56 acorns.

For Kuraku Yoshiyuki, this means that rice played a more important role in the diet than the other cereals. Moreover, many jars unearthed in the sites of Hyakkengawa-Harōjima and Jō (3rd century) in Okayama showed marks of rice grains. As for yields, Kuraku accepts the reasoning of the archaeologists at the Hyakkengawa site, which is as follows: the fossilized rice field on this site yielded on average 400 rice root-stocks per *tsubo* (3.3 sq. m). In the case of one ear per head and 12–50 grains per ear, the yield was 40.2–112.2 kg of hulled rice per *tan* (0.26 acres), or double if each head bore two ears. This corresponds to the single or double yields calculated by Sawada Gōichi and Terasawa Kaoru according to the norms of the 8th–9th centuries.[48]

Sahara Makoto thinks that it is pointless to compare the volumes of plant remains, owing to major differences in the size and fragility of cereal grains. While noting that the discoveries of grains do not reflect the accumulated stocks but the remains of stocks, he considers the site of Hashibara to be a representative indicator of household rice reserves. For Sahara, rice cultivation moved very rapidly from Kyūshū up to the north-east of the country (Aomori) in the space of one to two centuries, and therefore spread rapidly throughout the archipelago.[49]

Archaeologists have opposing views on the question. Some consider rice to be the principal source of calories; others think that it alone could not have fed the population. These opinions are based, as Negita Yoshio has pointed out, on the chance discoveries at four sites that have yielded statistical data. Generally speaking, rice attracts far more attention from archaeologists than the other cereals, owing (among other things) to the number of sites. Around two hundred rice-field sites and a dozen dry-field sites had been identified up to 1988.[50] In fact rice-field ridges and canals leave traces in the soil, whereas dry fields disappear, except when buried after a volcanic eruption. To the archaeological viewpoints outlined above have now been added more recent studies, as well as bio-botanical and ethnological research. We have presented these results in our Introduction.

Since the mid-20th century, historians too have shown a great interest in rice, about which there is a wealth of information in the 8th–12th century texts. Sawada Gōichi (1927) and Takigawa Masajirō (1943) did calculations and estimated that rice fields provided between one-half and three-fifths of the dietary needs of the Japanese population at that time. Naoki Kōjirō (1968) and others think that rice supplied 80–90% of them. They do not deduct taxes and seed. Quoting the figures given by Takigawa, Ishii Susumu thinks that rice supplied half the needs in early medieval Japan. However, all these scholars base their arguments on the cultivation of 100% of the land surfaces, without any natural disasters, and Naoki relies on very high yields of 400–500 sheaves per *chō*.[51] It is as a result of these traditional estimates that we have drawn up our own, which are presented below.

Yields and consumption of rice in the modern period

Before summarizing the estimates put forward by several historians for the modern period, we give the example of a holding of the Genroku era (1688–1704). It concerns a small farm, described by Ōhata Saizō (1642–?) in his work called *Saizōki* or *Jikata no kikigaki*. Appointed an intendant of the local administration in the village of Kamuro in Kii (Hashimoto-shi in Wakayama prefecture), Saizō reports the case of a family of ten people, who work 2 *chō* (2 ha) of rice fields and 5 *tan* (1.3 acres) of other fields. Located in the fertile region of Kii, this holding produces 45 *koku* (8,100 litres; 1 *koku* = 180 litres) of rice, 40 *koku* (7,200 litres) of wheat or barley and 4 *koku* (720 litres) of buckwheat. After payment of a yearly tax of 26.04 *koku* (4,687.2 litres) of rice, it is left with 18.96 *koku* (3,412.8 litres). The holding buys fertilizer for the rice at a cost of 250 *monme* of silver, fertilizer for the wheat and barley for 200 *monme*, tools and other materials for 107.4 *monme*, and outlays a further 1,409.1 *monme* for management. The costs total 1,966.5 *monme*. The harvests are therefore sold in exchange for cash. Since the harvest of the three cereals is worth 1,988 *monme* after tax deductions, the yearly profit amounts to 21.5 *monme* of silver or 4 *to* (72 litres) of rice. In other words, from a harvest of 45 *koku* of rice, 4 *to* or 0.8% is left. This household therefore had to feed itself with other crops. The situation described is comparable to that in the mountain villages of Aichi about which the *Hyakushō denki* (1683) states: "There is nothing to eat if one does not cultivate the millets *awa, kibi* and *hie*."[52]

In fact, the *Saizōki* mentions dry crops such as barley, wheat, foxtail millet *awa*, barnyard millet *hie*, common millet *kibi*, buckwheat, soybeans, red beans, yams, melons and other vegetables. This work also notes the diet of the farming family of ten people: in the morning and evening it eats dumplings made of common millet in a gruel, or an average quantity of 8 *shaku* (144 centilitres) of millet per person, and at midday each person eats 2.5 *gō* (0.45 litres) of "white wheat", i.e. barley. This family eats rice around twenty-six times a year, namely at New Year, during religious festivals and during festivals established by the government.

The household in Kii mentioned by the *Saizōki* survived thanks to dry cereal crops. This type of management originated in the land survey *taikō-kenchi* instituted by Toyotomi Hideyoshi in 1594, at the time of his agrarian reforms. The land survey provided for yields per *tan* (0.26 acres) fixed in hulled rice called "whole rice" *genmai*, according to the soil quality.

Land survey *taikōkenchi* of Hideyoshi in 1594 in hulled rice *genmai*:

- good rice field *jōden*   1.5 *koku*
- average rice field *chūden*   1.3 *koku*
- mediocre rice field *geden*   1.1 *koku*
- field of good quality *jōbatake*   1.3 *koku*
- field of average quality *chūbatake*   1.1 *koku*
- field of mediocre quality *gebatake*   0.9 *koku* (1 *koku* = 174 litres)

The tax paid to the authorities was equivalent to four-tenths or to half of the expected production and dry fields were also taxed in rice, even though these fields did not produce rice but other cereals. It was therefore from the production of the rice fields that farmers had to put aside an additional share to pay the dry-field tax. In fact, in the opinion of Furushima Toshio and other historians, most of the rice went in tax and the rest of the rice harvest was sold to manage the farm. Therefore little rice remained for household consumption.[53]

"Whole rice" and "white rice"

We may note that hulled rice or "whole rice" *genmai* in the Edo period was pounded in the same ratio as husked rice or "white rice" *hakumai* in the 8th century, that is, to 50%. The difference can be explained in the following way. In the 8th century, rice sheaves were threshed *dakkoku* to obtain unhulled rice ('whole rice') *momi*; the unhulled rice was then hulled to 50% and this product was called husked rice or 'white rice' *hakumai*. However, after 1624, this product was called hulled rice or 'whole rice' or 'brown rice' *genmai*. In the Edo period, rice hulled by more than 50% was called husked rice or 'white rice' *hakumai*. Therefore Edo-period hulled 'whole' rice roughly corresponds to 8th-century husked 'white' rice. In the 8th century, rice hulled by less than 50% was called 'black rice' *kokumai*. In the Edo period, rice hulled by 60% or more was called husked rice or 'white rice' *hakumai*.

Yields

By way of reference, the forecast yields of the 8th century[54] may be compared with those of the official land survey of 1594, taking account of the difference in the units of measure. Yields are expressed per *tan* (1 *tan* = 0.3 acres) in the 8th century and per *tan* (1 *tan* = 0.26 acres) in the modern period; 1 *koku* equals 85 litres in the 8th century and 180 litres in the modern period (except in 1594, when 1 *koku* = 174 litres); yields are expressed in sheaves in the 8th century and in hulled "whole" rice in the modern period. For comparative purposes, we have converted modern yields into unhulled rice, in the generally accepted ratio of 0.5 *koku* of hulled rice for 1 *koku* of unhulled rice. This gives Table 4.5, showing yields per *tan* in 1594.

We have also converted the 8th-century yields in unhulled rice (1 sheaf *soku* = 1 *to*; 1 *to* = 8.5 litres) and in modern *tan* (1 old *tan* = 1 modern *tan* x 0.84). The results are shown in Table 4.6 below.

It is clear that the forecast rice yields, in the 8th century, barely reached 56–68% of yields in the modern period, when technical advances enabled agriculture to make great progress. Yet it should be remembered that in both cases these are government forecasts and not actual production.

Rice production in the modern period is shown in two other documents. The "Register of agricultural products for the country in 1877", *Meiji jūnen*

Table 4.5 Forecast yields in unhulled rice per *tan* after the land survey of 1594

Rice field	Hulled rice	Unhulled rice per tan
good	1.5 *koku*	3 *koku* = 522 litres
average	1.3 *koku*	2.6 *koku* = 452.4 litres
mediocre	1.1 *koku*	2.2 *koku* = 382.8 litres

Table 4.6 Forecast yields in unhulled rice per *tan* in the 8th century

Rice field	Per *old* tan	Conversion in mod. tan	Comparison 1594: 8th c.
good	50 sheaves = 50 *to* = 425 litres	357 litres	68.4%
average	40 sheaves = 40 *to* = 340 litres	285.6 litres	63%
mediocre	30 sheaves = 30 *to* = 255 litres	214.2 litres	56%
poor	15 sheaves = 15 *to* = 127.5 litres	107.1 litres	

© Verschuer, *Rice, Agriculture, and the Food Supply in Premodern Japan*, Routledge, 2016.

zenkoku nōsan hyō, gives figures for each province and the *Aizu hiyō shūzei bugai kyūjo hō bassui*, a compendium of taxation for the Aizu domain dating from 1685, notes the yields according to the soil quality.[55] The 1594 land survey reflects the theoretical yields at the national level, the Aizu document shows the yields at the level of one region, and the 1877 register shows the average production at the level of each region. A map has been drawn up based on the 1877 register (see Map 4.1), enabling the regions with good, average and poor production to be identified. Converting the *koku* in the three documents to litres (1 *koku* = 180 litres) gives Table 4.7 below.

The document of 1877 shows regional differences, which are evident on Map 4.1. The best yields, ranging from 1.3–1.6 *koku* of rice, or more, were achieved only in the central region (that of the old capitals of Nara and Kyōto) and in four other provinces. The mountainous areas of Honshū, Shikoku and Kyūshū gave at best average yields of 1.1–1.3 *koku*. Kantō, Tōhoku and the south and west of Kyūshū were the regions where yields, often less than 1 *koku*, were the lowest. Regional differences between the maximum yields of more than 1.6 *koku* and the minimum yields of less than 0.9 *koku* are roughly double. Overall, the margin of 1–1.2 *koku* of rice per *tan* was geographically the most widespread. This figure corresponds to that of the mediocre rice fields in Hideyoshi's land survey of 1594. In other words, the survey shows a tendency to over-estimate yields, and the good and average rice fields were in reality far less

Table 4.7 Yields in hulled rice *genmai* per *tan* based on three documents

Yield	Survey of 1594	Regions in 1877	Aizu in 1685
good	1.5 *koku* = 261 litres	1.6 *koku* = 288 litres	1.95 *koku* = 351 litres
average	1.3 *koku* = 226.2 litres	1.2 *koku* = 216 litres	1.55 *koku* = 279 litres
mediocre	1.1 *koku* = 191.4 litres	0.9 *koku* = 162 litres	0.6 *koku* = 108 litres
poor			0.37 *koku* = 66.6 litres

© Verschuer, *Rice, Agriculture, and the Food Supply in Premodern Japan*, Routledge, 2016.

common than the mediocre ones. We have observed a similar situation concerning the forecast yields for rice fields in the 8th century.[56]

The variations were important both at the interregional level and within each area, taking the case of the Aizu region. We have found, for the year 1685, a margin ranging from 0.37–1.95 *koku*, i.e. up to five times as much. This significant variation reflects the appearance of the Aizu landscape with its fertile intramountainous basins and steep slopes ill suited to rice growing. This variation also takes account of the different qualities of rice. Good rice fields gave 3 *shō* (5.4 litres) of hulled rice *genmai* per sheaf, average rice fields 2.5 *shō* (4.5 litres) per sheaf, mediocre rice fields 1.7 *shō* (3 litres) per sheaf and poor rice fields 1.5 *shō* (2.7 litres) per sheaf. In other words, one could obtain on the same areas up to twice as much rice ready for consumption.[57] Let us recall, by way of comparison, that in the 8th century the norm was 0.5 *to* (4.25 litres) of husked rice per sheaf.[58] In this respect, the good-quality rice of the early medieval period is comparable to the rice of near-average quality in Aizu in the 17th century.

The diet in the Edo period

From the 17th to 19th century, the rural population ate little rice and many other cereals. Such, at least, is the general opinion based on certain texts and the tax system of the time. Earlier we gave an example of this situation noted in the *Saizōki*. However, according to the geographer-historian Arizono Shōichirō, this is a cliché that gained currency in Japanese national education after the Second World War, under the influence of the interpretations of certain historians who carried some weight at that time. For his part, Arizono gives estimates based on the figures known for 1721: a Japanese population of 29 million people and rice-growing areas of 1.64 million hectares.[59]

Arizono's calculations suggest that the population of tenant- farmers must have eaten at least 0.34 *koku* (51 kg or 61.2 litres) of whole rice *genmai* per year per person, and, in his view, even 0.5 *koku*. For the élites, namely the privileged classes of well-off farmers, high- and middle-ranking warriors and merchants, consumption amounted at that time, according to Arizono, to 1 *koku* (180 litres) of rice per year per person. Arizono studied the agricultural treatises in search of data relating to consumption and was able to find confirmation of his estimates in several documents. These are texts written by cultivators between the 18th and early 19th century, describing the conditions of life in their household. For instance, in 1739, the family of a wealthy Shinano landowner, who employed twenty-four farm workers, ate annually, per person, 1.3 *koku* of whole rice, 0.58 *koku* of crushed wheat and 0.6 *koku* of common millet. In this case and in five others identified by Arizono, rice accounted for 50–60% of the food, the rest being made up of barley, barnyard millet, foxtail millet, buckwheat and soybeans, with differences according to the regions. These examples, which show quite a high yearly consumption of around 1 *koku* of rice, concern a class of well-off cultivators.

Arizono mentions other cases concerning less privileged groups. For instance, during the first half of the 18th century, the inhabitants of a region in Ise ate in spring and summer (before the rice harvest), twice a day, a mixture of rice boiled with wheat or barley (*katemeshi*), accompanied by dumplings *dango* made with lily bulbs, taro, radishes, egg-plant, cowpeas, wheat or sorghum (*tōkibi, morokoshi*, Sorghum bicolor Moench). In autumn and winter, the villagers ate steamed rice of inferior quality (not good enough for taxes) in the morning and dumplings in the evening. The inhabitants of another village in Iyo (Ehime), ate rice only on the first three days of the year; from 1804 up to 1818, they lived all year long on a soup *zōsui* made with wheat or sorghum flour, accompanied by turnips and radishes cooked and mixed with *miso*. The poorest people mixed, in a soup, one part flour and three or four parts rice or wheat glumes. In another region, near Kawasaki, the mountain inhabitants did not have rice even at New Year. In 1721, they lived on gruels *kayu* of foxtail millet, barnyard millet, wheat and barley, supplemented by turnip and taro leaves, or leaves of soybeans and cowpeas. They "ate all kinds of other leaves and never saw cereals". In three other cases mentioned by Arizono, farming families lived on much the same foods. However, Arizono thinks that the postwar Japanese curriculum developers have downplayed the peasant diet. He quotes the "Farming Statistics", *Nōji tōkei hyō*, of 1861, and from it infers the following proportions concerning the average diet at the national level: rice 47%, wheat and barley 28%, other cereals 19%, sugar-cane 3%, other 3%. For Arizono, rice therefore supplied, at this time, half the dietary needs, even for the majority of the population that did not belong to the privileged classes.

Calorimetric estimates for the Edo period

We have calorimetric estimates for the regions of Hida, Ibaraki and the country as a whole. The calculations done by Matsuyama Toshio, based on information in the monograph *Hida gofudoki* dating from 1873, showed that the population of this region, on average, had a calorie intake in the following proportions: 35% rice (maximum figure), 36% acorns, horse chestnuts, sweet chestnuts, etc., 17% barnyard millet, 12% other cereals, tubers and soybeans. In this case, it was a mountainous area in present-day Gifu prefecture, with low rice production.[60]

Nakanishi Ryōtarō formulated estimates based on the statistics for the Ibaraki prefecture in the years 1909–1913.[61] He found significant regional variations within this prefecture. Depending on the locality, 0.74–1.2 *koku* of hulled rice *genmai* and 0.14–0.6 *koku* of barley (and a little wheat) were eaten per year per person. Among the cereals, the proportion of rice was higher than 60% in some villages and exceeded 70% in others. The average consumption for Ibaraki was 0.97 *koku* of rice per year per person, 0.35 *koku* of barley and 0.06 *koku* of other cereals. This represented a proportion, by volume, of 66.5% rice, 23.8% barley, 4.3% other cereals and tubers, and 5.4% other foods. In terms of calorific values, rice supplied 60.7% of calorie consumption, barley 16%, tubers 6.5% and beans 4.8%.

Kitō Hiroshi based his estimates on the "Register of agricultural products for the country in 1874", *Meiji nana nen fu ken bussan hyō.*[62] According to his calculations, the Japanese at that time ate 431 g of starches per day per person, which included 61% rice, 17% wheat and barley, 17% tubers, 4% other cereals and 1% beans. These foods provided in total 1,670 kcal per day per person, an insufficient amount, even if adult men were then only 156.5 cm tall on average.

Through these various examples from the modern period, two extremes are discernible. On the one hand there was a peasantry of fairly poor people, most of whom possibly lived in mountainous regions and did not often taste steamed rice, and on the other hand the better-off social groups who ate up to 1 *koku* of rice a year, which accounted for 50–60% of their diet. Individuals belonging to the well-off groups were perhaps more numerous than might be thought. In all cases, rice was always eaten with other foods, except on certain festival days. Thus, the well-off groups of the population in the modern period would have eaten per year per person 1 *koku* (180 litres) of hulled rice and 1 *koku* of other cereals, or 360 litres of cereals in total. Converted to a daily ration, this represents 0.274 *shō* (0.5 litres) of rice and 0.274 *shō* (0.5 litres) of other cereals.

How much rice did they eat?

The answer to this question lies between two extremes, if one leaves aside the aristocratic and military élites where rice was eaten all the year. Rice provided half the calorific needs of the well-off classes, or more when harvests were higher than the average; but the poorest farming households ate only 10% of rice in normal times, and even less in very bad years. As for early medieval Japan, with an average proportion of a quarter of rice in the diet, one may question the role of this cereal as the staple food. A quarter of the calories necessary for survival does not justify making rice a staple food, and we do not think that rice was the staple food of the Japanese between the 8th and 12th centuries, with the exception of the officials and employees in service at the imperial court. Nor does this justify the notion of a "structural hunger for rice" for the population that did not belong to the élite, a concept formulated by some ethnologists. As we have seen in Chapters 1 and 2, for the population of the country, rice growing was merely one form of agriculture and rice was only one of the cereals eaten.

As in the Neolithic on the one hand, and the modern period on the other, the gathering of the self-sown fruits of nature and cereal as well as vegetable polyculture made an important contribution to nutrition in the early medieval period. Neolithic people had access to around forty edible plants, among them starches such as nuts, seeds and roots. In some cases walnuts and acorns provided a third of the calorific needs; in other cases it was fish; in still others it was game. The differences lay in the natural environment: upland, coastal, etc. For the site of Kōsaku in Chiba, the proportion of plants in the diet has been estimated at 80%; for Sannai-Maruyama, it is thought that (sweet) chestnuts were a very important calorific resource. In certain regions, Neolithic people had to subsist exclusively on the products of the forests and rivers. Yayoi people also ate nuts and seeds, as shown by the acorn silos discovered in the rice-growing sites.

In premodern Japan, wooded mountains covered three-quarters of the land surface, and the consumption of tree fruits and nuts is attested by the written sources. Some nuts have a calorific value equal to or higher than that of rice. These are *kaya* (torreya), walnuts and horse chestnuts (see Table A.5 in the Appendix). Up to the modern period, foraging supplied food; starch was extracted from the roots of bracken and pueraria, and taro and yams were eaten. Even in the 19th century, chestnuts and acorns still provided, in certain regions such as the mountains of Hida and Ishikawa, between one-fifth and one-quarter of the food. Many other plants, discussed in Chapter 3, supplied the early medieval Japanese with vitamins, carbohydrates and minerals. The gathering of wild products was therefore an important element in their diet.

Crops were another important contribution. If, in the 18th–19th centuries, the well-off classes ate the same quantities of dry cereals as rice and the poorer classes survived mainly thanks to these cereals, the same can be said for the early medieval period. The calorific value of dry cereals is not very much lower than that of rice. Wheat, barley, buckwheat, common millet, foxtail millet and barnyard millet have almost 300 kcal per 100 g (compared to 351 kcal for rice). Other cereals can be added, such as wild rice *komo*, glyceria *mino* and wild barnyard millet, as well as graminae that are harvested and not cultivated, the calorific values of which are, however, unknown. One must also mention the oil seeds, such as sesame, perilla and hemp, with values close to 400 kcal. In the early medieval texts, beans (soybeans, red beans, cowpeas) come immediately after cereals. They are high not only in calories, but also in protein and calcium.

In the 8th century, dry fields were probably very numerous. Small tenant-farmers grew vegetable and cereal crops at the same time. As for the large temple holdings, the registers show that they were dotted with dry fields scattered among the rice fields. Dry crops were encouraged by the government, notably foxtail millet. The coexistence of dry and irrigated crops did not change throughout history. Statistical data from the Edo period has shown that dry crops occupied a significant amount of farmland and that the areas devoted to them even exceeded the irrigated rice fields in some regions (see Chapter 1, pp. 77–79).

However, the yields for dry cereals were much lower than those for rice, both in terms of seed and land areas. One could count on a rice harvest 7.5 to 25 times the seed, or an average of 15 times the seed in the case of average yields of 300 sheaves per *chō*. For dry cereals, the harvest was only two to five times the quantity of seed. As regards areas, a rice field produced four times, five times, or more calories than dry fields.[63] In other words, those who, in the early medieval period, wished to supply half their calorific needs by dry cereals, required, in addition to their rice fields, five times more land. There was a shortage of arable soils. The government did not, in principle, distribute lands intended for dry crops, nor did it levy taxes on them. Moreover, cereal cultivation was a time-consuming activity. Tillings, weedings and manurings required considerable labour. It should be remembered that the collection and transportation of green manure for 1 *chō* of land represented the work of an adult for one to two months. Except for irrigation works, the time and effort needed was the same for dry fields as for rice fields, whereas labour resources were limited for the small tenant-farmer plots.

Nonetheless, there was a solution that made it possible to offset the shortage both of arable soils and labour: swidden-field crops. These did not require tilling or fertilizing and few weedings. The Japanese landscape provided ample lands and natural conditions favourable to shifting cultivation. We think that swidden farming gave cultivators major possibilities, alongside permanent crops, and we also think that the Japanese could hardly have survived without it in the mountainous regions less well suited to rice growing, namely in three-quarters of the country.

One can therefore give an overall picture of the calorific resources of the Japanese showing rice cultivation, permanent dry crops, shifting dry crops and foraging, as well as hunting and fishing, emphasizing one or other of these practices, depending on the region. These resources, in normal times, should have been sufficient to feed the population. Throughout Japanese history, resources have fluctuated between the Neolithic practices, as regards foraging, hunting, fishing and swidden farming, and those of the Edo period as regards cultivation practices. As for the staple food, it is necessary to take account of the harvests of all permanent and shifting crops, i.e. all the dry and irrigated cereals and beans. These foods appear in the written sources under the name "five grains". It is therefore the "five grains" – and not exclusively rice – that constituted the staple food in premodern Japan.

Notes

1 Braudel (1979: 1: 120).
2 Tsuboi Hirofumi (1979); Tsuboi in Mori Kōichi, ed. (1986: 351–374); Tsuboi (1986: 296–341).
3 Amino Yoshihiko, "Shōen kōryōsei no keisei to kōzō", in *Taikei Nihonshi sōsho* vol. 6, 1973, reprinted in Amino (1991: 13–139). Kimura Shigemitsu, "Chūsei seiritsuki ni okeru hatasaku no seikaku to ryōyū kankei", *Nihonshi kenkyū* 180, August 1977, reprinted in Kimura (1992: 181–231).
4 Terasawa Kaoru *et al.* (1981: 1–130). See below, note 47.
5 Amino (1980, 1986: 17–28; Amino in Tanigawa Kenichi (1986: 60–73), and in Amino (1990: 44–59); Amino quotes Kuroda Hideo (1983: 50, 1984). Amino and Ishii (2000), with the documents quoted by Amino, are reproduced and corrected, pp. 34, 35, and see the opinion of Ishii Susumu (2000).
6 Harada Nobuo (1993) (a collection of articles by this author), esp. pp. 133–151.
7 Tanigawa Kenichi *et al.* (1993: 96–118); Amino (1986: 61, 1990: 45). Shiraishi Akiomi (1988; 1994); Nomoto Kanichi (1984).

8 Sahara Makoto (1995: 107–132); Sahara (1996: 227–233). He quotes Koyama Shūzō and Gotō Yoshiko (1985), but with chronological differences. Koyama (1985: 475–477) places the third wave of demographic growth beginning "from the Heian period", or in the year 1000. To explain the socio-economic changes from 1000, Koyama gives no explanation or reference. He concludes, p. 497, that rice was the staple food for two millennia (from Yayoi to the 20th century?). Koyama and Gotō rely on the population estimates of Kitō Hiroshi (1983), estimates strongly disputed by specialist medieval historians (Amino and Ishii (2000: 48, 49)). Sahara also quotes Ishige Naomichi (1986), according to whom the Japanese would have eaten more rice than other Asian peoples. For the consumption of rice in the Yayoi period and at Hida in 1873, see Chapter 3, pp. 222–223.

9 Sahara (1996: 232). In our view, there are ten to twelve centuries between the introduction of irrigated rice cultivation around 800–500 BC and the first tumuli in the 4th century AD; the speed of cultural changes owes a great deal to borrowings from the continent and this evolution was sustained both by rice and dry cereals. See the Introduction to this volume.

10 Harada Nobuo, "Han suiden chūshin shikan no mondaiten" ("The problem with the 'anti-rizicentric' view") (2006: 12–14).

11 Andō Hiromichi (2014: 410, 411). See also a dietary history by Yun So Sot (1995). For the estimates on rice, see below, notes 47, 51, 61–62, 34–35.

12 RSh, p. 367; RSK 15: 472, 481; SN Tenpyō 1.11.7. 729; SN Enryaku 10.5.29. 791; *Yamato no kuni Sōnoshimo-gun Keihoku handen zu; Yamashiro no kuni Kadono-gun handen zu;* see Takashige Susumu (1975: 72–83); Torao Toshiya (1969).

13 *Kōninshiki,* p. 18; ES 26: 654; SY 53: 317.

14 Hi 1: 63, 107–110, 111, 282–284; DNK 3: 334; DNK 7: 47, 48; TM book 4, map of Chimori; *Bungo fudoki* and *Izumo fudoki,* pp. 139, 193, 373. One finds nine grades in the *Harima fudoki.* Fertile or thin soils in RSK 8: 328, 19: 613; RGi, p. 274; *Kaden,* p. 883.

15 DNK 7: 45–48 (Sanuki in 735); for the land at Fujiwara-kyō, see *Fujiwarakyū shutsudo mokkan* 6, text reproduced by Fujii Kazutsugu (1986: 377) (we have calculated 478 and 420 sheaves); RSK 15: 435 (Kyūshū in 823); Kjg, p. 28 (Ise): the 2.4 *chō* produce 1,680 sheaves, thus 700 sheaves per *chō;* but converting these sheaves of 6 bundles to sheaves of 10 bundles gives 420 sheaves. See Hi 5: 1699 (Ise in 1121) and three other examples quoted by Ōishi in Toyoda (1964: 178). Fujii Kazutsugu (1986: 304, 327) (Kuwabara); RSK 15: 438 (Satsuma in 876).

16 *Hitachi fudoki,* p. 37; RSK 15: 433 (Izumi in 827); SJ Jōgan 6.1.28. 864 (Yamato and Yamashiro); RSK 15: 433 (Kawachi in 821); RK 83: 470–471 (Aki in 819).

17 This average established by Takigawa Masajirō using the 8th-century land surveys is also adopted by Naoki Kōjirō (1968: 69); Matsuo Hikaru (1994: 172); Ishii Susumu, in Amino and Ishii (2000: 49).

18 RR Denryō 3. On this section, see Veashua (2009a: 37–59).

19 Harashima Reiji (1962: 30). On the Tōdaiji holdings, in the 8th century, the proportion of uncultivated areas exceeds one-third of the total, see DNK 4: 375–391, 5: 476–690; also Chapter 1, pp. 56–59, and note 183 in that chapter.

20 See the land surveys for Kyūshū in DNK 1: 97–218; and the calculations by Matsuo Hikaru (1994: 173); see RR Denryō 3.

21 Regulations in RR Buyakuryō 9 and RGi; p. 119, RSh, pp. 396, 397; SY 60: 483–493. Reports in RK 173: 182 (insect attacks), RK 173: 183–188 (harvest losses); RK 83: 460–472 (exemptions, some for loss of harvest).

22 Harashima Reiji (1962: 25, 26); Murao Jirō (1961: 102–107): the 66 times gave rise 37 times to distributions of rice, 11 times to tax exemptions, 6 times to prayer sessions, the other cases being unknown.

23 Harashima Reiji (1962: 29).

24 RR Denryō 1 and p. 570.

25 Murao Jirō (1961: 392–439); Sonoda Kōyū (1981: 49–123). The ratio of 50% is confirmed by a document of 759, cf. *Nara ibun* chū, p. 700; it is reduced to 30% after the 8th century, with the evolution of the tax system. A register of deaths caused by indebtedness in Bitchū, in DNK 2: 247–252, records an average debt of 51 sheaves per person, but these debts do not result only from the *suiko*. For the loan amount, see Sonoda Kōyū (1981: 107–110); and RSK 14: 395, 402; RK 83: 457. Loans of 10–40 sheaves are attested in the *urushigami monjo* from the Kanoko C site; cf. Fujii Kazutsugu (1997: 20). See some examples of the management of public loans in Mikami Yoshitaka (2003).

26 DNK 1: 397, 399, 400, 411; SN Yōrō 7.2.14. (723). According to Otomasu Shigetaka (1992: 103), the seed-harvest return is 1/30 in the *Seikei zusetsu* and *Kōka shunjū*.

27 RSK 15: 435 (corvée labourers); for the court: ES 14: 407, 17: 455, 461 (with saké, fish, salt and seaweeds), 39: 868; for the sūtra-copying office, see DNK 6: 223 et seq.; see the Tōdaiji in Hi 1: 383 (with salt, *miso* and vinegar).

28 Ishiyamadera in DNK 5: 5, 10, 16, 27, 30, 32 and 88, 89, and 163 et seq. (with seaweeds, salt and *miso* conserves). For the complete menus of the Ishiyamadera, see Iyanaga Teizō (1980: 475–477). For the residence of Prince Nagaya, see Nara kokuritsu bunkazai kenkyūjo (1989: 21: 10–25). I thank Tateno Kazumi and Sugimoto Kazuki for having shared their interpretation of the daily rations as being more than the needs of one person, and including some reward. Other figures for workers' daily rations (that include the same amounts) are found in the tablet inscriptions *mokkan*, Shōsōin documents and *Engishiki*; see Yoshida Shūji (2010: 177–200).

29 0.8 *shō* of hulled rice gives 1,991 kcal. But with 0.8 *shō* for eight adults and 0.4 *shō* for two children, the average daily consumption for this household of ten people is 0.72 *shō*. See note 36.

30 See Chapter 1, "From ears to grains", pp. 48–53, and note 151 in that chapter.

31 DNK 1: 163, 164. The studies of Japanese historians on the household registers of the early medieval period *kodai no sekichō* have shown that some are completely or partially false, but the two registers that we have used for our study are not suspect documents.

32 DNK 1: 262, 263; RR Koryō 7 for handicapped people.

33 RR Buyakuryō 9; RGi, p. 119; RSh, p. 402. I thank Ishigami Eiichi and Kimura Shigemitsu for helping me to verify and correct the figures in my estimates. These two historians are not, however, in favour of these kinds of estimates.

34 See Braudel (1979: 104–107).

35 Ishii Susumu, in Amino and Ishii (2000: 49), quotes Takigawa's estimates; see below, note 51. See a not very successful attempt at calorific estimates by Toda Hidenori, in Haga and Ishikawa (1999: 49–80): he bases it on the foods in the dictionary *Wamyō ruijushō*, which does not reflect the diet of the Heian court (see Chapter 3 in this volume), and on the calorific values of an old scale of 1963, corrected by the Kagaku gijutsuchō after 1987. Toda's article dates from 1973 and was reprinted without revision in 1988 and 1999.

36 Koyama and Gotō (1985: 486). Braudel (1979: 104, 106) puts the average at 2,000 kcal in Paris, around 1780; p. 127, he quotes F. W. Mote, in K. C. Chang, *Food in Chinese Culture*, New Haven, CT, Yale University Press, pp. 198–200, for 2,000 kcal in China "for six or seven centuries". But Abel (1978: 27) counts 3,200 kcal during the late medieval period in northern Germany. See also Verschuer (2009b: 53–62).

37 Arizono (1997: 14–16, 161); thus 1 litre = 834 grammes of unhulled (whole) rice.

38 Kagaku gijutsuchō Shigen chōsakai ed. (2000: 22).

39 See Chapter 1, pp. 67–70.

40 See Chapter 1, pp. 61–63.

41 *Aizu nōsho*, p. 253, commentary by Shōji Kichinosuke. We have converted 1 *shō* = 1.8 litres; see below, note 56.

42 Kimura Shigemitsu (1996: 111, 112).
43 Arizono (1997: 164), and personal communication.
44 Sonoda Kōyū (1981: 100–103); and examples in DNK 1: 613, 2: 120.
45 Kimura Shigemitsu (1996: 115).
46 *Owari no gebumi*, art. 5, Hi 2: 475.
47 Terasawa Kaoru (1981; 1986: 344–350; 1992: 16–19); Kōmoto Masayuki (1987: 180). Terasawa quotes Andō Kōtarō, *Nihon kodai inasakushi kenkyū*, Tōkyō: Nōrin kyōkai, 1959, who suggests 100 kg per *tan*, which Terasawa takes to be *chūden*. Terasawa also quotes Sawada Gōichi (1927, 1972), and converts the figures given by Sawada to modern *koku* and to kilogrammes, based on 1 *koku* = 25.375 kg (a calculation incomprehensible to us). Terasawa's estimates do not take account of deductions for taxes or accidental losses (due to bad weather, etc.). Harada Nobuo (2006: 75–77) thinks that the yields of rice from the Toro site should be estimated at one-fifth of the modern yields and that this site could only have fed between 18 and 60 people.
48 Kuraku Yoshiyuki (1991: 38, 123–130).
49 Sahara Makoto (1987, 1994: 284–298, 309–311, 410–413).
50 Kōmoto Masayuki defends the theory of fairly high yields, and Negita Yoshio defends the opposing point of view, both in Sahara and Tsude (2000). See more details in the Introduction. See a list of the archaeological sites in Nōkō bunka kenkyū shinkōkai ed. (1988).
51 Naoki Kōjirō (1968: 69–71). Sawada, p. 614, estimates the daily consumption at 3.59 modern *gō* (64.6 cl) of unhulled rice per person. Here are the figures given by Takigawa: an average household of 10 people, with a holding of 1.266 *chō* and yields of 315 *soku* per *chō*, an actual harvest of 400 *soku* of rice, and deductions of 119 *soku* or 69 *soku*(?). There remains a yearly reserve of 331 *soku* or 281 *soku*(?). The household consumes per day 1.72 *soku* (73 cl) of rice per person or 629 *soku* a year. The reserves cover 163 days. See the presentation by Miyahara Takeo (1973: 272–275); cf. Table 4.1. Ishii Susumu quotes Takigawa's figures and thinks that production for the average household gave 2 modern *gō* (36 cl), or 300 g, or 1,100 kcal per day per person over the whole year, and that production therefore supplied half the needs; cf. Amino and Ishii (2000: 50). See above, note 36.
52 *Jikata no kikigaki, Saizōki*, in "Nihon nōsho zenshū", vol. 28, pp. 34, 35; interpretation by Hayama Teisaku, in Hayama Teisaku, ed. (1992: 191–194); Arizono Shōichirō (1997: 162) quotes the same document adding 5 *shō* of barley. See also Furushima Toshio (1975: 419). *Hyakushō denki*, p. 202.
53 Furushima (1975: 418). Before 1624, 1 *koku* is equivalent to 174 litres, cf. *Iwanami Nihonshi jiten*, 1999, p. 1342. Hayama Teisaku (1992: 192, 193).
54 See above, note 13.
55 Quoted after Tanaka Kōji, in Watabe Tadayo (1987: 300) (map of 1877). *Aizu nōsho*, p. 253, commentary by Shōji Kichinosuke, who quotes a document of 1685 entitled "Aizu hiyō shūzei bugai kyūjo hō bassui".
56 See above, note 23.
57 *Aizu nōsho, ibid.*
58 See Chapter 1, pp. 48–53, and note 151 in that chapter.
59 See Arizono Shōichirō (1997: 154–174), concerning everything that follows.
60 See Chapter 3, pp. 221–224, and note 118 in that chapter. The amount of 35% of rice is actually lower if the taxes deducted are taken into account.
61 Nakanishi Ryōtarō, in Haga Noboru (1998: 59–90). We do not give his calorific values, because he uses an old scale dating from 1963 that has since been corrected; see Kagaku gijutsuchō ed. (2000).
62 Kitō Hiroshi, in Haga Noboru (1998: 47–58).
63 Braudel (1979: 124); see above, note 1.

5 Polyculture in premodern Japanese traditions

At the end of our study on the technical and material aspects of premodern Japanese agriculture, we consider the cultural context of early medieval Japan, especially with respect to rice monoculture versus grain polyculture. We present a survey of the agricultural mythology and religious rituals, as well as the ritual foods in court ceremonies. We challenge the commonly accepted theory that sees rice growing as "the foundation of Japanese civilization" and as the "symbol of Japan's land and society". This theory, which associates "Japaneseness" and rice cultivation, is expressed in Japanese by the term "rice-growing culture" (or rice-growing civilization) *inasaku bunka*.[1] Yet the early medieval documents refer to cereals in general, that is rice and millets, as well as leguminous plants, by the expression "five grains" *gokoku*. We analyse in this chapter how the early medieval Japanese viewed rice and the "five grains" through the mythology, historiography, rituals and celebrations of the imperial court. Before giving our point of view, we present the debate concerning rice cultivation and Japaneseness.

Discussion: rice cultivation and Japaneseness, a concept re-examined

Many scholars give rice a predominant place in Japanese culture. Their argument focuses on three areas: diet, mythology and ritual, and is based on the fact that irrigated rice cultivation was introduced to Japan in the early 1st millennium BC. As regards diet, we saw in Chapter 4 that rice was not, in fact, the staple food for many centuries. The present chapter will be devoted to the mythology and rituals of the early medieval period, but we will first retrace the argument concerning the cultural primacy of rice in Japan, according to the historian Harada Nobuo. He presents the mythology of rice based on the early medieval texts, first and foremost the *Nihon shoki*.[2]

The mythology of rice

The *Nihon shoki* (720) relates the following episode concerning the birth of the "five grains": the goddess Ukemochi, whose ancestry may be attributed to the goddess Ukanomitama, divine personification of the rice spirit, turned her head towards the Central country of the reed plain and spat out a porridge of cereals; turning next towards the sea, she spat out the products of the sea; turning finally towards the mountain, she spat out game. These foods were laid out on an altar and presented as offerings to the deities. The same foods, namely the products of the cultivated areas, the sea and the mountains, later appear in the offering texts of the "Regulations of the Engi Era" *Engishiki* (927). According to Harada Nobuo, the foods thus represented in the mythology formed the basis of the early medieval Japanese diet, namely the products of agriculture, foraging, hunting and fishing.

In the *Nihon shoki* myth, Ukemochi is then killed by the brother of the sun goddess Amaterasu. Voided from her corpse are oxen, horses, foxtail millet, silkworms, barnyard millet, wheat or barley, beans and rice. Amaterasu collects these foods and orders their cultivation: cereals and beans in dry fields, rice in an irrigated rice field; this is how agriculture began in the divine countries. The rice field, according to the myth, gave abundant harvests in autumn. Harada Nobuo considers that this episode tells of the beginning of irrigated rice cultivation and thinks that "rice enjoys a special status, different from the other cereals". Harada also quotes another episode recorded in the *Kogo shūi* (807), according to which Amaterasu hands ears of cereals to her grandson who descends to earth, ears identified "in an early commentary" as "rice seed". For Harada, this is another indicator of the "predominant place of rice in agriculture" in Japanese mythology.

The goddess Ukemochi also appears in the *Kojiki* (712) under the name of Ōgetsuhime. This goddess, who gave birth to the various cereals, gave her name to the province of Awa. *Awa* means foxtail millet, a cereal that was grown on dry fields and swidden fields. Harada Nobuo thus distinguishes two traditions in the mythology: on the one hand the myths relating to dry cereal polyculture representing the farming traditions on swidden fields of the Neolithic that preceded the introduction of irrigated rice cultivation, and on the other hand the myths relating to rice that reflect "the special place accorded to rice by the annals" of the 8th century.

Harada also quotes the myths recorded in the geographical treatises *Fudoki* of the 8th century. The *Bungo fudoki* recounts that the abundant rice harvests allowed the peasants to make rice cakes *mochi* in the shape of (round) targets and that these cakes changed into white birds. According to another episode in the *Hyūga fudoki*, the peasants of a village plunged into darkness collected a thousand ears of rice, following the example of the divine grandson; they extracted the grains and scattered them in the four directions, after which the light of the sun and moon illuminated the land. According to Harada Nobuo, these myths give expression to the beginning of rice cultivation from the time when the divine grandson brought the rice seed to earth and they reflect "a rice-growing society centred on the emperor". In his view, rice cultivation played a role in the formation of societies controlled by an élite, for – he thinks – rice has a high nutritional value and keeps well, allowing it to be stored by the élites. It was conflicts between local groups involving rice that led to the emergence of a central authority. Harada also points out that these developments, as well as their mythological expression, are not the monopoly of Japan; they characterize a number of East-Asian societies. He stresses that rice on its own was not the staple food, but that the products of the sea and mountains, also symbolized by deities in the mythology, played just as important a role.

Rice in the early medieval rituals

In the area of offering rituals, the same applies. Harada notes that offerings were far more varied before the rituals were standardized at the beginning of the Meiji era. For the early medieval period, he again quotes the *Fudoki*. According to the *Bungo fudoki*, the rice cakes that changed into white birds were further changed into taro *imo*. Then the province of Bungo, which corresponds to present-day Ōita prefecture in Kyūshū, reported this auspicious sign to the throne. The emperor considered taro to be a sign of plenty and granted to the province the name "land of abundance" Bun(go) no kuni. The Bungo legend thus shows an interesting development: rice is made into cakes; these become white birds and end up as taro tubers. Another episode recorded in the *Hitachi fudoki* concerns the offering for the Harvest Festival *Niiname*, presented to the deities following the harvests. This offering consisted not only of rice, but also of foxtail millet. The Bungo and Hitachi myths therefore provide evidence of polyculture.

By contrast, rice was the main food in the Great Thanksgiving *Daijōsai*, the enthronement ritual for each new emperor. The object of this ritual was the collecting by hand *nukiho* of ears of rice in rice fields determined by divination. Here one sees, according to Harada Nobuo, the primacy of rice in the formation of the State and the intentional choice of rice by the early medieval Japanese government as a "social product of prime importance". "In the Great Thanksgiving, an essential State ritual, he says, rice was therefore the first offering for the deities; this shows the importance given to rice at the level of the State."[3]

Thus Harada Nobuo, who in principle is critical of the rizicentric argument of most writers and who argues for an agricultural plurality (also including animal products), in the final analysis does give a primary role to rice cultivation in the State rituals of early medieval Japan. It is clear that Harada separates rice in his notion of cereals and sees a chronological break between the polyculture characteristic of the Neolithic and the predominance of rice from the prehistoric period. Regarding the primacy of rice, Harada finally brings only three pieces of written evidence: a myth in the *Kogo shūi*, a myth in the *Hyūga fudoki* and the Daijōsai ritual, i.e. a tiny fraction of the panorama of the mythology and rituals of early medieval Japan. Other scholars, too numerous to mention, do not restrict themselves like Harada to the early written sources and add more speculation concerning the structure of the

State. One often finds in their arguments the link on which this conceptual chain rests: rice – central authority – and society.

Another scholar, the ethnologist Sasaki Kōmei, set out to analyse this argument, too simplistic in his view, in more detail. He became interested in the role of rice growing in Japanese culture after conducting research, in the late 1950s, into swidden-field crops in Japan and other Asian countries. His comparative approach and his critical attitude towards certain accepted ideas led him to discuss his viewpoint in a work called "The Structural Complexity of Japanese Culture", *Nihon bunka no tajū kōzō*, published in 1997. In this book, Sasaki summarizes the arguments formulated by historians and ethnologists for and against the predominant role of rice in Japanese society. In his analysis, he makes good use of the comparison with other Asian cultures. He conceives of a Japaneseness not limited to rice growing, but characterized by the complexity of its agricultural structure and by a cultural pluralism. Nevertheless, Sasaki in fact emphasizes the special role played by rice in Japan. We will now summarize his views on the evolution of Japanese rice cultivation (Sasaki, 1997: 243–318), trying to keep as closely as possible to the author's terminology.

Rice, power and society according to Sasaki Kōmei

In Sasaki's view, the "rice-growing culture" carved out a distinctive place from a common base dominant in East Asia, characterized by dry crops (*hatasaku*, denoting non-irrigated crops). In Japan, the special place accorded to rice gave rise to a particular symbolism. According to Sasaki, a political ideology of a "rice-growing country" *inasaku kokka* took shape from the early medieval period. This found a clearer expression in the Edo period, in association with the tax system tied to yields expressed in rice *kokudaka*.

The origin of the collective nature of Japanese behaviour has traditionally been attributed to rice cultivation, but Sasaki considers this to be a trait common to all 'hydraulic societies', i.e. characterized by irrigation management, since irrigation requires collaboration between neighbouring farmers. From the 1930s to 1950s, the ethnologist Yanagita Kunio (1875–1962) had stated his conviction that rice played a decisive role in the origin and originality of the Japanese people. He found that the rituals and beliefs surrounding rice were associated with the presence of a rice spirit and he especially drew attention to the role of rice in the mythology and the diet. However, according to Sasaki, beliefs concerning rice characterize most of the countries of South-East Asia and are not unique to Japan. The theories of Yanagita Kunio were, nonetheless, very influential in academic circles. Then they were called into question by Tsuboi Hirofumi. In his work "The Japanese and Taro", *Imo to Nihonjin* (1979), the writer discusses his theory of a dichotomy of irrigated and dry crops in Japanese society that he calls *inasaku bunka* and *hiinasaku bunka*. Sasaki adopts this concept while noting, in Tsuboi's work, the lack of a definition of "rice-growing culture" on the one hand, and of "dry cereal culture" on the other. He is therefore careful to define these two notions.

Sasaki Kōmei identifies the constituent elements of the Japanese "rice-growing culture" of the Yayoi period, referring to various agricultural techniques, methods of storing grains and the presence of ritual bronze implements. He notes that there is no fundamental difference (except for the use of animal traction) between these elements and the cultural characteristics of the entire laurel-forest zone of East Asia (Vietnam, Yunnan, northern Thailand) during the first half of the 1st millennium BC, a zone characterized precisely by the dual presence of irrigated and dry crops. Thus Japan shares a common base with these countries. This is explained by the fact that in Japan, as in the countries of East Asia, rice cultivation supplanted some of the existing dry crops. This process took place in Japan at the end of the late Neolithic (1000–500 BC); with rice cultivation, Japan imported bronze and its ritual context from the continent.

It is from this time on that Sasaki sees a more specifically Japanese "rice-growing culture" gradually taking shape, an evolution that he situates between the end of the Yayoi period (3rd century AD) and the 7th century, and of which he gives the following description. Irrigated rice was intentionally chosen for its qualities (taste, size of grains, culinary preparation, yields); it progressively took hold alongside dry cereal cultivation and then supplanted it. Rice later became culturally isolated from the other agricultural traditions (*hiinasaku bunka*). With the "ripening" of the "rice-growing society" (from here on Sasaki no longer speaks of agricultural society in general) and the management it involves, according to Sasaki, élites and centres of power were formed in the Japanese archipelago, hence the link between power and rice. Moreover, the ritual structure that had come from the continent with bronze gave rise, once it was applied to rice cultivation, to a symbolism of rice (*ine no shinborizumu*) expressed in the rituals *matsuri* (unspecified). Sasaki points out that the ritual of rice also existed in the other countries of East Asia, but he thinks that the cultural isolation of rice in relation to the other cereals was more evident in Japan. Over the centuries, the symbolism of rice became more pronounced as the role of rice in the economy gradually increased. Rice symbolized political and social unity (*tōgō*) and resulted in a political ideology of a "rice-growing country" *inasaku kokka* around the 7th century, when the Yamato clan put in place an agrarian and tax system based on rice cultivation. This ideology is clear, for instance, in the mythology, which was written down at this time. According to Sasaki, it expresses the coexistence of the rice-growing and cereal (non-irrigated) traditions, through the myth of the birth of the "five grains" (of South-East Asian origin) on the one hand, and through the myth of the divine descent that gives prominence to the "ear of rice", as well as the mirror and sword (these two regalia being of Korean origin) on the other hand. This is how we understand the interpretation of a Japanese "rice-growing culture" proposed by Sasaki. We have tried to give an accurate summary of his text, which is not altogether clear in places. What is more, Sasaki's explanations in the second half of his discussion lack concrete examples.

Overall, Sasaki Kōmei corrects certain received ideas relating to the original characteristics of Japanese culture. He highlights the superposition of rice-growing and non-rice-growing factors, of indigenous and external elements. He thus puts forward what he calls the structural pluralism of Japanese culture. This sets him apart from the prevailing opinion that sees rice cultivation as the main constituent element of Japanese society. In this sense, Sasaki's concept is innovative. The aim of his analysis, however, is not to question whether a "rice-growing culture" even existed, but to redefine it. Thus he treats rice cultivation as a monoculture, separate from dry crops. His hypothesis of a cultural isolation of rice is not supported by any arguments; nor is his theory of the emergence of a political ideology concerning rice. For the latter, he mentions only the tax system based on rice as a temporal reference for the 7th century. As the sole example of the expression of this political ideology, he quotes the myth of the divine descent that mentions the ear "of rice" of the sun goddess, but without examining this myth (the original text of which does not actually mention rice, as we will see later).

On the material level, Sasaki also argues for the coexistence, even after the 7th century, of two cultures: one rice-growing, the other non-rice-growing. Geographically he situates the first on the lowlands, the second in the mountains. He considers that foxtail millet played an important role, at least up to the 9th century, and that rice yields in fact remained relatively low until the 15th century. Sasaki thinks he can discern a predominance of rice from the 15th century, even if the population depended to some extent on dry crops up to the modern period (see the 1874 census, according to which rice supplied only 60% of dietary needs). In the 18th century, according to Sasaki, following land clearance campaigns undertaken by the military government, there developed a "structural preference" for japonica white rice. This concept links up with the idea of a "structural hunger" of rice attributed to the early medieval Japanese by Tanigawa Kenichi *et al.* (1993) and other ethnologists and historians. Sasaki goes on to say that the "structural preference" for rice has characterized, since the modern period, the "typically Japanese rice society" (*wagakuni dokutoku/nihonteki kome shakai*). In his view, this evolution went hand in hand with an ideological renewal, supported by the school of nationalist thought (Kokugaku), whose arguments concerning rice are still to be analysed.[4]

To sum up, Sasaki sees a coexistence of rice-growing and non-rice-growing traditions, but he makes no temporal distinction between the appearance of a primacy of rice in the 7th-century political ideology and an intellectual preference for rice from the 18th century. Sasaki goes against the opinion dominant since the 1960s that attributes to rice both an ideological and material supremacy from the end of the Neolithic. Yet he in no way calls into question the principle of a primacy of rice in Japanese culture; he merely situates it differently in time. In short, Sasaki's views show certain elements of rizicentric thinking, even though he is a scholar well known for his research into swidden-field crops. Thus we cannot accept his interpretations. We will now examine the theory of a political ideology of rice by pointing out the elements relating to agriculture that can be found in the mythology, historiography, rituals and celebrations of the imperial court of Nara and Heian.

Mythology and agriculture

Japanese mythology is recorded in the *Kojiki* (712), a chronicle edited by Ō no Yasumaro (?–723), the *Nihon shoki* (720), another chronicle edited by Prince Toneri (676–735) and the regional monographs *Fudoki*, completed in the first half of the 8th century. All these works are compilations of 7th-century written and oral traditions. The mythology traces the events from the founding of the heavenly universe up to the creation of the earthly country. It introduces deities, the actors in many myths, and ends with the descent to earth of the god Hononinigi, ancestor of the first legendary emperor of Japan. There are different versions of these episodes, presented in the works cited and recorded under distinct variants in the *Nihon shoki*. We will now survey the principal myths relating to agriculture.

The birth of the five grains

In the *Nihon shoki*, the origin of the "five grains" *gokoku* as follows: before dying from burns, after the birth of her son Kagutsuchi, god of fire, Izanami, "lying down gave birth again to the goddess of earth Haniyama and the goddess of water Mizuhanome. Kagutsuchi then took Haniyama as his wife and she gave birth to Wakumusubi. From the head of this deity came forth silkworms and mulberry [branches/leaves], and from her navel were born the five grains."[5]

A few paragraphs further on, the *Nihon shoki* gives more details about the "five grains" in an episode devoted to the goddess Ukemochi, the content being as follows: the god Tsukiyomi, by order of Amaterasu, came down to see Ukemochi. She offered him a feast consisting of spat-out rice, fish and game meat. Tsukiyomi, affronted by this reception, drew his sword and slew Ukemochi. Amaterasu dispatched to the place a messenger who witnessed the following scene:

> By this time Ukemochi had already expired. But from the top of her skull came forth an ox and a horse; from her forehead came forth foxtail millet; from her eyebrows appeared silk cocoons; from her eyes sprouted barnyard millet; from her stomach sprouted rice; from her vagina sprouted wheat/barley, as well as soybeans and red beans. The messenger took away all these foods and offered them to Amaterasu. She said delightedly: 'These will be foods for all of humankind', and she named foxtail millet, barnyard millet, wheat/barley and beans as the fruits of dry (non-irrigated) fields *hata(ke)tsumono* and rice as the fruit of irrigated fields *tanatsumono*. She chose an official and gave him grains of rice to be sown for the first time in the long and narrow heavenly rice fields. In autumn, the long and heavy ears grew thickly. It was a delight. In addition, the silkworms that had piled up behind the mouth [of the deity] enabled silk to be spun. This was the beginning of the way of the silkworm.[6]

The *Kojiki* gives a slightly different version of this myth, in these words:

> The eight hundred myriads of deities, assembled in council, decided to impose on Susanoo a punishment of a thousand meals, to make him shave off his beard and cut his fingernails and toenails by way of purification and to banish him from their society. Susanoo went to Ōgetsuhime seeking food. She prepared various foods from gifts drawn from her nose, mouth and posterior to offer [to the deities]. Susanoo, having watched her and indignant that she dared to offer polluted foods, killed Ōgetsuhime. From the body of the defunct deity there then came forth various things: from the head were born silkworms, from the two eyes sprouted rice, from the two ears sprouted foxtail millet, from the nose red beans, from the vagina wheat/barley and from the posterior soybeans. Whereupon the god Kamimusubi had these foods collected and extracted their seeds.[7]

These episodes show the early 8th-century Japanese notions about the origin of agriculture. Millets, wheat or barley, beans, rice and silk are named. The first legend introduces the god Wakumusubi, son of the earth and fire, as the procreator of the five grains which, according to Ōbayashi Taryō, makes him the god of swidden-field crops. Likewise, Ukemochi and Ōgetsuhime are the procreators of cereals and beans, which were the main crops grown on swidden fields up to the 20th century. Ōgetsuhime was the wife and mother of the mountain gods. Ōbayashi also points out that, according to the *Kojiki*, Ōgetsuhime gives one of the provinces in Shikoku the name "Province of millet", Awa, when the Japanese provinces are created. At the same time are born the province of Kibi, "Common millet", and that of Azuki, "Red beans".[8]

These legends follow the usual model of myths concerning the excretion of agricultural products from a divine corpse, of which there are various expressions in the countries of South-East Asia. J. E. Jansen identified a prototype in the mythology of Hainuwele, a goddess celebrated by the Wemale tribe in Indonesia; but the coconut of the Indonesian myth was replaced in Japan by the crops of this country. The Japanese episodes therefore allow us to infer the practice of polyculture and the existence of swidden farming in mountain areas.[9]

The divine descent

Another myth is more directly associated with rice by historians and ethnologists such as Sasaki Kōmei. We give it here, even though the role of rice in this myth actually seems to us to be debatable. It is the myth of the "descent to earth of the divine grandson" *tenson kōrin*, according to which a deity, grandson of the sun goddess Amaterasu, descends to earth where he becomes the ancestor of the first Japanese emperor. The episode, recorded in the *Nihon shoki*, says:

Amaterasu took the mirror – a divine treasure – and gave it to [her son] Amanooshi-homimi saying (to him): 'My son, when you look in the mirror, you will see me...'. Then she said: 'I also give my son ears [of rice] from my purified garden on the heavenly plain.' She married Amanooshi-homimi to the princess Yorozuhatahime, daughter of Takamimusubi, and ordered him to descend [into the world]. Finding themselves in the [inter-mediate] country of Ōzora, they had a son and called him Amatsuhiko-hononinigi. The [couple] wished the divine grandson to descend [from heaven] in their place. [Amaterasu] therefore entrusted him with [this mission] with the gods Amanokoyane, Futotama and the other [gods of his retinue]. The goddess gave him, as agreed, the [imperial] garments and regalia. Then [his father] Amanooshi-homimi returned to heaven, and Amatsuhiko-hononinigi descended onto the peak of the Thousand piled-up ears [of rice] Takachiho at Kushihi in the country of Himuka (southern Kyūshū). From this country he later went and settled in the country of Ata (in Satsuma, Kagoshima).[10]

We have translated this episode according to the annotations (in square brackets) in the *Nihon shoki* editions published in 1967 (edited by Inoue Mitsusada) and in 1994 (edited by Kojima Noriyuki). This myth retraces, according to these commentators and Japanese historians in general, the descent to earth of the divine grandson of Amaterasu who becomes the ancestor of the emperors and passes on to the people rice growing as the basis of sustenance. This myth explains, in the eyes of the Japanese, the divine lineage of the emperor and the relationship between the imperial family and rice cultivation. However, these interpretations reveal certain problems.

Since this myth, in the annals, is placed after the episode of the birth of the five grains (including rice) designated by Amaterasu as "foods for all of humankind", the myth of the divine descent with "ears of rice" is given a different role, that of the origin of the Great Thanksgiving, the enthronement ritual of the Japanese emperors. Yet ears are mentioned in only one of the five versions of this episode in the *Nihon shoki* and are absent from that in the *Kojiki*, according to which the grandson of Amaterasu is entrusted by the goddess with the three regalia: sword, mirror and jewel *magatama*.[11] Also, the *Nihon shoki* version mentions ears, but not ears of rice. In fact, rice is totally absent from the myth of the divine descent to earth (*tenson kōrin*), as we will see.

The ear of rice that wasn't

We propose below a different interpretation of the myth of the divine descent, beginning with the meaning of 'ears'. The editors of the *Nihon shoki*[12] gloss *inanoho/inaho* (ear of rice) for the single character *ho* (ears), even though the reading *inaho* is not attested by the early texts. In the 8th century, the *Man'yōshū* writes *inaho* with the two characters of rice and ear and refers to ears of other plants such as eulalia. The *Hitachi fudoki* mentions ears of reeds.

In the 9th century, the annals several times refer to ears of foxtail millet. The dictionaries of the 10th–12th centuries define the ear *ho* as the "tip of a cereal". Ears occur in the prayer texts of the *Engishiki* with the reading *ho*, or *wo* in a gloss of the 12th-century Kujōkebon. The poetry of the 10th–12th centuries speaks not only of ears of rice, but also of ears of eulalia *susuki* or *kaya*, of barley or wheat *mugi*, etc. The term 'ear' *ho* was therefore not reserved for rice. In our legend, it rather refers to ears of unspecified graminae.[13]

These come from the "sacred garden" *yuniwa* of the goddess, a term interpreted by the editors as "a sacred and forbidden rice field where the rice intended for the offerings is cultivated", an explanation that transforms the garden into a rice field. But *niwa* was not an agricultural space. This term had three meanings in the 8th century: a wide expanse of water and sea, an area reserved for household use, and a garden near a dwelling that could be planted with trees or include some self-sown vegetation. The *Man'yōshū* refers to "the grass of the garden", "the hemp of the garden" and "the plum trees of the garden", and the poetry of the 10th–12th centuries speaks of the gardens of aristocratic residences with a vegetation of grasses *kusa*, eulalia *kaya*, flowers *hana* and trees including flowering cherries. In our account, they are therefore ears of unspecified graminae from the garden of the goddess. If they were ears of rice, the goddess would be expected to have collected them not in her garden, but in the divine rice fields where she had previously had rice sown (in an earlier episode). However, another interpretation defines *yuniwa* as a ritual place for offerings for the Great Thanksgiving (of the enthronement of the emperor). This is a homonym of *niwa* meaning 'place', which is written not with the character for garden but with that of place *ba* (as in *basho*). In this case, *yuniwa* denotes a "ritual place" for offerings. This notion is attested by the regulation on the Great Thanksgiving and by a prayer text composed by the Nakatomi family in 1142. However, in the context of *yuniwa* located on the heavenly plain, this interpretation is difficult to understand. In fact, the Nakatomi prayer text actually distinguishes between the garden *yuniwa* of Amaterasu and the physical ritual place *yuniwa* at Heian, where the offerings were celebrated on the occasion of the enthronement of Emperor Konoe in 1142. In short, the meanings of "rice field" and "ritual place" remain unclear. The textual meaning of "sacred garden" *yuniwa* of the goddess appears to be the simplest and most coherent.[14]

The *Nihon shoki* editors have put forward other interpretations. They consider the spouse Yorozuhatahime to be the "princess of myriads of looms" *hata* or many woven fabrics; but *hata*, with this character, also means 'swidden field', for instance in the 8th-century toponyms and in the dictionary *Wamyō ruijushō*. This term is found precisely in the legend relating to the "five grains", mentioned above. In this legend the editors gloss *hataketsumono* meaning "(dry) field harvests" instead of *hatatsumono*, "swidden-field harvests", a reading that is, however, attested for the 9th–10th centuries. So Princess Yorozuhata could also be called "Myriad of swidden fields".[15] The son of the divine couple is called Amatsuhiko-hononinigi, interpreted by the editors as "divine son furnished with full ears of rice", which still assimilates the ear *ho* to the ear of rice; but this name contains the character for fire *ho* (a homonym of ear *ho*). Later, this child will have a son called Ameno-hoakari, "Firelight" (a name that in turn is interpreted by the editors as "Ripening of rice"). Thus Hononinigi is rather the god of "Auspicious fire".

The grandson of Amaterasu, Hononinigi, finally descends to the summit of Takachiho in Himuka. The *Nihon shoki* editors interpret the toponym Takachiho as "a thousand ears of rice piled up high". Himuka was southern Kyūshū, which from 713 included the three provinces of Hyūga, Satsuma and Ōsumi, a very mountainous region. The toponym Himuka is explained in the *Kojiki* version as being a good place *itoyoki tokoro* that enjoys sunshine morning and evening. The dictionary *Iroha jiruishō* (12th century) reads *tō* (east) for *himukashi* and explains this term by the morning sun. A place with a precise orientation was probably located on a mountain slope. We know, thanks to ethnology, the crucial importance of the orientation of lands, especially the easterly exposure of mountain swidden fields. By contrast, on a slope facing west, harvests were greatly reduced. The notion of the slope facing the morning sun *asahi no himukai* is also expressed in the prayer text of the Tatsuta ritual, a ritual devoted to prayers for the ripening, not of rice but of the "five grains". It may therefore be supposed that the summit of Takachiho had nothing to do with ears from irrigated rice cultivation.[16]

The localization of the summit of Takachiho also it makes possible to link it with cereal crops on swidden fields. There are two places with this name in Kyūshū: the summit Takachiho 1,574 m above sea level in the mountains of Kirishima straddling Kagoshima and Miyazaki; and the summit Futakami (Kunimi-dake, 1,739 m) in the old village of [Taka]chiho, in the mountains on the border between the prefectures of Miyazaki and Kumamoto. The mountainous region of Kyūshū has been well known for swidden-field crops since Sasaki Kōmei conducted research there in the 1960s. Such a location seems unsuitable for rice cultivation; in fact, historians have long expressed doubts as to the choice of the place, and ask why the divine descent occurred in the mountains of Kyūshū and not on the Yamato plain. More recently, the historian Senda Minoru (1999) has cast doubt on the presence of rice in the toponym Takachiho in a book called "The Illusion of Takachiho".[17]

Taking account of the various elements discussed above, we now propose a different reading of the myth of the divine descent: Amaterasu gives the mirror and the ears of cereals (not rice) to her son, whose name, Amanooshi-homimi, refers to the tip of the ears. She marries him to the princess called "Myriad of swidden fields" (or "Myriad of looms") and the couple produce the "Heavenly man of auspicious fire" (of swidden fields?). He descends to earth and alights on a slope facing east, in the mountains of the Hyūga region, near the summit of Takachiho. From this place, he propagates in the realm the cereal crops that are to feed the people. In other words, the myth of the divine descent does not mention rice and, in our view, constitutes an agrarian myth relating to cereal polyculture, just like the myth of the birth of the five grains.

It remains to be seen whether these episodes refer to cereal crops in general, or more especially to swidden-field crops. The second interpretation corresponds to the ethnobotanical point of view, according to which swidden farming is the first form of agriculture. One can imagine that the grandson of Amaterasu brings to earth ears of domesticated cereals, or self-sown cereals such as Echinochloa crus-galli L. (*inubie*), a wild barnyard millet that has left numerous traces in the Neolithic archaeological sites, including that of Sannai-Maruyama; this graminae was later domesticated. Let us recall that some versions of the same myth attribute to the deity who descends to conquer the country the sword called Kusanagi, meaning "Grass-cutter"; this clearing operation precedes the burning of land and the cultivation of swidden-field crops.[18] In this sense, our myth of the divine descent would correspond to an agricultural-type logic that would prevent philologists from transforming cereal ears into ears of rice, the garden into a rice field, and interpreting as 'rice ear' the semantic element of fire *ho* in the names of the god Hononinigi and the semantic element *ho* of the toponym Takachiho.

The country of vigorous ears

Philologists also see rice in another context of the same myth. At the beginning of the legend, the realm to be conquered by the gods, i.e. Japan, is called "Country of vigorous ears of one thousand five hundred autumns of the plain of luxuriant reeds" Toyo ashihara no chiiho aki no Mizuho no kuni, and later "Central country of the reed plain" Ashihara no nakatsu kuni. In this episode, the way is being prepared for the divine descent, by driving away the evil spirits *ashimono*, a term assimilated by commentators to reeds *ashi*.[19]

The "Country of vigorous ears" Mizuho no kuni is today a poetic name for Japan that is interpreted as the "Land where the vigorous ears of rice ripen" and understood as the "land of abundant rice". This is the meaning assigned to it by modern dictionaries based on the mythology. However, in our myth, the text has "Land of the vigorous ears of reeds" (and not of rice). The 1994 edition of the *Nihon shoki* explains this contradiction as follows: "After the evil spirits had been driven from the reed plain, rice was able to grow on this plain. The autumns *aki* mean both 'years' and 'rice harvests', and the vigorous ears are those of rice. The country *kuni* (with the character for *tokoro*) is the land. [Mizuho no kuni] is therefore the land where they have been harvesting for a very long time the abundant ears of rice on the plain where the reeds grow thickly." In their explanation, the editors transform the reed plain into a land of rice (even though the latter does not appear in the text). This kind of interpretation is common in current publications on mythology.[20] Yet in the 8th century, the *Man'yōshū* poets presented the divine descent into the "land of the vigorous ears of reeds" with no mention at all of rice cultivation. One of the poems in this anthology refers to the burning of a clearing aimed at promoting the regrowth of plants; this quite widespread fire technique has nothing to do with rice cultivation. Moreover, the prayer texts that mention the "plain of the vigorous ears" accompany rituals that have no direct connection with the growth of rice.[21]

Plains of reeds *ashi*, an aquatic graminae that corresponds to Phragmites communis Trinius, the common reed, even today characterize certain landscapes in Japan although the Meiji government, from 1910, undertook many campaigns to drain wetlands. Before these works, the soil was much wetter and the reed plains were probably far more numerous. This is why reed plains are often mentioned in the early medieval poetry and have given Japan another poetical name, that of "Land of the reed plains" Ashihara no kuni. It therefore seems to us more logical to interpret literally the divine descent into the land of the reed plains than into a land where there are already abundant harvests of rice.[22]

According to our hypothesis, rice disappears entirely from the myth of the divine descent (in which it does not occur anywhere in the text); the first Japanese ruler does not play the role of transmitter of rice cultivation in the empire and the name Mizuho no kuni by which this empire is designated does not mean 'rice-growing land'.

Rice, one of the "five grains"

Our hypothesis does not deny the existence of other myths that actually do concern rice. A legend devoted especially to rice cultivation is recorded in the *Izumo fudoki*. It retraces the origin of the name of the village of Tane ("Seed") in these words: "When the great deity created the world under heaven, the gods Ōnamochi and Sukunahikona visited the countries under heaven, and, when they arrived in this place, they dropped rice seeds *inatane*. This is the origin of the name Tane, but the character was changed in Jingi 3 (726)." This brief account is similar to another type of myth found in several Asian countries: the ear dropped on the ground in a precise spot. It is also reminiscent of the creation myth of the Kanaks in New Caledonia, according to which the moon throws into the sea a yam cake wrapped in taro leaves and from it are born the first human beings. This myth reflects the importance of the yam and taro both in agriculture and in the offering and exchange rituals of this Pacific island. While the Asiatic myths are known to us mainly through oral tradition, there is also a Chinese written account of the 8th century, the sūtra *Wubaifanzhi jing* (Jap. *Gohyakubonshi kyō*) that mentions "the human body born of the five grains". Returning to Japanese mythology, in the myths of several provinces one comes across a crane, another kind of bird, or a mythical being, who drops from on high ears of rice, ears of other cereals or the "five grains", but these legends mostly date from late medieval times or the Edo period.[23]

For example, according to an undated tradition from the village Ōse-machi near Miyazaki city (the old region of Hyūga), the local tutelary deity descends from heaven bringing the "five grains" at the time of the spring equinox.

Then she takes away the seeds of the same cereals harvested at the autumn equinox. For these two rituals, the local shrine is decorated with a portrait depicting the tutelary deity holding a sheaf of rice. Here rice and the "five grains" merge in the popular tradition. It was the same, apparently, in the early medieval period: as Ōbayashi Taryō has pointed out, the god Sukunahikona, who dropped the grains of rice in Izumo, also sowed foxtail millet in the province of Hōki.[24]

Thus, even the myths devoted to rice contain elements of non-rice-growing cereal cultivation. There is nothing contradictory in this if it is remembered that rice is offered as one of the "five cereals" in the first Japanese agricultural myth, written down in the 8th century. However, 20th-century historians and ethnologists see a contradiction here. They isolate the rice-growing tradition from a general cereal tradition and attribute a cultural primacy to rice. Everything began with the ears "of rice" in the myth of the divine descent. While these ears of rice are not found in the 8th-century text, neither are they an invention of 20th-century philologists. We may therefore ask how this interpretation came about. A study of the *Nihon shoki* manuscripts has given us the key to answering this question. We retrace the transmission of the myth of the divine descent and attempt to go back to the origin of the 20th-century interpretations.

The ear of rice: the emergence of an idea

The *Nihon shoki* has come down to us in a number of manuscripts. The first two books of this 8th-century chronicle, devoted to mythology, were called "The Age of the Gods" *Jindai no maki* from the late medieval period. The earliest copies date from the 13th century: the Kateibon of 1236 that we owe to an anonymous copyist and the Kanekatabon of 1286 copied by Urabe Kanekata (?–?). These manuscripts offer the particularity of glossing in *katakana* the original text written in Chinese characters.[25] Here we have found the readings of the terms that interest us. The three earliest copies gloss the character for ear *ho* by "rice ear" *inanoho* and other later copies also gloss it by "rice ear" *inaho*. However one copy, the Enkakubon of 1306, gives the two readings *inaho* and *ho*.[26] The author of this copy therefore gives the meaning of ear (and not of rice). This copy belongs to an earlier tradition of the Heian period, whereas the reading that gives the meaning of "rice ear" dates from the 13th century and was proposed by the Urabe family, who held posts as diviners and middle-ranking officials in the Ministry of Religious Affairs at the imperial court. Urabe Kanekata, active ca. 1278–1306, studied the *Nihon shoki* under his father Kanefumi (?–?), who gave lectures on the *Nihon shoki* to the regent Ichijō Sanetsune (1223–1284). Kanekata also wrote the *Shaku Nihongi* in twenty books, a methodical thematic study that quotes earlier works such as the 8th-century *Fudoki* and the *Nihongi shiki*, a compilation of the 9th–10th centuries.

In the section on mythology in the *Shaku Nihongi*, Kanekata explains that the garden *yuniwa* of Amaterasu is the province of Yuki that delivers the rice for the Great Thanksgiving. The section on the readings of the terms in the *Shaku Nihongi* is fragmentary and the part that should have included the term *ho* (ear) is lost. By contrast, book 8 gives the following explanation of the "Country of vigorous ears of one thousand five hundred autumns of the plain of luxuriant reeds", Toyo ashihara no chiiho aki no Mizuho no kuni: "It is the fertile land of abundance where reeds grow thickly; the one thousand five hundred autumns signify that, during autumn, over a very long period of time, there will inevitably be splendid ears of rice, 'one thousand five hundred' being a term of the Ancients meaning 'a great number'. It was in autumn that the divine grandson descended to earth."[27] This is how Urabe Kanekata introduced the notion of the ear of rice into the reed plain.

Urabe Kanekata was able to consult the *Kogo shūi*, presented to the throne in 807 by Inbe no Hironari (?–?). The Inbe family was in charge of the religious (Shintō) rituals at the court, and also took on the role of imperial messengers to Ise. The *Kogo shūi* summarizes the myth of the divine descent and notes that the goddess gave her grandson "ears from the sacred garden on the heavenly plain", without mentioning rice.[28] However, the earliest copy that has come down to us adds a commentary on the subject: this is the Karokubon of 1225, a copy annotated and glossed by Urabe Kanenao (?–?), poet and priest of the Yoshida Shrine in Yamashiro. Kanenao adds this note about ears: "These are grains of rice" (*kore inadane nari*). It is possible that Kanenao is referring to the legend in the *Izumo fudoki*, according to which a deity dropped grains of rice in the village of Tane in Izumo.[29] Be that as it may, this interpretation may have inspired Urabe Kanekata to gloss 'ear' by *inanoho* in the *Nihon shoki*. Hence it was under the Urabe that the following three changes occurred in the 13th century: ears became ears of rice; the purified garden from which they came was transformed into the province of Yuki that provided the rice offered at the Great Thanksgiving; and autumn became the season that produced fine ears of rice. At the time of the Great Thanksgiving for each reign, it was in fact the Urabe who had the task of designating by divination the two provinces that were to supply the rice for the offerings and were called for the occasion Yuki and Suki. It is therefore probable that the Urabe attempted in this way to strengthen the reputation of their tradition, as had previously, in their works, the Inbe and Mononobe families whom the Urabe had succeeded as the family responsible for the imperial rituals. In this way, ears of rice came into the myth of the divine descent, five centuries after they first appeared in writing.

A journey of five centuries

The Urabe founded a tradition of Shintoist scholarship that was handed down by the school of Yoshida Shintō in the late medieval period and later inspired the philologists of the Edo period. Among them was Tamaki Masahide (1670–1736), who discussed mythology in his work *Jindai no maki moshiokusa*, printed in 1739. He was quoted and followed by Tanikawa Kotosuga (1709–1776), known as the author of the monumental encyclopedia *Wakun no shiori.* In Tanikawa's book *Nihon shoki tsūshō* (1751), ears become ears of rice, and the garden *yuniwa* becomes a "sacred rice field" *yuta* where the rice for the Harvest Festival offerings is collected.[30]

From the 18th century on, it was mainly the Kokugaku school that devoted itself to the study of the Japanese classics. Kokugaku scholarship, which emerged as a reaction to Western studies (Rangaku) and Chinese Confucianist studies, advocated a return to the ancient Japanese traditions and the veneration of the imperial family. Among its representatives, Motoori Norinaga (1730–1801) left to posterity the *Kojikiden* (1798) and Hirata Atsutane (1776–1843), more strongly Shintoist, wrote, with others, the *Nihon shoki den* (1853), a commentary on the *Nihon shoki*.[31] One finds in all these works a reformulation of the notions concerning rice put forward by the Urabe family in the 13th century. The ground was thus prepared for the political ideology of the Meiji government.

Rice for the emperor: an invented tradition

After the enthronement of Emperor Meiji in 1868, the imperial court celebrated the first Great Thanksgiving of the new reign *Daijōsai* in the 11th month (December). On this occasion the government issued the following proclamation:

> On the eighteenth (day) of this month, the Great Thanksgiving will be celebrated in Kyōto, and the sovereign will pay his respects from afar (from Tōkyō). Here is the meaning of this celebration: in accordance with the will of the great goddess Amaterasu Ōmikami, rice, in our Empire, is a food that comes from a fine green plant. The rice that the goddess had planted in her long and narrow heavenly rice fields she gave to the divine grandson when he descended to earth. We shall never forget this divine favour. Moreover, in order to ward off droughts and tempests, the emperors have continued, from generation to generation since Emperor Jinmu, to celebrate the offering of the new grain harvest to the heavenly and earthly gods, on the second day of the hare in the eleventh month. This has continued for three thousand years. From the first day of this eleventh month, the emperor observed a prohibition for three days, in order to prepare himself for this ritual, to show his compassion for the people. Let everyone high and low show their gratitude with sincerity.

The text then enjoins the populace to show its respect for the deities and to pray for the growth of the "five grains" and for peace in the country. It explains that it is the duty of each person to participate in the national religion. The text goes on: "Let everyone know that rice, our daily food, was originally a gift of the divine ancestor by which he honoured our land, in order to distribute it (to everyone)."[32] The interpretations of the myth of the divine descent, which started only in the 13th century, strongly influenced the political ideology of the Meiji era.

The celebration of the Great Thanksgiving in 1868 inaugurated the restoration of thirteen "ancient" rituals, eleven of them in fact being new creations that retraced the history of the imperial line; these rituals celebrated certain episodes drawn from the *Kojiki* and *Nihon shoki*. To support imperial authority, Shintō was made the State religion. The Ise Shrine was assigned to the worship of Amaterasu and installed as the chief organ at the head of all the shrines in the country. The emperor, invested with supreme political, military and religious power, was gradually elevated to god-like status, by means of ritual celebrations and a propaganda campaign through the national education system. Rice became a link in the chain of ultra-nationalist thinking. The scholarly works of the 1920s to 1940s pointed out the primacy of rice and presented rice cultivation as an activity dedicated to the sun goddess Amaterasu.[33] They thus consolidated the rizicentric notions established by the philologists of the Edo period. The government propaganda effort continued until 1945.

After 1945, the divine status of the emperor was abolished, and mythology was no longer part of the secondary education curriculum. It was from this time on that the scholarly community was able to give prominence to the theories of Tsuda Sōkichi (1873–1961) who, since the 1940s, had denied mythology any historical basis. Contrary to the tradition established by Edo-period philology, the mythological accounts, according to Tsuda, were not based on historical facts, but resulted from the wish of the 8th-century authors to consolidate the imperial power of the Yamato line.[34] Post-war research on mythology also made use of comparative studies. Yet the philological analysis of the texts dealing with mythology is still partly dependent, it seems, on the 18th-century theories, and some historical and ethnological studies continue to be influenced by the Meiji legacy. This explains why one finds evidence of interpretations similar to those in the 1868 declarations by the Meiji government in the Iwanami edition of the *Nihon shoki*, published in 1967. The editors say so themselves: in the introduction they quote as references the studies of the *Nihon shoki* mentioned above, which date from the Edo and Meiji periods. This is because, in 1967, they did not do a philological analysis of the myth of the divine descent based on the 8th-century literature. The Iwanami edition has been so influential that its interpretations have not been questioned. It still forms the basis of the more recent editions of the *Nihon shoki* and of studies relating to mythology.

If the interpretations of the myth of the divine descent go back to the 13th century, what was the situation before that time? We saw earlier that this myth was not associated with rice cultivation in the literature and poetry of the 8th–12th centuries. The same applies to the early philology of the *Nihon shoki*. The *Kogo shūi* (807), which derives from the tradition of the Inbe family, is a compilation of the most important myths for the imperial court. It gives summaries of them and reworks the episodes. This text includes the myth of the divine descent and mentions ears *ho*. A note in small characters adds that these are "rice ears", citing another myth, that of the birth of the "five grains". The sentence in the *Kogo shūi* is repeated by the *Sendai kuji hongi*, dating from the 9th century and based on the traditions of the Mononobe family, whose members were often responsible for reading prayer texts, notably on the occasion of the harvest festivals.[35] However, the interpretation of the *Kogo shūi* and *Sendai kuji hongi* was later ignored, and the traditions of the Inbe and Mononobe did not interest the nobles of the Heian court when they were reading the mythology.

In fact, the *Nihon shoki* was the subject of readings at the court on a number of occasions. The notes for seven sessions that took place in 712, 813, 839, 878, 904, 936 and 965 were compiled by an anonymous author in the 9th–10th centuries in a work called *Nihongi shiki* that simply gives the reading *yuniwa* for the sacred garden and makes no mention at all of the ear of rice, or even of ears in general, in the sections relating to the divine descent.[36]

If the authors of the 9th–10th centuries, those closest to the imperial family, ignored rice in the myth of the divine descent, it is because it did not occur there. In all likelihood, the image of the transmission of rice cultivation to Japan by the grandson of the sun goddess Amaterasu did not form part of the conceptual parameters of the authors of the 8th-century mythology, or of the Heian nobility. This idea seems to have entered the myth of the divine descent five centuries after the memory of it was fixed in the annals. In our view, the nobility of early medieval Japan did not attribute a symbolic value to rice any more than it thought of rice in isolation from the other cereals, but it had an overall view of agricultural products. We will examine this point by analysing the early medieval Japanese perception of rice and the other cereals in the historiography, rituals and annual celebrations of the imperial courts of Nara and Heian.

Cereals in the historiography of the early medieval period

The compilation of the *Nihon shoki* was the initiative of Emperor Tenmu (r. 673–686). In 681, he summoned the princes and his ministers and ordered them to draft an administrative code; he then named Prince Kusakabe (662–689) heir to the throne. One month later, the sovereign ordered the compilation of an imperial genealogy *teiki* and an account of the events of antiquity *jōko shoji* (*kuji*), all based on oral tradition, but also on writings of the 6th–7th centuries no longer extant. Emperor Tenmu gave Hieda no Are (?–?) the task of reading these writings and Ō no Yasumaro (?–723) that of compiling the annals. This work, the *Kojiki*, was only completed in 712, twenty-five years after Tenmu's death. The *Kojiki*, in three books, covers the mythology, the legendary age and the history up to the reign of Empress Suiko (r. 592–628). The *Nihon shoki*, completed in 720, was also the initiative of Emperor Tenmu who entrusted its compilation to Prince Toneri (676–735). It is more detailed than the *Kojiki* and continues up to 697. The two works reveal the concern of the Yamato clan to consolidate its power and lay the foundations of an empire.

They therefore reflect the political ideology of the late 7th and early 8th century. Agricultural products are mentioned according to the notion of them held at the imperial court.

In the *Nihon shoki*, Prince Toneri attributes to the legendary sovereigns (called emperors) the virtue of having taken care of cereal cultivation. The following episodes may be quoted. In 91 BC, Emperor Sujin established in his country many shrines and ritual places. Afterwards, epidemics disappeared, "the five grains *gokoku* ripened and the people were happy". Five years later, this sovereign took the first census of the population in order to levy taxes; the deities approved his action, "the rain and wind came at the right time and the hundred grains ripened", the people were prosperous and peace reigned in the land.[37] In 316 AD, Emperor Nintoku proclaimed in a solemn edict that, having observed his country from the top of a hill, he had seen no smoke rising from dwellings and poverty everywhere among the people; thus he did not hear the people singing his praises as in previous reigns. "Know, he said to his ministers, that poverty rules where the five grains do not ripen", this being true for our country and even more so for foreign lands. Nintoku had a tax exemption proclaimed for three years. Afterwards, the wind and rain came at the right time, the five cereals *gokoku* ripened, and after three harvests prosperity returned to the people. They began to sing the sovereign's praises and thick smoke rose from all the dwellings. In 406, Emperor Hanzei established his court at Tajihi in Kawachi. At that time, the wind and rain came at the right time, the five grains ripened, the people were prosperous and peace reigned under heaven. Emperor Ninken also had the satisfaction, in 495, of seeing "the five cereals ripen, the rearing of silkworms and the harvest of wheat/barley increase", peace reigning in the near and distant regions, and the population growing.[38]

Later, the legendary emperors Keitai and Ankan also observed that the people lived without anxiety thanks to good grain harvests. Then Senka proclaimed in 536 that food was the most valuable asset under heaven, and that it was therefore necessary to fill the granaries with grain in order to guard against bad years, this being the first principle for maintaining peace in the land. However, in the 28th year of the reign of Kinmei (567), the land was hit by floods, to the point where hunger drove "people to eat each other". Kinmei then requested assistance from the neighbouring country in the form of a gift in grain.[39]

It is clear that the *Nihon shoki* speaks of cereals in general and not of rice in particular. The chronicle uses the terms "five cereals", an expression of Chinese origin and, in the latter cases, grains *koku* in general. The modern editors of the *Nihon shoki* gloss *tanatsumono* or *momi* for grains *koku*, and *itsutsuno tanatsumono* for the five grains *gokoku*. *Tanatsumono* has two meanings: "rice-field harvests" *ta no mono* defined by Amaterasu in the myth quoted above, and "fruits from seeds" *tane no mono* in general. The early medieval dictionaries attest the readings *koku, momi* and *tanatsumono*.[40] The modern editors of the *Nihon shoki* think that rice is meant, even though *tanatsumono* occurs, in the references given above, in the same context as the five cereals *gokoku*.[41]

The five grains and prosperity

For the historical period, the *Nihon shoki* again uses the expression "five grains" from the reign of Suiko, in connection with a good harvest in 617 and poor harvests in 623 and 628. For 642, this work relates a royal progress by Kōgyoku (r. 642–645) in her country, stating that the sovereign's prayers brought rain, wind and the growth of the "nine grains"; after which the people rejoiced and praised the supreme virtue of the sovereign.[42] This account is one example among others of the cycle of causality linking the following elements: imperial virtue; ripening of the (five) grains; peace in the country; and prosperity of the people. According to this notion, good cereal harvests result from the imperial virtue. The image of the imperial virtue being responsible for the growth of the "five grains" (and not exclusively rice) is also the subject of a number of prayer texts.[43]

Cereals also symbolize a high level of civilization, in the following articles relating to foreign countries. During a mission sent to China by Empress Saimei in 659, accompanied by men of the Emishi tribe who were presented as a tributary gift to the Chinese court, the Japanese envoy was questioned by the Chinese emperor who wished to know whether this tribe cultivated the five grains; to which the envoy replied that the Emishi were meat-eaters (and therefore barbarians). Then Tenji (r. 668–671) sent a mission to Paekche in 661, at the time of the wars on the Korean peninsula, a mission carrying arms and a quantity of the five grains by way of a special allocation. The following year, Tenji handed over seeds of the five cereals to an envoy from Tanra (Chejudo island in southern Korea).[44] A reading of these episodes gives the impression that the five cereals (and not just rice) were considered by the early Japanese as the source of the country's prosperity.

Rice: one of the five cereals

For the 8th–9th centuries, the annals record many food shortages, but also some good harvests. They mostly use the terms "five grains" and "cereals of the year" *nen koku*.[45] One finds for instance this text of 757:

> Imperial edict: From what we have heard, the five grains did not ripen last year owing to prolonged rains, or this year owing to drought. For this reason, the price of rice has soared. The people are beset by suffering and famine compounded by epidemics that have caused the death of a great many people. We deeply sympathize with their suffering. An exemption from the rice taxes for this year must be proclaimed for the capital, the central region and all the provinces of the seven circuits.[46]

In this kind of report, poor harvests follow rain, wind, floods and droughts. The measures taken by the government include full or partial tax exemptions, distributions of food, offerings in the shrines and readings of sūtras in the temples.[47] In 874, the three Inari deities in Yamashiro (Fushimi) were rewarded with court ranks after a very good harvest of the "five cereals", because these gods personify both the spirit of rice Ukanomitama and that of the five grains Ukemochi. In 739, there was a reading of the/a "sūtra for the ripening of the five grains" *gokoku jōjuku kyō*, an expression designating either the sūtra *Gokokuse kyō*, known by the Chinese catalogue *Kaiyuanlu* (8th century) but no longer extant, or one of the sūtras read on these occasions such as the *Konkōmyō saishōō kyō*. In 741, the foundation of the provincial temples was associated with prayers for "the abundance of the five grains" *gokoku hōjō*.[48]

At first sight, the accounts in the annals give the reader the impression that *gokoku* is a generic term referring in practice, however, almost exclusively to rice, since, depending on the situation, bad years result in an increase in the price of rice, an exemption from the rice tax or a shortage of rice seed. Moreover, prayers take place several times in the 7th month, i.e. before the rice harvest. One may also mention a temporary reduction in the rice allowances for officials in the wake of variations in the harvest of the five grains in 820; and a decree issued in 841 explains the importance of the five grains while at the same time referring to the drying of rice.[49]

However, it seems to us that the meaning of *gokoku* was not limited to rice, as the reports of poor harvests refer to the damage caused "in the rice fields *ta* and dry fields *sono*", *sono* being the area where they cultivated "the five grains, hemp and the mulberry tree", according to the *Izumo fudoki*. When the annals speak of famines, this refers in our view not to the lack of rice for the people, but to all cereals used for food. Thus the "Regulations of the Jōgan Era" (859–877) provided for the loan of rice, foxtail millet and wheat or barley during famines. Distributions of food probably involved all kinds of cereals, including foxtail millet from the granaries *gisō*, intended for aid during famines, as well as all the cereals whose cultivation was encouraged from the early 8th century: the millets, wheat, etc. When harvests were poor, only the officials received allocations of rice.[50]

The various cereals served at the palace

The annals therefore appear to us to show the simultaneous interest of the court for all the cereals, including rice. It is hardly surprising that the court should be concerned for the survival of the rice-growing population, who formed the tax base on which it depended for its rice resources. But in reality, while depending on rice as the basis of its food supply, the court also stocked other grains. To meet its needs, the Ministry of Religious Affairs had fields of foxtail millet cultivated in the province of Yamashiro. The Grains Office in fact prepared both rice and foxtail millet for the daily meals of the imperial family. In addition, the Bureau of the Imperial Table had wheat, barley and beans cultivated. Each year, the court received deliveries, not only of rice, but also of various other cereals and beans to be distributed to the offices responsible for the meals of the officials.[51] Thus rice was not the only cereal found on the palace tables.

The high-ranking officials also had different cereals prepared in the kitchens of their residences, according to the account of the move of Fujiwara no Morozane (1042–1101) to the Kazan-in residence in 1063. On his official entry into the new pavilion, he had in his retinue ten servants who symbolically carried the essential objects for daily use. Among them the fourth carried a large pot filled with the "five grains" and the ninth a steaming strainer filled with "steamed five grains" *gokokuii*, probably meaning in both cases cereals including rice. The two containers were then placed in the kitchen of the new residence and were the object of prayers for the two days following the move.[52]

What are the five grains?

Whether in the administrative terminology of the 8th–9th centuries or in the dietary customs of the Heian nobility, rice was not isolated from the five cereals, but had its place among them. The same applied in the mythology. The myths variously name as the "five grains": foxtail millet *awa*, barnyard millet *hie*, rice *ine*, wheat or barley *mugi*, soybeans *mame* and red beans *azuki*. It can be seen that beans belonged with grains in Japan, as in China. By way of comparison, the granaries intended for food aid during famines stocked mainly foxtail millet, with smaller quantities of soybeans, wheat and barley. There is a known case of an order for the "five grains" from the sūtra-copying office at Nara in 770, in which however only soybeans and red beans are mentioned. If, in the 9th century, the commentaries on the "Code" of 701 explain that "the rice field *ta* is the area where the five grains are cultivated", this is an error, as Kimura Shigemitsu points out. In fact these early medieval commentaries do not refer to Japanese reality, but quote the Chinese classics in which the term *ta* means both rice field and dry field. They also list the five grains according to the Chinese sources, namely common millet *kimi*, foxtail millet, beans, wheat or barley and rice. This definition and its variants, with or without rice and hemp, were later reproduced in the early medieval Japanese dictionaries.[53]

For the annual offerings of the "five grains", on the 15th day of the 6th month, offerings prepared in the residence of the regent Fujiwara no Tadamichi (1097–1164), the five grains are specified as: rice, soybeans, red beans, barley and wheat. The *Shinsarugakuki* paints a picture of an ideal agricultural holding in the 11th century, where the intendant "delights in seeing the ripening of the five grains and reaping an abundant harvest of them". This man cultivated rice, wheat or barley, soybeans, cowpeas, red beans, foxtail and barnyard millet, buckwheat *sobamugi*, perilla *e* and sesame *goma*. The expression "five grains", originally borrowed from China, in Japan therefore means variously cereals and beans, with a preponderance of rice and foxtail millet and a relative scarcity of common millet and buckwheat. This notion remained in place until the Edo period. For the agronomists of this time, rice was "the best *among* the five cereals", probably from the point of view of yields, and possibly also from the point of view of taste.[54]

Thus the imperial court took an interest in all the cereals and had an overall view of them. We therefore do not follow the reasoning of Kimura Shigemitsu, for whom the term "cereal" *koku* meant only rice in the 7th–8th centuries. Kimura states: "from the 8th century, the government thought essentially of rice when speaking of the five grains" and "it was at this time that the image of cereals *koku* meaning rice took shape".[55] Having closely examined the annals and protocol regulations of the 8th–12th centuries, we can say that rice is not treated there in isolation from the other cereals.

Cereals in the religious rituals

The imperial court celebrated many religious rituals intended to ward off disasters or evil spirits, to ask for good harvests and good fortune and to mark important events. The "Regulations of the Engi Era" list the "seasonal rituals", celebrated regularly each year on fixed days, the "special rituals" devoted to certain offerings that did not necessarily take place each year, and other offering rituals celebrated regularly in a dozen palace offices, including the Bureau of the Imperial Table and the Office of Divination. Shibusawa Keizō, who has counted the rituals and offerings, gives some interesting figures concerning the role of rice in the offerings that may be summarized as follows: the *Engishiki* lists 49 seasonal rituals, 80 special rituals and 48 office rituals, or 177 rituals a year, including parallel celebrations in different places (for example, a ritual conducted at the same time by the Ministry of Religious Affairs and in the residence of the Crown Prince is counted twice). Each ritual includes one or more offerings, for instance one presented to the deity (Amaterasu), another by way of purification and a third to close the celebration. Each offering consists in turn of several kinds of foods. There are three distinct categories: artisanal objects, including textiles; marine products, including sea-ear *awabi*, bonito *katsuo*, seaweeds, salt; and plant foods. These include: rice in ears *ine*, rice in husked grains *hakumai*, grains of glutinous rice *mochiyone*, boiled rice *ii*, saké, and more rarely beans, tree fruits, nuts, foxtail millet, a soup of barnyard millet, a food fermented in soy *hishio*, etc. Shibusawa Keizō counted the foods used in offerings (leaving aside their respective quantities), presented in total in all the rituals for the year. He identified 1,706 foods for the seasonal rituals, 1,315 foods for the special rituals and 424 foods for the palace office rituals (number of occurrences of each food in each ritual); these figures do not include the artisanal products, but do include all the foods, as shown in Table 5.1.

In the seasonal rituals, marine products account for 65% and plant products for 35% of the foods presented as offerings. Rice makes up 47% of the plant foods, or 16% of all the foods. The figures are similar for the special rituals and office rituals. In other words, marine products occupy the most important

Table 5.1 Offerings celebrated by the imperial court (number of foods used in offerings)

Types of rituals	Numbers of rituals	Offerings	Products (total)	Marine products	Plant products	Rice
seasonal	49	266	1,706	1,110	596	280
% of total				65%	35%	16%
special	80	175	1,315	862	453	216
% of total				65.5%	35.3%	16%
of offices	48	49	424	224	200	75
% of total				53%	47%	18%

Notes: Figures: the number of occurrences each year as mentioned in the *Engishiki* (927).
Read: in 49 seasonal rituals each year, there are 266 offerings that include a total of 1,706 mentions of foods: 1,110 marine products (or 65% of food offerings) and 596 plant products (or 35% of food offerings); and among the latter, rice occurs 280 times; rice accounts for 16% of all food offerings in the seasonal rituals.
Sources: *Engishiki*; after Shibusawa Keizō (1954).

place among the offerings, and rice appears in only one-fifth to one-sixth of the number of foods used as offerings in the rituals celebrated at the court. From a statistical point of view, the presence of rice is therefore relatively low.[56] However, we will now look at the role of rice in a number of special rituals.

Rituals for the growth of the five grains

At New Year, i.e. the 1st day of the monkey in the 1st month, the imperial court and all the local administrations of each village throughout the country celebrated an exorcism ritual to drive away evil spirits and ask for good harvests. A ditch nearly a metre deep was dug and 255 litres of muddy water was siphoned into it from an easterly watercourse. Twenty-five litres of unrefined saké was then poured in and the whole area was covered over with earth. The celebrants stamped on the surface to compact the earth. Then they hit the surface again twenty-seven times with a pestle while uttering magic formulas saying: "May disasters disappear; may everyone be spared from illness; may the five grains ripen."[57]

This ritual was celebrated in two places at the palace: inside and outside one of the palace gates, and also by each regional government. Another ritual, that of "fire calming" Hi-shizume, took place in the four corners of the palace, in the 6th and 12th months, with the aim of preventing fires. The prayer text refers to the myth of the five grains voided by the dead goddess burned during the birth of her son who became the god of fire. According to the regulation, the offering at that time consisted of rice; but at some time in 762 a Nara office prepared the five grains to celebrate this festival.[58]

For cereal growth, the hygrometric conditions were of the utmost importance. Thus the imperial court organized prayers to cause or to stop rain, sometimes on particular occasions, and at regular times every year.[59] The "Regulations of the Engi Era" list eighty-five shrines devoted to prayers for rain *amagoi*, the most important being those of Matsunoo, Inari, Kibune, Isonokami, Hirose, Tatsuta and Kamo.[60] The annual festival at Kamo, a shrine located north of the imperial palace in Yamashiro, goes back to the time of Emperor Kinmei (6th century). While this sovereign was offering up prayers, on an auspicious day in the 4th month, men of the region wearing a wild boar or lion mask chased after horses decked with small bells. This ritual ensured the growth of the "five grains" and the prosperity of the country.[61]

The festivals of Hirose and Tatsuta go back to the reign of Tenmu. In 676, in the 4th month, this sovereign sent messengers to offer prayers to the wind god Kazekami of the Tatsuta Shrine (at Heguri-gun, Yamato) and to the pro-hibition god Ōimikami at the Hirose Shrine (at Hirose-gun, Yamato). He did this the day after setting up the loan of public rice to farmers to enable them to proceed with the sowing of rice.[62] The two rituals of Hirose and Tatsuta were later celebrated twice a year, on the 4th day of the 4th and 7th months. This time corresponds to the growing period of rice, first in the seedbeds (sown in the 3rd month), then in the rice fields (transplanted in the 5th month). The god Ōimikami saw to it that the mountain and lowland streams watered the seedlings at the right time, and the god Kazekami caused the wind to blow and the rain to fall at the right time. At first sight, these two rituals appear to give priority to rice, but the prayer texts specify that these rituals are devoted to the good growth of "everything that the people cultivate under heaven". In the case of Tatsuta, this means: "everything, beginning with the five grains adorned with vermilion-red ears down to the smallest blade of grass and the smallest leaf."[63] Thus, the watering of the seedlings, a task undertaken by the god of Hirose, was linked rather to rice, whereas the deity of rain and wind of Tatsuta also influenced the growth of the other cereals.

It may be noted that the 4th month also coincided with the growing period of soybeans (sown in the 3rd month) and barnyard millet (sown in the 2nd month), and that the 7th month coincided with the growing period of buck-wheat (sown in the 7th month and harvested in the 9th month) and foxtail millet, which is a summer crop like rice. Thus, the rituals celebrated in the 4th and 7th months were not necessarily reserved for rice cultivation. In fact, the prayers at Tatsuta were intended to make "the five grains ripen".[64] Other rituals may not have been limited to rice cultivation either. It would be necessary to analyse systematically for each ritual recorded in the *Engishiki* the time of the celebration (in relation to agriculture), the prayer text and the products pre-sented as offerings. If, at Hirose and Tatsuta, rice was the only graminae among the food offerings (consisting of marine products, plants, textiles and artisanal objects), it was absent from the offerings in some other rituals.[65]

Rice and foxtail millet

Rice is absent from the "Monthly field tasks" ritual, Tsukinami, among the offerings presented in the 198 shrines in the country responsible for this celebration.[66] The Tsukinami ritual took place twice a year on the 11th day of the 6th and 12th months, from the beginning of the 8th century. Unlike the festivals of Hirose and Tatsuta, for which messengers from the court presented the offerings in the two shrines, the Tsukinami ritual was celebrated at the same time throughout the country, at Ise and at the imperial court in the Ministry of Religious Affairs.[67] It was intended to promote the growth of the "five grains" and was of popular origin; this ritual went back to the offerings presented by peasant families to their tutelary deities.[68] At the court, the emperor took a ritual meal, Jinkonjiki, at dawn, the day after the Tsukinami offerings. At that time he was served not only rice, but also foxtail millet, steamed and in a gruel, accompanied by fish, seaweeds and nuts.[69]

Rice and foxtail millet were found together every day on the emperor's table. The Grains Office prepared the daily rations of both foods for the Bureau of the Imperial Table, responsible for the daily meals of the sovereign and his principal consort.[70]

Apart from the Tsukinami, the Harvest Festival Niiname was another important ritual celebrated annually at the court and throughout the country. The Niiname took place on the 2nd day of the hare in the 11th month. The 9th-century commentaries explain that they gave thanks to the gods for the harvest of "new grains" *shinkoku*, without naming the cereal(s). Cereals are not found among the offerings; but, as pointed out by Araki Toshio, the Niiname was by no means limited to rice: at that time they prepared at the palace rice and foxtail millet, steamed *ii* and also in a gruel *kayu*, as for the Tsukinami festival. It may be asked whether the court chose these two cereals because they represented irrigated and dry cereals, or because they were both grown in the fields *kanden* managed by the government. Next day at dawn, the Niiname celebration ended with a ritual meal during which the emperor ate the "gruel of seven cereals ending the prohibition".[71] Elsewhere, the Niiname ritual was devoted solely to foxtail millet, as for example in the district of Tsukuba in Hitachi (Ibaraki) and probably in Awa (Tokushima). In the case of the office of the Ise priestess, the list of foods prepared for the Niiname celebration (and not the offerings) is extremely detailed. Here we find: rice, foxtail millet, soybeans, beans, wheat, sesame, glyceria, common millet, vegetables and nuts. The identification of glyceria *mino* is uncertain. It was a self-sown graminae, perhaps also cultivated, though attested only during the 8th–10th centuries.[72]

It is clear from the above that the Kamo and Tatsuta festivals and the Tsukinami and Niiname rituals concerned not only rice, but also the other cereals. These rituals headed the religious ceremonies at the court. Among the high- and middle-level rituals, only the Daijōsai, Kanname and Toshigoi are missing. The Daijōsai or Great Thanksgiving celebrated at the start of each reign, and the Kanname, celebrated annually at Ise in place of the Niiname, included the stripping of ears of rice *nukiho* in the sacred rice fields. The Toshigoi, celebrated at the court and at each of the 3,132 shrines in the country, concerned the prayers intended to ward off disasters.[73] This is another ritual where rice is not found among the offerings. In other rituals, rice went hand in hand with beans *mame*: the offerings at the Kasuga, Hirano and Hiraoka shrines, and the Yomo no mikado, Mikawa-mizu and Kanname rituals. Let us recall that, generally speaking, marine products played a more important role in the offerings than plant foods, and that rice, normally, did not have pride of place among the plant products. On the contrary, it is clear that rice is absent even in some regular rituals that were part of the major annual religious celebrations at the court.[74] In fact, the content of the offerings was quite varied and it is necessary, in short, to reconsider the place of rice in the rituals.

Cereals in the court banquets

Alongside the rituals of the Japanese religion (Shintō), the imperial court calendar was dotted with official banquets organized in parallel with the auspicious and seasonal occasions of the year. Kimura Shigemitsu has pointed out that many plant foods are found in the menus for these banquets. Beginning in the 1st month, the emperor was served, on the 15th day of the New Year, a "seven-cereal gruel" *nanakusa kayu*, the ingredients being rice, foxtail millet, common millet, barnyard millet, glyceria, sesame and red beans.[75] Certain ingredients were occasionally replaced by soybeans, (sweet) chestnuts, cowpeas and persimmons.[76] With a variation in the quantities, the recipe was the same for the dawn gruel, the day after the Niiname.[77] One has the impression that the court saw to it that all the cereals were offered up on important occasions.

At New Year, there was also a celebration associated with Buddhism. From the 8th to the 14th day, the court organized banquets, Gosaie, on the occasion of readings of the sūtra *Konkōmyō saishōō kyō*, intended to ensure the protection of the State. During these readings, the monks were served rice, wheat noodles and beans, as well as other vegetables, fruits, nuts and marine products. When it was over, they were thanked with foods consisting of rice, common millet, barnyard millet, sesame, barley, wheat, soybeans and red beans.[78] These sessions ended on the 14th day with the offering of the "five grains", presented on a dais installed in the Daigokuden (Great Hall of State). It may be assumed that this offering symbolized the fact that readings of the sūtra *Konkōmyō saishōō kyō* were intended to ensure not only the protection of the State, but also the growth of the "five grains". The monk Kūkai (774–835) had actually given the court an explanation along these lines in 834.[79]

Regarding the other seasonal banquets of the year, Kimura Shigemitsu has drawn attention to a text by Nakahara Moromitsu, secretary in the Ministry of Religious Affairs in 1264. In his protocol manual for the court ceremonies, Moromitsu refers to the journal of Emperor Uda (r. 887–897) concerning the seven-cereal gruel in these words:

The *[Uda tennō] gyoki* notes this in the 2nd year of Kanpyō (890), 2nd month, 30th day: I gave this instruction to Yoshi: the seven-cereal gruel on the 15th day of the 1st month, 'peach-flower cakes' *chōkahei* on the 3rd day of the 3rd month, rice cakes wrapped in leaves *chimaki* on the 5th day of the 5th month, noodles *sakuhei* on the 7th day of the 7th month and rice cakes *mochi* on the day of the wild boar in the 11th month are a custom of the common people on festival days. Starting today, each of these [specialities] must be served at our banquets. I gave this instruction when Yoshi was in charge of my personal affairs.[80]

This custom was in fact introduced to the court. In the 3rd month, there were "grass cakes" *kusamochi*, i.e. made with mugwort *yomogi* among other things. In the 5th month, there were glutinous rice cakes *chimaki* cooked in their wrapping of wild rice *komo* leaves and containing, according to the *Engishiki*, cowpeas and crushed (sweet) chestnuts; also in the 5th month were served early melons *wasauri*. Finally in the 7th month, there were noodles made with wheat flour *muginawa*.[81]

The above-mentioned specialities were served at seasonal banquets that accompanied certain court celebrations. The regulations had fixed several official banquets during the New Year (the 1st, 7th, 16th and around the 20th day of the 1st month) and still others during the year, including those mentioned above. Refreshments were also provided at all administrative (nominations, etc.) and ritual (offerings, etc.) events. On each occasion, the offices in charge of the food prepared for the emperor and officials the following foods: rice, foxtail millet, wheat, perilla, sesame, soybeans, red beans, saké, fish and seaweeds, as well as common millet in the 7th month.[82] Thus it was very unusual to find only rice on the palace tables. Rather, one has the impression that the court made a point of including the various cereals in ceremonial refreshments.

An ideology of the "five grains"

It is clear, in short, that rice and the other cereals often appear together in the mythology, in the administrative terminology, on the tables of the emperor and nobility, in the rituals and court banquets. This suggests that for the early medieval Japanese administrators there was no dichotomy between rice and the other grains. One sees no evidence in the ceremonial protocol and ritual of a conceptual isolation of rice. The nobility of the early medieval period seems rather to have had a broad view of all the cereals, one of them being rice. Certainly rice was the most important item in the public administration as well as in commodity exchanges, and exceeded in volume the other cereals in the foods eaten at the imperial court; this does not mean, however, that in the early medieval period it had a mythical or sacred value superior to that of the other cereals. One does not see in the early medieval texts a conceptual particularity of rice or a distinctive symbolism for this food. The legislators of the 7th–8th centuries do not seem to have felt the need to assign to rice a special place in the mythology. If rice was the object of special treatment, this must have come about in the late medieval or in the Edo period. Rather, the texts of the 8th–12th centuries reveal the concern of the imperial court to include all the cereals and beans in the most important celebrations, and it called them the "five grains" in the official terminology. For the annalists of the 8th century, the prosperity and peace of the country depended on the ripening of the five cereals (and not just of rice). In this sense, one may speak of an ideology of the "five grains" in early medieval Japan.

Notes

1 Berque, "Fūdo", "Riz", and Joseph Kreiner, "Japonais", in Berque, ed. (1994: 216–220, 236, 437–439). These authors summarize the arguments in the Japanese publications.

2 Harada Nobuo (1996: 44–58).

3 End of summary. See the texts quoted by Harada in *Hitachi fudoki*, p. 39, *Hyūga fudoki*, pp. 523, 524, *Bungo fudoki*, p. 357, *Kogo shūi*, p. 28; for *Nihon shoki* and *Kojiki*, see below, notes 5–12.

4 End of our summary of Sasaki Kōmei (1997: 243–318).

5 *Nihon shoki* (N), book 1 Jindai jō, beginning of 5th dan, in "Nihon koten bungaku taikei" (NKBT), Iwanami, 1967, vol. 1 p. 89.

6 N (NKBT), end of 5th dan, p. 102. See below, note 15, on the readings of *hata(ke) tsumono* and *tanatsumono*.

7 *Kojiki* (K), book jō, episodes on Amaterasu and Susanoo, in "Shinchō Nihon koten shūsei" (SNKS below), Shinchōsha (1979, 1991), ed. Nishimiya Kazutami (1979: 53); and K (NKBT), p. 85. See Kimura Shigemitsu (1996: 31–43).

8 Ōbayashi Taryō (1973: 7–9, 31–34, 48–50); K (SNKS), p. 31 (also pp. 34, 76, 77); K (NKBT), p. 57.

9 Ōbayashi (1973: 5–14); Yoshida Atsuhiko (1990: 81–116); Ōkuma Kiichirō *et al.*, eds (1993: 29–31).

10 N (NKBT) Jindai ge, 9th dan, 2nd version, pp. 152, 153; and N, in "Shinpen Nihon koten bungaku zenshū" (SNKBZ below), Kojima Noriyuki, ed. (1994: 2: 139). On rice in the myth of the divine descent, see Veashua (2012: 165–182).

11 In the *Nihon shoki* (NKBT, 1967), the Jindai section is annotated by Ōno Susumu, Ienaga Saburō and Sakamoto Tarō. In the *Nihon shoki* (SNKBZ, 1994), the Jindai section is annotated by Kuranaka Susumu and Mōri Masanori; these notes differ only slightly from those of the Iwanami edition thirty years earlier. See Yoshida Atsuhiko (1990: 88–90), for the Great Thanksgiving; see the divine descent in K (SNKS), pp. 88–90; K (NKBT), pp. 111, 127–129, and a version similar to that of the *Kojiki*, in N Jindai ge, 9th dan, 1st version. See Mishina Akihide (1971).

12 We quote the commentaries of N (NKBT) and N (SNKBZ), as well as those of the *Kojiki* (SNKS), pp. 88–92; Yoshida Atsuhiko (1990: 88–90); and Ōkuma Kiichirō (1993: 63–67).

13 See the details of what follows in Veashua (2012). See the reading *ho* in WR 17: 1; IJ entry under *ho* (shokubutsu); and RMS. See M 1: 6 (*inaho* written with two characters), 2: 88 (*ho*), 8: 1601 (*susuki ho*); *Hitachi fudoki*, p. 37. See the ears of millet *awa* in NK Kōnin 5.8.18., MJ Ninju 1.8.2., SJ Jōgan 11.8.13., Gangyō 8.11.5. See the *Engishiki* book 8, and the gloss *wo*, p. 164. *Kokin waka rokujō* 2: 4–3701, 3703; *Fuboku wakashō* 2: 16–4452; *Yoshitada shū*, no. 135. But according to another version of this myth, recorded in the *Hyūga fudoki itsubun*, p. 523, there is stripping of the rice ears that are later found as part of the Great

Thanksgiving, Daijōsai. Ōbayashi Taryō (1973: 17, 135) mentions the "ears of rice" of our episode, and Ōbayashi Taryō (1984: 183–186) points out that the character *ho* is read *inaho*, without giving any explanation.

14 See *niwa* in *Nihon kokugo daijiten*, *Jidai betsu kokugo daijiten* (*Jōdai* hen), etc.; M 10: 2160 *niwa kusa*, 4: 521 *niwa ni tatsu asa*, 19: 4140 *sumomo*. WR 10: 12, records *niwa* as a garden of a dwelling. *Yoshitada shū*, no. 67, 96, 144, 376; *Goshūi wakashū* 1: 4–224, 225, 122, 148, 398, 1207; *Shinsen waka rokujō* 2: 14–816 to 820; *Kokin waka rokujō* 2: 4–1354 to 1356; *Fuboku wakashō* 2: 16–4446, 13563, 14887–14903. In the *Nihongi shiki* one finds *yuniwa*, p. 103, but nothing on *ho*, pp. 17, 18. See the synonym of *yuniwa* in ES 7: 144, and *Kojiki Norito*, pp. 459–463. See Kakubayashi Fumio (1999: 426, 427); Kakubayashi does not offer a new interpretation for the myth of the divine descent.

15 Princess Yorozuhata, called by different names depending on the versions; it is rather the loom in the other versions of the *Nihon shoki* and according to that of the *Kojiki*; and according to *Hitachi fudoki*, p. 85, the princess, called Kamuhata (written like the loom), introduces weaving to the Hitachi region. See *hata* in WR 1: 11, and the *Dōbuntsūkō* book 4, by Arai Hakuseki. Kuroda Hideo (1984: 141–146) drew attention to the meaning of *hata* 'swidden field' in his article "Chūsei no hata to hatake", written in 1980. See Chapter 2, pp. 154–156, for the toponyms with *hata*. The editions of N (NKBT), 1967, p. 102, and N (SNKBZ), 1994, p. 61, gloss *hataketsumono*, fruits of permanent fields, in the first episode mentioned above. One finds *hatatsumono* in *Nihongi shiki*, Otsubon (9th–10th century), p. 66, but *hataketsumono* in IJ (12th century), a version in three books, entry under *ha* zatsubutsu (and the term *hata* 'swidden field' in the two versions of IJ *ha* chigi), and *hataketsumono* in WR, a version in ten books, quoted in *Nihon kokugo daijiten*, art. "hata" and "hatake" (which, however, confuses the meaning of the two terms).

16 See Himuka in K (SNKS), pp. 91, 92; and (NKBT), p. 129. The *Hyūga fudoki itsubun*, p. 523, and the *Shaku Nihongi* 8: 116, explain that the place faces the direction where the sun rises, from the morphology of the terrain. See ES 8: 163, and *Honchō gatsuryō*, p. 265, for Tatsuta. For the importance of the direction of slopes for swidden-field crops, see Tachibana Reikichi (1995: 80, 296, 612); and Masuda Shōko (1990: 130, 148).

17 On the question of Hyūga, see Inoue Mitsusada (1960); Tsuda Sōkichi (1948); Hidaka Masaharu (1993). The *Satsuma fudoki itsubun*, p. 527, mentions the move of the deity to Ata. See Senda Minoru (1999: 36–40).

18 See the sword Kusanagi, in N Jindai ge 9th dan 1st version, p. 127; and in K (SNKS), p. 90; and K (NKBT), p. 127. See Chapter 2.

19 See Mizuho no kuni in N Jindai jō 4th dan and Jindai ge 9th dan, NKBT, pp. 82, 146; SNKBZ, pp. 29, 111; and K (SNKS), pp. 77, 89, and K (NKBT), pp. 111, 127. *Kuni* is written *chi/tokoro*, the reading *kuni* being attested by WR 2: 1.

20 See the quotation in N (SNKBZ), p. 28, note 6. This more or less corresponds to the explanation of the editor of K (SNKZ), p. 77. According to Saigō Nobutsuna (1967, 1984: 15–29, 137–140), the plain *ashihara* was transformed into a land of abundant rice by exorcism. Modern dictionaries explain Mizuho no kuni as "the country where the wet ears of rice ripen". Kawasoe Taketane disputes this definition and gives that of "reed plain", in his article "Toyoashi no Mizuho no kuni" in the *Kokushi daijiten*. Torao Toshiya, *Engishiki* vol. 1, p. 471, follows this last interpretation in his note relating to the Ōtono ritual in ES book 8.

21 See *ashihara no mizuho no kuni* in M 18: 4094 (in the Kōdansha edition, Nakanishi Susumu notes that it refers to the land where the reeds grow thickly), M 13: 3227 and 3253, 2: 167, and M 2: 199 (the burning of a moor). See the *norito* in ES 8: 167, 170, 172, 175.

22 See wetlands, Chapter 2, note 79. This did not prevent the clearing of wetlands for rice cultivation, including reed plains, see *Hitachi fudoki*, p. 55. See Kanai Tenbi,

Shitsugen saishi, who nonetheless follows the interpretation of N (NKBT) concerning the myth of the divine descent.

23 *Izumo fudoki*, p. 217. Ōbayashi Taryō (1973: 105–210). Kasarhérou *et al.* (1989: 7–58). See the sūtra catalogue *Kaiyuan shijinglu* book 18, *Taishō daizōkyō* no. 2154, vol. 55, p. 671.

24 Ōbayashi (1973: 136) (Miyazaki), p. 133 (the five grains in Hida), p. 116 (the five grains in Harima), p. 119 (Sukunahiko); see *Hōki fudoki itsubun*, p. 480.

25 For the manuscripts, see *Nihon shoki* (NKBT), editors' introduction; and Kokugakuin daigaku Nihon bunka kenkyūjo ed. (1989: 17–19).

26 For what follows, see Veashua (2012); Kokugakuin daigaku (1989: 480).

27 *Shaku Nihongi* 8: 117–119, *yuniwa*, 5: 77.

28 *Kogo shūi*, p. 25–30, 126–129.

29 *Kogo shūi*, p. 28; *Izumo fudoki*, p. 217.

30 *Nihon shoki tsūshō*, pp. 84, 195, 228.

31 Verschuer (2004).

32 Murakami Shigeyoshi (1977: 68, 69). See also E. J. Hobsbawm, *The Invention of Tradition*, Cambridge University Press, 1997, first published as: *The Construction of Nationhood: Ethnicity, Religion and Nationalism*, Adrian Hastings, 1983.

33 See Murakami Shigeyoshi (1977: 75–106), for the newly created rituals. For the propaganda about rice, one may mention *Kokutai shikan* (1929), *Kokutai no hongi kaisetsu taisei* (1940) and *Kokutai no hongi to nōdō* (1942).

34 Tsuda Sōkichi (1948).

35 *Kogo shūi*, p. 128; *Sendai kuji hongi*, 3: 35, divine descent in Himuka.

36 *Nihongi shiki* (Kōhon, dating from 813), pp. 17, 18, nothing on *ho*.

37 N Sujin 7.11.13. (91 BC) and 12.9. (86 BC). The expression "hundred grains" is quoted from the *Hanshu* (1st century AD).

38 N Hanzei 1.10. (406), Nintoku 4.2. and 3. (316), Ninken 8 (495). The expression relating to the five grains is quoted from the *Houhanshu* (5th century AD).

39 N Keitai 24.2. (530), Ankan 2.1. (535), Senka 1.5. (536), Kinmei 28 (567).

40 See *itsutsuno tanatsumono* in *Nihongi shiki*, Otsubon (9th–10th century), p. 59; WR 17: 2; ES 8: 163, 173 (Kujōkebon); and RMS. *Tanatsumono* is found in RMS with the meaning "fruits of the rice fields"; but in *Horikawa-in hyakushu* 4: 26–1498, with the meaning "fruits of the fields" *(mi)sono*. Grain is glossed *koku* in IJ, *momi* in WR and RMS, and also *yashinau*, 'to feed', in RMS.

41 See the glosses and notes in N Keitai 24.2., Senka 1.5. (the editors read *momiine* instead of *momitane*). See Veashua (2009a: 37–59).

42 N Suiko 25 (617), Suiko 31.11. (623), 36.9.20. (628); N Kōgyoku 1.8.1. (642); the expression 'nine grains', taken from the *Zhouli*, is a variant of 'five grains'.

43 See the imperial virtue in RK book 173, p. 183 (in 705); SN Tenpyō 13.3.24. 741, Jingo keiun 1.1.8. 767; see *norito* in *Chōya gunsai* 12: 306; RK 11: 100 (Seiwa 8.7.6. 866); ES 8: 173; and "Cereals in the religious rituals", pp. 289–291.

44 N Saimei 5.7. (659), Tenji sokui 7.8. (668).

45 In SN (below, note 47), and in RK book 173, pp. 183–187, one finds *gokoku* 21 times, *nenkoku* 11 times, autumn harvests *aki no nariwai* 3 times and other expressions 15 times. The three terms are also found in RK 11: 91–102. These texts do not mention rice harvests.

46 SN Tenpyō hōji 7.8.1. 757.

47 N Tenmu 11.7.27. 683; SN Tenpyō 4.7.5. 732, 5.2.16. 733, 5.3 intercal. 2. 733, 8.10.22. 736, 11.7.14. 739, 13.3.24. 741, Tenpyō shōhō 1.1.4. 749, Tenpyō hōji 7.1.15. 757, 7.8.1. 757, Jingo keiun 1.1.8. 767, 1.2.22., 1.12.16., Hōki 6.4.7. 775, Enryaku 4.10.10. 785, 8.4.19. 789; RK vol. 1, pp. 93, 94, 100–102, 119, 468, vol. 2, p. 152 (dates of Jōwa 1.2.10. 834, 3.7.15. 836, Seiwa 8.7.6. 866, 9.9.11. 867, 10.12 intercal. 10. 868, 16.4 intercal. 7. 874, Enryaku 18.7.23. 799, Jōgan 17.6.8. 875); and the same articles as in SN, in RK vol. 1, pp. 449, 463, 466, 468, 480, vol.

2, pp. 183–185, 191, 230. See RK vol. 2, book 173, pp. 183–187, twenty-six other articles relating to food shortages dating from 705 to 841. And later references: *Azuma kagami* Kennin 1.8.23. 1201, Kangi 2.6.16. 1230, Bun'ei 3.3.5. 1266.

48 RK 16: 119 (Inari in 874); SN Tenpyō 11.7.14. 739; Tenpyō 13.3.24. 741 and RSK 3: 107; Tenpyō 13.2.14. 741.

49 See the price of rice in SN Tenpyō hōji 7.8.1. 757 and RK book 173, p. 185 (in 803). In case of loss with silkworm rearing and/or that of mulberry trees and hemp, there is an exemption from the tax in kind *chō* and from that replacing corvée labour *yō*, cf. RK vol. 2, p. 183 Yōrō 5.3.7. 721, Tenpyō 18.10.5. 746, p. 185 Hōki 6.11.7. 775. See a shortage of seed in SN Jingo keiun 1.2.22. 767; see the references to the 7th month above, note 47. See the allocations of the officials and the drying of rice in RSK 3: 125 and 8: 324.

50 See *den-en* in RK book 173, p. 186, 11: 91, 93, dates of 806, 816, 834; *sono* in *Izumo fudoki*, p. 193. See the loan *Jōgan gishiki*, p. 25. See the granaries *gisō* in RR Buyakuryō 6; ES 26: 655; RGi, p. 337; RSh, p. 395. See the decrees for the encouragement of cereal cultivation in RSK 8: 326–329. The annals mention hunger *ue, kikin* 13 times and the distribution of food *shinkyū* 11 times (see the references, note 45). In 749 there is a distribution of rice to officials, cf. SN Tenpyō shōhō 1.1.4. 749, in an amount of 6 *to* per person per month, i.e. 2 *shō* a day.

51 ES 39: 878, 879 (Bureau of the Imperial Table); ES 35: 803, 805 (Grains Office); ES 22: 578, and RSK 10: 359 (fields of *ine* and *awa*); RSK 14: 411; see the delivery of *shōmai* and *zakkoku* in RR Shikiinryō 42; RGi, p. 50; RSh, p. 126 (with one 9th-century commentary that protests: "Not at all: the meals [of officials] consist only of rice"). See also Satō Masatoshi (2008).

52 *Ruiju zatsuyōshō* 2: 538.

53 DNK 6: 53 for the order in 770; RSh, p. 345, commentary in the Denryō. See WR 17: 2; IJ under entry *ko inshoku*; nothing in ShJ; see *Shūgaishō* ge inshoku for *kyūkoku* and four definitions of *gokoku* (after *Kojiruien* Shokubutsuhen 1: 752); see *gokoku* in the Chinese classics, in the *Daikanwa jiten*.

54 *Shisseishoshō* ZGR 10 jō, p. 461; *Shinsarugakuki*, section on the third son-in-law Toyomasu. For Edo, see *Seiryōki*, p. 48, *Nōgyō zensho* 2, *Kōka shunjū* 5: 214.

55 See Kimura Shigemitsu (1996: 36, 38, 39); and above, notes 45 and 47.

56 Shibusawa Keizō (1954: 264–280).

57 ES 16: 435, ES 50: 993.

58 DNK 16: 292, purchase of *gogoku* for Shizume sai, which may be Hanashizume, Tamashizume or Hishizume. We opt for the last one. See Hishizume in ES 1: 9, 28, 8: 170, 171; RGi, p. 77.

59 See examples of prayers for the growth of *gokoku*: RSK 2: 67, annual readings of sūtras requested by Kūkai; SJ Gangyō 1.7.19. 877, *norito* concerning the change of era and prayer to safeguard the harvest; Kjg, p. 32, annual prayers at Ise from the beginning of the 7th month to the end of the 8th month; Hi 1: 16, register of the Tado Jingūji of 801; an episode in *Konjaku monogatari* 13: 33.

60 See Chapter 1, note 90; ES 3: 55, 56.

61 *Yamashiro fudoki itsubun*, p. 416. The Kamo festival, first celebrated in the 4th month, later also took place in the 11th month, cf. ES 2: 32.

62 N Tenmu 5.4.10. 676; ES 9: 188; SN Hōki 9.6.26. 778.

63 RR Jingiryō 4 and 6; RSh, pp. 195, 196; RGi, p. 77; and the *norito* in ES 8: 162–164, and in *Honchō gatsuryō*, pp. 264–265. On rituals and prayers, see Veashua (2009a).

64 For the dates of crops, see Chapter 1, "Harvesting", pp. 42–43. See *gokoku seijuku* in *Honchō gatsuryō*, p. 265.

65 For Hirose and Tatsuta, see ES 1: 19, 20 and ES 8: 163, 164; *Honchō gatsuryō*, pp. 264–265; the celebrations are mentioned in *Moromitsu nenjūgyōji*, pp. 348,

357; *Nenjūgyōji*, pp. 167, 172; *Moromoto nenjūgyōji*, pp. 243, 247. See the annual rituals with the lists of offerings in ES book 1. Rice is absent from certain rituals (see below, note 74).

66 ES 1: 24, 8: 166–167; but rice is present in the same ritual at Ise, cf. ES 4: 79.

67 RR Jingiryō 5 and 9; ES 1: 9. For Hirose and Tatsuta, ES 1: 20 mentions the messengers; ES 1: 9 also lists them among the festivals celebrated at the court.

68 *Nenjūgyōji shō*, p. 308; RSh, p. 196; RGi, p. 77; RR commentary p. 533; ES 8: 173 and Kjg, p. 33 record the *norito* with *gokoku*.

69 ES 1: 25, 35: 799, 39: 865; *Nenjūgyōji shō*, p. 308, *Nenjūgyōji*, p. 171, *Moromoto nenjūgyōji*, p. 246, *Moromitsu nenjūgyōji*, p. 355; *Honchō gatsuryō*, pp. 280–282. This is a "communion" of the emperor with the gods, similar to the Niiname, the difference being that old rice, i.e. of the previous year, is served (*Kokushi daijiten*).

70 ES 35: 803.

71 See *shinkoku* in RGi, p. 78, and *Shaku Nihongi*, p. 97; for rice and millets, see ES 2: 45, 35: 799, 800, 31: 749; and *Nenjūgyōji*, pp. 177, 180–183. See the other foods in ES 39: 866, and the offerings in ES 2: 44 (without rice, as for the Tsukinami). The Niiname originally took place in the 9th month (ES 1: 9), and was then replaced by the Kanname at Ise, celebrated on the 16th and 17th days of the 9th month (Kjg, pp. 39, 40; ES 4: 80, 81, 8: 174); and the Niiname was moved to the 2nd day of the hare in the 11th month, cf. *Moromoto nenjūgyōji*, p. 252, and RR Jingiryō 8 (Ainame, Ōname). M 14: 3386 mentions an offering of rice. *Nenjūgyōji*, pp. 180–183 (detailed description of the Niiname); *Moromoto nenjūgyōji*, p. 252; *Nenjūgyōji shō*, pp. 322, 323; *Moromitsu nenjūgyōji*, p. 369. See the gruel recipe in ES 40: 897, 898; see below, note 73.

72 *Hitachi fudoki*, pp. 39, 40. The *Awa fudoki* has not come down to us. For the priestess of Ise (not the shrine), see ES 5: 119; for *mino*, see Chapter 3, note 71.

73 ES 1: 9 lists the following annual rituals: Daijōsai (major ritual); Toshigoi, Tsukinami, Kanname, Niiname, Kamo (intermediate rituals); and Ōimi, Kazekami, Hanashizume, Saikusa, Ainame, Tamashizume, Hishizume, Michiae, Sonokarakami, Matsunoo, Hirano, Kasuga, Ōharano (minor rituals). The ES book 3 lists the other special rituals. See Daijōsai in ES book 7, Kanname in Kjg, pp. 39–42, Toshigoi in ES 1: 9, 8: 159–161, and RGi, p. 77.

74 Araki Toshio (1986). Rice is absent from the offerings for Toshigoi (ES 1: 9, 10, and 8: 159), as well as those for Tsukinami (ES 1: 24, 8: 166–167), for Niiname (ES 2: 44) and the other rituals of the 11th month (ES 2: 32–45). Okada Seishi (1970: 161), attributes the absence of rice from these offerings to the negligence of the shrine priests. At Kamo, rice is absent from the offerings, in the list after cloth, but the rice intended for making saké is there (ES 2: 33), as is the case for nearly all the offerings of the 11th month (ES 2: 32–42). Rice and beans are found together in the rituals of Kasuga (ES 1: 12), Hirano (ES 1: 21), Hiraoka (ES 1: 16), Kanname (ES 2: 31), Yomo no mikado (ES 1: 22), Mikawa-mizu (ES 1: 22). See the special rituals in ES book 3.

75 Kimura Shigemitsu (1996: 85–109); ES 40: 899; this gruel is also mentioned in *Ononomiya nenjūgyōji*, p. 378; *Nenjūgyōji hishō*, pp. 481, 482; and DNK 6: 140–142.

76 *Nenjūgyōji*, p. 159; *Moromoto nenjūgyōji*, p. 239; *Moromitsu nenjūgyōji*, p. 336; *Shisseishoshō*, p. 432; *Nenjūgyōji hishō*, p. 482.

77 ES 40: 897–899 and RR Shikiinryō 53 (regulations of the Water Bureau); *Nenjūgyōji*, pp. 171, 181 (gruel prepared by this office); *Moromoto nenjūgyōji*, p. 252; *Nenjūgyōji shō*, p. 323; *Moromitsu nenjūgyōji*, p. 369.

78 See the Gosaie banquets in ES 33: 767, 768, and the allocations in ES 35: 802; see gruels in ES 40: 899; and other references to Gosaie: *Kenmu nenjūgyōji*, p. 443; *Ononomiya nenjūgyōji*, p. 376; *Nenjūgyōji*, pp. 157, 158; *Moromoto nenjūgyōji*,

p. 238; *Shisseishoshō*, pp. 430, 431; *Nenjūgyōji shō*, pp. 267, 268; according to these texts, meals take place in the Daigokuden, readings in the Shingon-in.

79 *Gōke shidai* book 3 art. "Gosaie kyōjitsu"; *Chōya gunsai* 3: 50; *Gishiki* book 5 (*zakkoku* instead of *gokoku*, after *Kojiruien* Shokubutsuhen 1: 756); see Kūkai in RSK 2: 67, and *Mikkyō jiten*, p. 245.

80 *Moromitsu nenjūgyōji*, p. 336; and also *Nenjūgyōji hishō*, p. 482. Kimura Shige-mitsu (1996: 99, 100). Yoshi (the first character of his name is missing) is an unidentified person.

81 See *kusamochi* in WR 16: 15, *Moromitsu nenjūgyōji*, p. 344, and *Nenjūgyōji hishō*, p. 303; *chimaki* in WR 16: 13, ES 30: 775, *Nenjūgyōji hishō*, p. 303; melons *wasauri* in *Nenjūgyōji*, p. 169, *Morotō nenjūgyōji*, p. 214; *Moromoto nenjūgyōji*, p. 245; *Nenjūgyōji hishō*, p. 518; see *muginawa* in ES 30: 775, *Nenjūgyōji shō*, p. 314; *Moromitsu nenjūgyōji*, p. 357; *Nenjūgyōji hishō*, p. 528, and p. 529 (*mochi*).

82 RR Zōryō 40; they are: 1–1 audience, 7–1 festival of grey horses *aoba*, 16–1 dances *tōka*, 3–3 watercourse banquet *gokusui-en*, 5–5 *tango*, 7–7 Sumō, and the Niiname of the 11th month; and many other banquets that were added later. See the list of banquet foods in ES 39: 867–870 Naizen (imperial family), and ES 32: 761–765 Daizen (officials). There are many early protocol manuals on the court celebrations; see Yamanaka Yutaka (1972); Fujii Kazutsugu (1997).

Conclusion

The main purpose of this work is to situate irrigated rice cultivation in the overall context of the crops grown in premodern Japan. We have put forward a number of facts regarding the coexistence of rice growing and dry cereal cultivation, the practice of swidden farming, the gathering of plant foods, the relative proportion of cultivated and wild plants in the diet, and finally the cultural portrayal of rice and the other cereals. To remind readers of how our conclusions have been arrived at, we have included discussion sections where we summarize the history of research. This debate is governed by two factors: on the one hand, the scarcity or even the absence of written sources, and on the other, the difficulty of evaluating a historical reality that is not mentioned in the texts. The debate comes down to a problem of methodology and to one question, namely: what relative importance should be given to information recorded in the written sources as compared with that provided by the geographical and archaeological context?

The silence of the sources

"In archaeology, presence is significant, absence not always." This principle was stated by Jean-Marie Pesez during his seminars at the École des Hautes Études in Paris in 1997. Archaeologists are used to the disappearance of objects in the ground, by reason of the fragility of the materials, and they do not, *a priori*, draw any conclusions from the absence of certain categories of objects in archaeological sites. Historians are less familiar with the idea that "in written sources, presence is significant, absence not always". They are confronted by the absence of certain subjects in the sources. The scarcity of written sources is well known, but is managed differently depending on the subject. Do we not sometimes equate the occurrence in texts with the spread of evidence in history? Do we not tend to consider as widespread what is often mentioned, and as marginal what is found only sporadically in the texts? Japanese rice cultivation is a subject that in our view has produced disproportionate conclusions, because of the abundance of information about rice and the scarcity of sources on the other agricultural products. Thus, the importance of rice growing and the other crops has sometimes been assessed more by their appearance in the texts than from the historical context.

On the subject of agriculture and the history of the Japanese economy, one can distinguish different levels of silence in the sources. The historical texts show on the one hand an objective silence reflecting the absence of facts, and on the other hand a conjunctural silence, due to the loss of texts. There is also a structural silence specific to the premodern period (7th–16th centuries) that depends on the social milieu, context and sector being considered: the sources reflect the social milieu of the aristocratic and military élites, the economic context of public management, the political context of the government, and in general give priority to the public rather than the private sector. They deal only indirectly with the illiterate population and the private economy at the individual level. As we have considered agriculture from a private point of view, namely that of production and the subsistence economy, we have sometimes been led to use the historical sources deductively; for example, to evaluate foraging in the provinces from the fiscal regulations and to identify certain cultivation practices with the aid of toponyms. We have also drawn on literary texts (narrative and poetical) to fill in the gaps in the administrative texts. The poetry has given us an idea of the natural setting and the agricultural environment of the early medieval period. To support our arguments, we have drawn on the perspectives offered by archaeology and ethnology that show the practices and technical realities in prehistoric and premodern Japan. The approach we have described has enabled us to arrive at the following conclusions regarding plant products in early medieval Japan.

Mythology and cultural tradition

The random nature of the written sources available to us has resulted in the role of rice being overestimated in the mythology and the historiographical tradition. In fact, through our philological study of the myth of the divine descent, we have seen that contrary to the generally accepted interpretation, the ears brought from heaven to earth were not those of rice but of cereal polyculture. The presence of the various cereals is far more important than was previously thought. In early medieval Japan, successive emperors considered the "five grains" (not rice) to be the source of the country's prosperity. In the offerings organized by the court, rice occupied a minor place behind the other plant foods and marine products. The object of the rituals celebrated by the court was more often the growth of the "five grains" than that of rice. And in the official banquets and court celebrations, the "five cereals" played an important role, including in the emperor's New Year meal. There is no symbolism of rice in the court rituals and celebrations, or even in the nomenclature of the early texts. Modern theories that privilege a dichotomy opposing rice and dry cereals are therefore not based on the traditions of early medieval Japan, in which the different products of Japanese polyculture appear to coexist in harmony. For this period, it therefore seems more correct to speak of a "culture of the five cereals" than of a rice-growing culture. Only from the late 17th century did the Kokugaku school of nativist thought develop the idea of a mythological symbolism and a cultural supremacy of rice.

The fruits of polyculture were designated by the term "five grains" by the élite and annalists of the early medieval period. In their eyes, the "five grains" included cereals (millets, rice, wheat, barley) and beans (soybeans, red beans). The birth of the "five grains" is recounted in the country's mythology. Born from the corpse of the goddess Ukemochi, the cereals were then brought to earth, in Japan, by the grandson of the sun goddess. Whether in the annals or in the administrative texts, historians have sometimes translated by "rice" generic terms (grains, ears, etc.) that in fact refer to cereals in general.

Polyculture

From an agronomical point of view, the advantages of certain plants (at the nutritional and technical level) combined with the heavy tax burden on rice may have led the medieval Japanese to choose crops other than rice for their subsistence. Certainly rice occupied an important place in their everyday lives from the point of view of the investment in labour, but it did not always play a dominant role as regards the volume of production and consumption. Japanese agriculture was in fact characterized by a polyculture that was to continue throughout the medieval period and up to the 20th century. In our view, the conditions allowing rice to be assigned a predominant place in the crop production of the country – namely the technical innovations, a notion of productivity (defined in the agricultural treatises), the tax system established by Toyotomi Hideyoshi that subjected the dry cereals to the tax régime of rice, and the land clearances undertaken at the instigation of the Edo shogunal government – only came together from the 17th century.

In premodern Japan, irrigated rice did not occupy the most important place in the diet of the country's population, but it had its place among the cereals. Rice followed a historical path different from the other cereals. Arriving in the archipelago in the early 1st millennium, it made its way among the other graminae that had been cultivated from earlier times and were more widespread. Rice was later chosen by certain élites wishing to impose their authority. Used as an instrument of power, rice later spread to the territory controlled by the élite of the Yamato clan in the 7th century. It served from this time as the value standard for tax obligations and public accounting. Therefore, rice growing was forced onto the population beyond the favourable areas of the alluvial plains and intramountainous basins that cover no more than a quarter of the total land area of the archipelago. By contrast, the other cereal crops were able to spread across the entire country. Rice was therefore less omnipresent than is sometimes thought.

Plant tripolarity in agriculture and the food supply

The premodern Japanese were guided by the natural conditions, the country's rich biodiversity and their concern for food security to adopt three methods of acquisition: (1) permanent crops including irrigated rice cultivation, dry cereal crops and horticulture; (2) shifting crops including swidden fields and agri-sylviculture; (3) foraging for wild edible plants, fruits, nuts and tubers. Calorimetric estimates suggest that the medieval Japanese must have subsisted to a large extent thanks to dry crops and the gathering of plants, with the contribution of irrigated rice averaging not more than 25% in the diet of the common people. In fact, non-irrigated graminae, as well as beans, nuts, seeds, acorns and tubers must have formed an important part of the diet throughout premodern Japan. One may therefore speak of a threefold method of plant acquisition that is not limited to a coincidental triad, namely permanent agriculture, shifting cultivation and the harvesting of wild plants.

Appendix

Table A.1 Catalogue of edible and industrial plants

List of plants (excluding medicinal plants) drawn up from the following chapters of the *Engishiki* (927): 15 Kuraryō, 23 Minbu ge, 24 Shukei jō, 31 Kunaishō, 33 Daizen, 39 Naizenshi. These plants were eaten or used at the imperial court and delivered by the court services or by the provinces.

The following information is given for each plant:

- Written form according to the *Engishiki*, Shōsōin monjo documents and *Fudoki*;
- Old name according to the Shōsōin monjo, *Engishiki, Man'yōshū* and/or *Wamyō ruijushō* (in this order);
- Modern Japanese name according to the identification of Aoba Takashi 1991, Kimura Yōjirō 1991, Sekine Shinryū 1969, or Nihon gakushiin 1980 (in this order);
- Family according to Makino Tomitarō 1974, unless otherwise specified;
- Latin name according to Makino Tomitarō 1974, unless otherwise specified;
- English name according to *Zander* 2000, sometimes according to GP or BJ or GD; in brackets: another name. The English names in *Zander* are from Charles Quest-Ritson and are based on David J. Mabberley, *The Plant Book* (1997) etc.; in **bold** are the names used in the present work;
- French name according to Métailié, 1997, and/or *Le Bon Jardinier* and/or Cobbi 1978 (in this order, unless otherwise specified); and Péronny 1993 (PM);
- German name according to *Zander*; otherwise, no name.

Notes: In most cases, plants are identified by their family; and in English, French and German only by their genus and not by their species or variety.

We thank Georges Métailié for his advice regarding identification. See Métailié, 1999.

We have added *sobamugi* and *zakuro*, which are not found in the *Engishiki*.

Edible plants mentioned in sources other than the *Engishiki* are listed in Table 3.1.

Abbreviations

BJ: *Le Bon Jardinier*, 3 vols, ed. Jean-Noël Burte, Paris: La Maison Rustique, 1992

GD (general dictionary: *Harrap's Standard French and English Dictionary*, 2 vols, 1947, 1948; *Britannica World Language Edition of the Oxford Dictionary*, 2 vols, 1962; *Kenkyūsha's New Japanese–English Dictionary*, 1978)

GP: Michael Wright, *The Complete Handbook of Garden Plants*, London: Michael Joseph/Rainbird, 1984

Mé: Métailié, personal communication

PM: Claude Péronny, *Les plantes du Man'yōshū*, 1993

Zander: W. Erhardt, E. Götz, N. Bödeker, S. Seybold, *Zander Handwörterbuch der Pflanzennamen, Dictionary of plant names, Dictionnaire des noms de plantes*, 16th edition, Stuttgart: Ulmer, 2000 (1st edition 1927)

Zhongyao dacidian, 3 vols, Jiangsu Xinyixueyuan ed., Shanghai: Renmin chubanshe, 1977

***

Ai 藍 / ai, tadeai / Polygonaceae / Polygonum tinctorium Lour. / **Japanese indigo** (GD); knotweed (GP) / indigo d'Extrême-Orient (Cobbi) / ein Knöterich (see *itadori, tade*)

Akane 茜・茜根 / akane / Rubiaceae / Rubia cordifolia L. / **madder**; Indian Madder / garance (à feuilles cordiformes) (PM) / Krapp

Akebi 蔔子 / **akebi** / Lardizabalaceae / Akebia quinata Houtt. Decne. / akebia, five-leaf akebia (GP); Chocolate Vine / akebi, akébie / Fingerblättrige Akebie

Amazura, amakazura 甘葛・甘葛煎 / (a) tsuta / Vitaceae / Parthenocissus tricuspidata (Sieb. et Zucc.) Planch. / **(Japanese) ivy** (GD); Boston ivy (GP) / vigne vierge de Veitch, lierre du Japon (Mé) / Dreilappige Jungfernrebe / (b) amachazuru (14th century) / (c) kansho (17th century) / sugar-cane / canne à sucre

Aoi, afuhi 葵 / (a) aoi / Malvaceae / (b) fuyuaoi, okanori / Malvaceae / Malva verticillata L. / **mallow**; Curled Mallow / mauve (verticillée) (PM) / Quirl-Malve

Aona, kabura 菁・菁奈・蕪菁・蔓菁・菁根 / kabu / Cruciferae (Brassicaceae) / Brassica campestris L.; Brassica rapa L., according to *Zander* / **turnip**; Field Mustard (Rapa Turnip) / chou-navet; navet (Mé and Cobbi); rave (PM) (see *ochi*)

Aonori 青苔・青海菜・青乃利 / aosa / Ulvaceae / Ulva pertusa Kjellman / **ulva** (seaweed)

Aouri 青瓜 / (a) makuwauri, see *uri* / (b) shirouri / Cucurbitaceae / Cucumis melo L. var. conomon Makino / melon / **melon** conomon / Melone, Zuckermelone

Arame 滑海藻・荒海藻 / arame / Laminariaceae / Eisenia bicyclis Setchell / **eisenia** (seaweed)

Araragi 蘭・阿良々伎 / (a) undetermined / (b) nobiru, see *hiru*

Asa, o 麻 / asa, ōasa / Cannabinaceae / Cannabis sativa L. / **hemp** / chanvre (PM) / Hanf

Asatsuki 嶋蒜 / asatsuki / Liliaceae / Allium schoenoprasum L. / **chives** / civette, ciboulette / Schnittlauch

Ashi 葦 / ashi, yoshi / Graminae / Phragmites communis Trinius / **reed**; Common Reed / roseau, roseau à balai; roseau (PM) / Gewöhnliches Schilf

Awa 粟・禾 / awa / Graminae / Setaria italica Beauv. / **foxtail millet** (GD); Italian Millet, Foxtail Bristle Grass / millet à grappes, millet des oiseaux; millet d'Italie, millet à grappes (PM) / Kolbenhirse

Ayame, ayamekusa 菖蒲 / shōbu / Araceae / Acorus calamus L. var. asiaticus Pers. / **sweet flag** (GP); Calamus, Flag Root, Sweet Myrtle / jonc odorant (Mé); acore (Cobbi and PM) / Kalmus

Azami 葉薊・莇・薊 / (a) variety of Cirsium with many species / (b) sawaazami / Compositae / Cirsium hilgendorfi Makino / **thistle** (GD) / chardon à fleurs violettes / Kratzdistel

Azuki 小豆 / azuki / Leguminosae / Vigna angularis (Willd.) Ohwi et Ohashi, according to BJ / red bean / haricot rouge

Biwa 枇杷・枇杷子 / biwa / Rosaceae / Eriobotrya japonica (Thunb.) Lindl. / **loquat**; Japanese Medlar / néflier du Japon, bibacier / eine Wollmispel

Boke, see *shitomi*

Chi, kaya, tsubana 茅・茅花 / chigaya / Graminae / Imperata cylindrica Beauv. / **imperata**; Japanese Blood Grass, (Lalong Grass) (Lalonggrass) / impérata, herbe à paillottes (Mé); impérata (PM)

Chisa 萵苣・苣 / chisha, chisa / Asteraceae / Lactuca scariola L. / **lettuce** / laitue; styrax (du Japon) (PM) / ein Lattich, (Blattsalat)

Daizu, see *mame*

E, enomi 荏・荏子 / egoma / Labiatae (shiso-ka) / Perilla frutescens (L.) Britton var. japonica Hara / perilla (GP); Beefsteak Plant / **perilla**, pérille / eine Schwarznessel

Fufuki 蕗 / fuki / Compositae / Petasites japonicus Miq. / **butterbur** / chapelière / ein Pestwurz

Funori 布乃利・不乃里 / funori, fukurofunori / Endocladiaceae / Gloiopeltis furcata Postels et Ruprecht / **gloiopeltis** (seaweed)

Futomugi, mugi 大麦・麦 / ōmugi / Graminae / Hordeum vulgare L. / **barley** / orge (cultivée) (PM) / Mehrzeilige Gerste, Saat-Gerste

Gama, see *kama*

Gishigishi, see *shibukusa*

Goma 胡麻 / goma / Pedaliaceae / Sesamum indicum L. / **sesame** / sésame / (Sesam)

Hachisu, hasune, kaefu 蓮・蓮根・荷葉 / renkon, hasu / Nymphaeaceae / Nelumbo nucifera Gaertn. / lotus; Sacred Indian Lotus / lotus, lotus des Indes; **lotus** (PM) / Lotusblume

Hahako 波波古 / hahakogusa / Compositae / Gnaphalium multiceps Wall. / **cudweed** / immortelle (Mé), gnaphale / Ruhrkraut

Hajikami, see *kurenohajikami* and *naruhajikami*

Hashibami 榛子 / hashibami / Betulaceae / Corylus heterophylla Fisch. / **hazel tree, hazelnut** (GD); Hazel / noisetier, aveline / Mongolische Haselnuss

Hasune, see *hachisu*

Hemi 閂美・閂弥 / yabudemari / Caprifoliaceae / Viburnum tomentosum Thunb. / **Japanese snowball** (GP); snowball-tree (GD); Lace Cup Viburnum; lacecap viburnum (GP) / boule de neige du Japon, viorne / Japanischer Schneeball

Hie 稗・稗子・稗 / hie / Graminae / Echinochloa crus-galli (L.) Beauv., according to BJ / **barnyard millet**; Barnyard Grass, Cockspur / panic pied-de-coq (PM) / Gewöhnliche Hühnerhirse

Himegurumi 姫胡桃子, see *kurumi*

Hiru, nuhiru, nobiru, nebiru 蒜•野蒜•沢蒜 / (a) Liliaceae, (b) nobiru / Liliaceae / Allium grayi Regel / **wild chives** (from French term); ciboulette sauvage; ail (de Gray) (PM) / ein Lauch (see *ki, mira*)

Hisago 瓠•瓠•匏 / hisago / (a) Cucurbitaceae of the genus Lagenaria with many species / **gourd** / courge / Flaschenfrucht / (b) yūgao / Cucurbitaceae / Lagenaria leucantha Rusby var. clavata Makino / **calabash**; calebasse / (c) hyōtan / Lagenaria leucantha Rusby var. Gourda Makino / white-flowered gourd (BJ) / gourde / Flaschenfrucht-Kürbis (BJ) (see *uri*)

Hishi 菱•菱子 / hishi / Trapaceae (Hydrocaryaceae) / Trapa natans L. / **water chestnut**; caltrop / mâcre, châtaigne d'eau (PM) / Gewöhnliche Wassernuss

Hiyu 莧 / hiyu / Amaranthaceae / Amaranthus inamoenus Willd. / **amaranth** (GP) / amarante / ein Fuchsschwanz

Hosoki 樏椒•樏椒, inuzanshō 犬山椒 / Rutaceae / Fagara mantchuria Honda / **Fagara**

Hozochi, see *uri*

I 藺 / i, igusa / Juncaceae / Juncus effusus L. / **rush**; Common Rush, Soft Rush; corkscrew rush (GP) / jonc / Flatter-Binse

Ichigo 覆瓫子 / (a) Rosaceae / (b) e.g. momijiichigo or kiichigo / Rosaceae / Rubus palmatus Thunb. **blackberry** (GD), brambleberry; brambles (GP) / ronce (mûre de la ronce) / eine Himbeere, Brombeere

Ichii 伊智比•伊知比古•櫟子 / ichiigashi / Fagaceae / Quercus gilva Blume / **oak** / chêne; chêne (jaune pâle), chêne premier (PM) / eine Eiche

Igisu 小凝菜•伊伎須 / igisu, egonori / Ceramiaceae / Campylaephora hypnaeoides J. Agardh / **campylaephora** (seaweed)

Imo, yamanoimo, yamatsuimo 暑預•薯預•署豫 / yamanoimo / (a) Dioscoreaceae / (b) Dioscoreaceae / Dioscorea japonica Thunb., according to Makino Tomitarō; Dioscorea alata Linn., Dioscorea batatas, according to Hashimoto Seiji / **yam**; Japanese Yam / igname / eine Yamswurzel (see *tokoro*)

Imo, umo 芋•家芋 / satoimo, taroimo / (a) Araceae / (b) Araceae / Colocasia antiquorum Schott et Engl., according to Zhongyao dacidian; Colocasia esculenta (L.) Schott, according to Hashimoto Seiji / **taro**; Dasheen / colocase, taro (PM) / eine Taro

Ine, yone (grains) 稲・米 / ine / Graminae / Oryza sativa L. / **rice** / riz (PM) / Reis, see *mochiyone, uruchi(no)yone*

Inuhōzuki, see *konasubi*

Itabi 木蓮子 / itabikazura / Moraceae / Ficus foveolata Wall. / **fig** / figuier / ein Feigenbaum

Itadori 虎杖・唐丈 / itadori / Polygonaceae / Polygonum cuspidatum Sieb. et Zucc. / polygonum; Japanese Knotweed / **polygonum**, une renouée (Mé) / ein Knöterich (see *tade, ai*)

Itajihajikami, see *hosoki*

Kaba 樺 / (a) kabanoki-ka / Betulaceae / (b) shirakaba / Betulaceae / Betula platyphylla Sukatchev var. japonica Hara / **birch**; Japanese White Birch / bouleau / Mandschurische Birke

Kabura, see *aona*

Kaefu, see *hachisu*

Kaki 柿・柿子・柿 / kaki / Ebenaceae / Diospyros Kaki Thunb. / **persimmon** / kaki, plaqueminier / Kakipflaume, Kaki

Kama 蒲・莞 / gama / Typhaceae / Typha latifolia L. / **bulrush**, Cat Tail; reedmace (GP) / massette, queue de renard / Breitblättriger Rohrkolben

Kamouri 冬瓜・鴨瓜 / tōga / Cucurbitaceae / Benincasa hispida (Thunb.) Cogn., according to *Zhongyao dacidian* / **wax gourd** / courge à la cire, benincasa / Wachskürbis (see *uri*)

Kanshi, kōji 甘子・柑子 / (1) variety of mikan / Rutaceae / citrus, **mandarin** (generic term), mandarine, see *tachibana* / (2) kōji, karatachibana (Edo period) / Myrsinaceae / Ardisia crispa DC. / Japanese Holly/ardisia, mandarine

Karamomo 杏 / anzu / Rosaceae / Prunus armeniaca L. var. Ansu Maxim. / **apricot**; ornamental cherry (GP) / abricotier anzu / Aprikose, Marille (see *ume*)

Karamushi, o 紵・苧・絳 / karamushi / Urticaceae / Boehmeria nivea Gaud. / ramie; China Grass, (False Nettle) / **ramie**, ortie de Chine / Rami

Karashi 芥・芥子 / karashina / Cruciferae / Brassica juncea Czern. et Coss. / **leaf mustard** (GD); Chinese Mustard, Indian Mustard / moutarde à feuilles de chou, moutarde / Ruten-Kohl, Sarepta-Senf (see *takana*)

Kariyasu, kariyasukusa 苅安草 / (a) kariyasu / Graminae / Miscanthus tinctorius Hackel / **miscanthus**, silver grass (GP) / miscanthe *kariyasu* / (b) kobunagusa / Arthraxon hispidus Makino (see *kaya*)

Kashiwa, kae 柏 / (a) Euphorbiaceae / (b) akamegashiwa / Euphorbiaceae / Mallotus japonicus Muell.-Arg. / (c) mitsunagashiwa / Polypodiaceae / Neottopteris antiqua Masamune / (d) kashiwa / Fagaceae

Kashiwa 槲・柏・栢 / (a) Fagaceae / (b) buna / Fagaceae / Fagus crenata Blume / **beech** / hêtre / Japanische Buche

Kawahone 荊根・川骨 / (a) kōhone / Nymphaeaceae / Nuphar japonica DC. / **water-lily**; nenuphar (Yellow Pond Lily, Cow Lily) / nénuphar du Japon, lis d'eau jaune / Teichrose / (b) onibasu / Nymphaeaceae / Euryale ferox Salisb. / foxnut / euryale (see *mizufufuki*)

Kaya, kae 榧 / kaya / Taxaceae / Torreya nucifera Sieb. et Zucc. / *kaya*, Japanese Torreya; nutmeg tree (GP) / Japanese nutmeg-yew seed (Makino) / **kaya**, torreya / Japanische Nusseibe

Kaya 草・茅・萱 / (1) kaya / thatch / chaume, generic term / (2) susuki / Graminae / Miscanthus sinensis (Thunb.) Anderss. / **eulalia**; miscanthus; Chinese Silver Grass, Tiger Grass; silver grass (GP) / miscanthe; eulalia (Mé); miscanthe (de Chine) (PM) / Eulalie / (3) see *chi* / (see *kariyasu*)

Kemushi 枲 / (a) asa / (b) karamushi

Ki, negi 葱・蔥 / negi / Liliaceae / Allium fistulosum L. / **spring onion** (GD); Welsh Onion, Japanese Bunching Onion / ciboule / Winter-Zwiebel (see *mira, hiru*)

Kibi, see *kimi*

Kihada 黄蘗 / kihada / Rutaceae / Phellodendron amurense Rupr. / **Amur cork tree**, phellodendron (GP) / liège d'Amour (Mé), phellodendron / Amur-Korkbaum

Kimi, kibi 黍子・黍 / kibi / Graminae / Panicum miliaceum L. / **common millet**, broomcorn millet, (Indian Millet) / millet de Bordeaux, millet à balai, millet commun, panic millet, millet à panicule; panic millet, millet de Chine (PM) / Echte Hirse, (Rispenhirse)

Kiuri 生瓜 / kyūri / Cucurbitaceae / Cucumis sativus L. / **cucumber** / concombre / Gurke (see *uri*)

Kobu, see *konbu*

Kōji, see *kanshi*

Kokorofuto, see *korumoha*

Komira, see *mira*

Komo (komozono) 蒋•薦子•蒋子 / makomo / Graminae / Zizania latifolia Turcz. / **wild rice**; American wild rice; Water Rice / zizanie américaine; zizanie (à larges feuilles) (PM) / Wilder Reis

Komugi, mugi 小麦•麦 / komugi / Graminae / Triticum aestivum L. / **wheat** / blé / Weizen

Konagi, see *nagi*

Konasubi 龍葵子•蘢葵 / inuhōzuki / Solanaceae / Solanum nigrum L. / **black nightshade**; Common Nightshade / morelle noire, amourette / Schwarzer Nachtschatten

Konbu, hirome, ebisume 昆布 / makonbu / Laminariaceae / Laminaria japonica Areschoug / **laminaria** (seaweed)

Konishi 胡荽 / (a) karashi (8th century) / a crucifer; **coriander** / une crucifère / (b) koendoro (17th century) / Coriandrum sativum / coriandre

Korumoha, kokorofuto 凝海藻•大凝菜•凝菜•心太 / tengusa, makusa / Gelidiaceae / Gelidium Amansii Lamouroux / **gelidium** (seaweed)

Koshiabura 金漆 / koshiabura / Araliaceae / Acanthopanax sciadophylloides Franch. et Sav. / **acanthopanax** / – / eine Fingeraralie

Kōzo, taku 穀皮•楮•栲•梓 / kōzo / Moraceae / Broussonetia papyrifera (L.) Vent., according to Zhongyao dacidian / **paper mulberry** (GP) / mûrier à papier (PM) / Papiermaulbeerbaum

Kuchinashi 支子 / kuchinashi / Rubiaceae / Gardenia jasminoides Ellis f. grandiflora Makino / **gardenia**, Cape jasmine (GP) / jasmin du cap, gardénia

Kukutachi 茎立・莖立 / kukutachi, generic term / **crucifer stems** / tiges de Cruciferae

Kurenai, kurenoai 紅花 / benibana / Compositae; Asteraceae according to Zander / Carthamus tinctorius L. / **safflower** (GD); Fake Saffron, Dyer's Saffron / carthame, safran bâtard; carthame des teinturiers, safran bâtard (PM) / Färber-Distel, Färber-Saflor

Kurenohajikami, hajikami 薑・生薑 / shōga / Zingiberaceae / Zingiber officinale Rosc. / **ginger** / gingembre / Ingwer (see *mega*)

Kuri 栗子・栗 / kuri / Fagaceae / Castanea crenata Sieb. et Zucc. / **(sweet) chestnut** (GD); Japanese Chestnut; chestnut (GP) / châtaignier du Japon, kuri; châtaignier (PM) / Japanische Kastanie

Kurumi 胡桃子・胡桃・呉桃子 / (a) Juglandaceae / (b) onigurumi / Juglandaceae / Juglans mandshurica Maxim. / (c) himegurumi / Juglans mandshurica Maxim. var. cordiformis Kitam. / **walnut** (GP); Manchurian Walnut / noyer de Manchourie (Mé) / Mandschurische Walnuss

Kusabira, see *take*

Kuwa 桑 / kuwa / Moraceae / Morus bombycis Koidz. / **mulberry** (GP) / mûrier; mûrier (du bombyx) (PM) / Maulbeerbaum

Kuzu, kuzukazura 葛・葛葉 / kuzu / Leguminosae; Fabaceae according to Zander / Pueraria Thunbergiana Benth. / **pueraria**, kudzu vine (GP) / pueraria, kudzu; puéraire (PM)

Makomo, see *komo*

Makuwauri, see *uri*

Mame 大豆 / daizu / Leguminosae / Glycine max (L.) Merrill / **soybean**; Soya Bean / soja; soja sauvage, soja grimpant (PM) / Sojabohne

Me, see *wakame*

Mega 蘘荷・売我・女我 / myōga / Zingiberaceae / Zingiber Mioga (Thunb.) Rosc. / **Japanese ginger** (GD); Mioga Ginger / gingembre / Ingwer (see *kurenohajikami*)

Mino 菫子・菫 / (a) mutsuoregusa? or (b) kazunokogusa? / Graminae / Glyceria acutiflora Torrey or Beckmannia erucaeformis Host / glyceria?; variegated manna-grass (GP)?, Sweet Grass? / **glycérie**? / ein Schwaden?

Mira, komira 韭 / nira / Liliaceae / Allium tuberosum Rottl. / **Chinese chives**, Oriental Garlic / ciboulette chinoise, ail sauvage; ail (tubéreux) (PM) (see *hiru, ki*)

Miru 海松 / miru / Codiaceae / Codium fragile Suringar / **codium** (seaweed)

Mizufufuki 鶏頭子 / onibasu / Nymphaeaceae / Euryale ferox Salisb. / **foxnut** / euryale (see *kawahone*)

Mochiyone, mochi(no)yone, mochiine 糯米・粳米・糯稲 / mochiine / Graminae / Oryza sativa L. var. glutinosa / **glutinous rice** (GD); waxy rice, sticky rice / riz gluant

Moke, see *shitomi*

Momo 桃・桃子 / momo / Rosaceae / Prunus persica (L.) Batsch. / **peach** / pêcher (PM) / Pfirsich

Mozuku 毛都久 / mozuku / Chordariaceae / Nemacystus decipiens Kuckuck / **nemacystus** (seaweed)

Mube 郁子 / mube / Lardizabalaceae / Stauntonia hexaphylla (Thunb.) Decne. / **stauntonia** / stauntonia

Mugi, see *komugi, futomugi*

Mume, see *ume*

Murasaki 紫草・紫 / murasaki / Boraginaceae / Lithospermum erythrorhizon Sieb. et Zucc. / **gromwell** (GD); Lithospermum (GP) / grémil du Japon; grémil (à racines rouges) (PM) / Steinsame

Murasakinori, see *nori*

Myōga, see *mega*

Nagi, konagi 水葱 / (a) mizuaoi / Ponderaceae / Monochoria Korsakowii Reg. et Maack / monochorie, poireau d'eau, mauve d'eau, all (PM) / (b) konagi / Ponderaceae / Monochoria vaginalis Presl **water-leek** (from French term)/ monochoria, poireau d'eau

Namai 沢寫 / omodaka / Alismataceae / Sagittaria trifolia L. / **arrowhead** (GP) / sagittaire / Pfeilkraut

Naruhajikami 蜀椒・檔椒・椒枡・椒檔・椒 / sanshō / Rutaceae / Zanthoxylum piperitum DC., according to BJ / **Japanese pepper** / clavalier, poivrier du Japon / Japanischer Pfeffer

Nashi 梨子・梨 / nashi / Rosaceae / Pyrus serotina Rehder / **pear**; Sand Pear / poirier du Japon (PM) / China-Birne

Nasubi, nasu 茄・茄子 / nasu / Solanaceae / Solanum melongena L. / **egg-plant** / aubergine / Aubergine

Natsume 棗 / natsume / Rhamnaceae / Ziziphus jujuba Mill. / **jujube** / jujubier (PM)

Nazuna 薺 / nazuna / Cruciferae / Capsella bursa-pastoris (L.) Medikus / **shepherd's purse** / bourse-à-pasteur, capselle / Gewöhnliches Hirtentäschel

Nigime, see *wakame*

Nira, see *mira*

Nire 楡 / akinire / Ulmaceae / Ulmus parvifolia Jacq. / **elm**; Chinese Elm / orme; orme (à petites feuilles) (PM) / Japanische Ulme

Nobiru, nuhiru, nebiru, see *hiru*

Nori, murasakinori 紫菜 / asakusanori / Bangiaceae / Porphyra tenera Kjellman / **porphyra** (seaweed)

Nunawa 根蓴・蓴・奴縄・蓴 / junsai / Nymphaeaceae / Brasenia Schreberi J. F. Gmel. / **water shield** / brasenia, brasénie; brasénie (PM) / Schleimhaut

O, see *asa, karamushi*

Ochi, uchi 蕓薹 / aburana (cf. aona) / Cruciferae (Brassicaceae) / Brassica campestris L. subsp. Napus Hook. fil. et Anders. / **colza**; a Field Mustard / colza / ein Kohl (see *aona*)

Ōdochi, ōtsuchi 茶 / nogeshi / Compositae (Asteraceae) / Sonchus oleraceus L. / **sow-thistle**, Milk Thistle / laiteron (maraîcher) / Kohl-Gänsedistel

Ogonori 於期菜 / ogonori (genus) / Gracilariaceae / genus **gracilaria** (seaweed)

Ohagi, see *uhagi*

Okera, see *ukera*

Ōmugi, see *futomugi*

Ōne, *suzushiro* 蘿蔔・蘿蔔・蘿蔔子・大根 / daikon / Cruciferae / Raphanus sativus L. / **radish** / radis, radis géant / Garten-Rettich

Sasage 大角豆 / sasage / Leguminosae / Vigna Catiang Endl. / **cowpea**; (Asparagus Bean) / dolique

Seri 芹・茎芹・葉芹 / seri / Umbelliferae / Oenanthe stolonifera DC. / **water-celery** (from French term); Japanese parsley (GD) / oenanthe; céleri d'eau (PM)

Shibukusa, *shi* 羊蹄 / gishigishi / Polygonaceae / Rumex japonicus Houttuyn / **sorrel**, Dock / oseille du Japon (Mé), rumex / Ampfer

Shii 椎 / (a) shiinoki / Fagaceae / genus castanopsis / (b) tsuburajii / Fagaceae / Shiia cuspidata Makino / **castanopsis**, shiia; shii or castanopsis (PM)

Shitomi, moke, boke 木瓜実・白花木瓜・枹 / kusaboke / Rosaceae / Chaeno-meles japonica (Thunb.) Lindl. / **wild quince** (GD); Japanese Quince; Maule's quince (GP) / cognassier du Japon / Japanische Scheinquitte

Shōmokkō 青木香 / umanosuzukusa / Aristolochiaceae / Aristolochia debilis Sieb. et Zucc. / **aristolochia**; birthwort (BJ) / aristoloche / Pfeifenblume

Sobamugi, somamugi 蕎麦 / soba / Polygonaceae / Fagopyrum esculentum Moench / **buckwheat** / sarrasin, blé noir (Mé) / Echter Buchweizen

Soraji 蘇良自・蘇羅自 / (a) undetermined / (b) kasamochi / Umbelliferae / Nothosmyrnium japonicum Miq. / **nothosmyrnium**

Suge 菅 / (a) Cyperaceae / laîche (PM) / (b) karasuge / Cyperaceae / Carex dispalata Boott / **sedge** (GP) / carex, laîche / eine Segge

Sumomo 李子・李 / sumomo / Rosaceae / Prunus salicina Lindl. / **plum**; Japanese Plum / prunier (Mé); prunier-saule, prunier trifolié (PM) / Pflaume (see *ume*)

Susuki, see *kaya*

Suzushiro, see *ōne*

Tachibana 橘子・橘 / (a) citrus, generic term (see *kanshi*) / (b) tachibana / Rutaceae / Citrus Tachibana C. Tanaka / **orange** / oranger; oranger (tachibana) (PM) / Orange (see *yuzu*)

Tade 蓼 / yanagitade, hontade / Polygonaceae / Polygonum hydropiper L. / **water-pepper**; Red Knees / poivre d'eau (PM) / Wasserpfeffer-Knöterich (see *itadori, aî*)

Takamuna 笋・竹子・筍 / takenoko, generic term / **bamboo shoot** / pousse de bambou / Bambussprosse

Takana, takuna 菘・辛芥 / takana / Cruciferae (Brassicaceae) / Brassica juncea (L.) Czern. et Coss. var. integrifolia Sinsk. / **leaf mustard** (GD); Chinese Mustard, Indian Mustard / moutarde takana, moutarde à feuilles de chou (Mé), navette / Ruten-Kohl, Sarepta-Senf (see *karashi*)

Take, kusabira 菌・茸 / kinoko, generic term / **mushroom** / champignon / Pilz

Take 竹 / take, generic term / **bamboo** / bambou / Bambus

Taku, see *kōzo*

Tara 太羅・多羅 / taranoki / Araliaceae / Aralia elata Seem. / **Japanese angelica tree** / aralia élévée, according to BJ / Japanischer Angelikabaum (see *tsuchitara*)

Tochi, tochinoki 栃・橡 / (a) tochinoki-ka / Hippocastanaceae / (b) tochinoki / Hippocastanaceae / Aesculus turbinata Blume / **horse chestnut** (GP); Japanese Horse Chestnut / marronnier d'Inde / Japanische Rosskastanie

Tokoro, tokorozura 莟・薢・蘭蔙 /   tokoro, onidokoro / Dioscoreaceae / Dioscorea Tokoro Makino / **yam** / igname tokoro / eine Yamwurzel (see *imo*)

Tokusa 木賊 / tokusa / Equisetaceae / Equisetum hyemale L. / **scouring rush**, Dutch Rush / prêle d'hiver / Winter-Schachtelhalm

Tsubaki 椿木・椿・海石榴 / tsubaki / Theaceae / Camellia japonica L. / **camellia** / camélia du Japon; camellia ou camélia (PM) / Japanische Kamelie

Tsuchitara 獨活 / udo / Araliaceae / Aralia cordata Thunb. / **angelica tree** / aralia cordiforme (see *tara*)

Tsuge 黄楊 / tsuge / Buxaceae / Buxus microphylla Sieb. et Zucc. / **boxwood**; Japanese Boxwood; little-leaf boxwood (GP) / buis à petites feuilles (Mé); buis (PM) / Kleinblättriger Buchsbaum

Tsunomata 鹿角菜 / kotojitsunomata / Gigartinaceae / Chondrus elatus Holmes / **chondrus** (seaweed)

Tsunomata 角俣菜・角俣 / tsunomata / Gigartinaceae / Chondrus ocellatus Holmes / **chondrus** (seaweed)

Tsurubami 橡 / (a) see tochinoki / **horse chestnut** / marronnier d'Inde / (b) buna-ka / Fagaceae / (c) kunugi / Fagaceae / Quercus acutissima Carruth. / **sawtooth oak** / chêne à dents de scie; chêne-charbon (PM) / Seidenraupen-Eiche

Tsuzura 葛・黒葛 / (a) Menispermaceae / (b) tsuzurafuji / Menispermaceae / **Sinomenium** diversifolium Diels / Sinomenium / cocculus (trilobé) (PM) (see *kuzu*)

Uchi, see *ochi*

Uhagi, ohagi 莪蒿・莪・莪頭蒿 / yomena / Compositae / Aster Yomena Makino / aster (GP) / aster à fleurs blanches; **aster** (yomena) (PM) / eine Aster

Ukera, okera 白朮・宇家良 / okera / Compositae / Atractylis ovata Thunb. / atractylis / atractyle (PM)

Ume, mume 梅・梅子 / ume / Rosaceae / Prunus mume Sieb. et Zucc. / **Japanese apricot** / abricotier du Japon; prunier du Japon (PM) / Japanische Aprikose (see *karamomo*)

Umo, see *imo*

Uri, hozochi 瓜・孰瓜 / makuwauri / Cucurbitaceae / variety of Cucumis melo L. / **melon** / melon (PM) / Melone (see *aouri, kiuri, hisago, kamouri*)

Uruchi(no)yone, urushine 粳米・粳 / uruchine, ine / Graminae / Oryza sativa L. / **non-glutinous rice** (GD) / riz non gluant

Urushi 漆 / urushi / Anacardiaceae / Rhus verniciflua Stokes / **varnish tree** / arbre à laque (Mé), sumac, sumac à laque / Lack-Sumach

Wakame, nigime, me 稚海藻・海藻・和布 / wakame / Laminariaceae / Undaria pinnatifida Suringar / **undaria** (seaweed)

Warabi 蕨・薇 / warabi / Polypodiaceae / Pteridium aquilinum (L.) Kuhn / **bracken** / fougère aigle ou grand-aigle (PM) / Adlerfarn

Wasabi 山薑・山葵 / wasabi / Cruciferae / Wasabia japonica Matsum. / **Japanese horseradish** / raifort japonais, wasabi

Yamaararagi 山蘭 / kobushi / Magnoliaceae / Magnolia kobus DC. / **magnolia kobus**; Kobus Magnolia; magnolia (GP) / un magnolia / Kobushi-Magnolie

Yamamomo 楊梅子・楊梅・山桃 / yamamomo / Myricaceae / Myrica rubra Sieb. et Zucc. / **red myrica**; wax myrtle (GP) / arbousier d'Extrême-Orient / (Wachsmyrte)

Yamanoimo, see *imo*

Yomena, see *uhagi*

Yomogi 艾 / yomogi / Compositae / Artemisia vulgaris L. var. indica Maxim. / **mugwort**; artemisia (GP) / armoise des Indes (Mé); armoise commune (PM) / Gewöhnlicher Beifuss

Yuri 百合 / (a) variety of yuri / Liliaceae / variety of Lilium L. / **lily** (GP) / lis (PM) / Lilie / (b) e.g. yamayuri / Liliaceae / Lilium auratum Lindl. / Golden Rayed Lily; gold-rayed lily (GP) / lis, lis doré, lis sauvage / Goldband-Lilie

Yuzu 柚子・柚 / yuzu / Rutaceae / Citrus Junos Tanaka / **citron**; Meyer's Citrus / orange amère, cédrat (see *tachibana*)

Zakuro 拓榴 / zakuro / Punicaceae / Punica Granatum L. / **pomegranate** / grenadier, grenadier commun / Granatapfelbaum

Table A.2 **Measures**

8th century

1 *chō* = 1.13 hectares;	1 *tan* = 0.3 acres
1 foot *shaku* = 29.6 centimetres;	1 inch *sun* = 2.96 centimetres
1 *koku* = 85 litres;	1 *to* = 8.5 litres; 1 *shō* = 0.85 litres

1 sheaf *soku* = 10 bundles *ha*

1 *soku* produces 0.1 *koku* or 1 *to* (8.5 litres) of unhulled rice *momi, kome*.

1 *to* (8.5 litres) of unhulled rice gives 0.5 *to* or 5 *shō* (4.25 litres) of pounded hulled rice *shōmai* or husked rice *hakumai* (at a pounding rate of 50%; see Chapter 1, note 151).

1 sheaf *soku* gives 5 *shō* (4.25 litres) of husked rice *hakumai*.

1 *koku* (85 litres) of unhulled rice gives 5 *to* (42.5 litres) of husked rice *hakumai* pounded at 50%.

1 litre of husked rice weighs 834 grammes.

100 grammes (0.12 litres) of husked rice has 351 kcal.

1 *shō* of husked rice = 0.85 litres = 709 grammes = 2,489 kcal.

1 *koku* of husked rice = 85 litres = 70.9 kg = 248,900 kcal.

1 sheaf *soku* of rice produces 12,445 kcal of edible rice pounded at 50%.

Modern period (after 1624)

1 *chō* = 0.992 hectares;	1 *tan* = 0.26 acres
1 *koku* = 180 litres;	1 *to* = 18 litres; 1 *shō* = 1.8 litres

1 *to* (18 litres) of unhulled rice gives 0.5 *to* or 5 *shō* (9 litres) of hulled rice or 'whole' rice *genmai* (at a pounding rate of 50%; see Chapter 4); or 0.4 *to* or 4 *shō* (7.2 litres) of husked rice *hakumai* pounded at 60%.

1 *koku* (180 litres) of unhulled rice gives 4 *to* (72 litres) of husked rice *hakumai* pounded at 60%.

1 *koku* (180 litres) of hulled or 'whole' rice weighs 150 kilogrammes.

1 litre of hulled or 'whole' rice weighs 834 grammes.

100 grammes (0.12 litres) of 'whole' rice has 351 kcal.

1 *koku* of 'whole' rice = 180 litres = 150 kg = 526,500 kcal.

Note: These are the measures used in the present work.

Sources: *Iwanami Nihonshi jiten* (1999) (measures); Kagaku gijutsuchō Shigen chōsakai, *Shinpen shokuhin seibun hyō* (2000) (calories); Arizono Shōichirō, *Zairai nōkō no chiiki kenkyū* (1997) (figures in kg and *genmai* pounding rate after 1624).

Table A.3 **Chronological table**

	Prehistoric period
Neolithic *Jōmon*	10000 to mid-2nd or mid-1st millennium BC*
Yayoi	mid-2nd or mid-1st millennium BC to late 3rd century AD*
Tumulus *Kofun*	late 3rd to 6th century AD
	Early medieval period
Asuka	592–710 AD
Nara	710–784 AD
Heian	794–1185 AD
	Late medieval period
Kamakura	1185–1333 AD
Nanbokuchō	1336–1392 AD
Muromachi	1392–1573 AD
	Early modern period
Edo	1603–1867 AD
	Modern period
Meiji	1868–1912 AD
Taishō	1912–1926 AD
Shōwa	1926–1989 AD

Note: *These dates have been revised since the 1990s as a result of archaeological discoveries. Shitara Hiromi (2014: 465) situates the beginning of Yayoi culture at the time of the transition from a foraging to a farming society (before the introduction of irrigated rice growing), with a significant difference between the dates for the north-east and south-west of the archipelago. However, works published up to the 1990s place the end of the Neolithic and the beginning of the Yayoi period around 500 or 300 BC. In this book, we give the older dates when quoting the works of earlier authors.
The division into three main periods (prehistoric/ancient, medieval, modern) adopted in this table is based, for convenience, on European usage. For Japanese periodization, see Minamikawa Takashi (2010) and Sakamoto Shōzō (2012).

Table A.4 Agricultural calendar (Kaga region in 1717)

Month / Crops	1	2	3	4	5	6	7	8	9	10	11	12
rice *ine*			S........		T........			H........	H			
millet *awa*			S........			H....H						
millet *awa* (swidden fields)				S........				H?				
barnyard millet *hie*			S........		T........		H					
barnyard millet *hie* (swidden fields)				S........					H?			
buckwheat *soba* (swidden fields)						S........		H........	H			
wheat *komugi*	....				H			S.....S........				
barley *ōmugi*	....				H			S.....S........				
soybeans (early) *daizu*			S........		H							
soybeans (late, after wheat)					S........			H				
soybeans (swidden fields)				S........				H?....H?				

Month / Crops	1	2	3	4	5	6	7	8	9	10	11	12
taro *satoimo, taroimo*			S................................				H?....H?....H?					
perilla *egoma*				S..............................H								
colza *natane*			...........................H					S...............T after rice..........				
hemp *asa*			S.....................H									
ramie *kara-mushi*			S....................H?									
indigo *ai*		S..........................H										
cotton *momen*			S.....................H?									
radish *daikon* (swidden fields)					S................................H							
radish after wheat					S................................H							
eggplant *nasu*		S?...........................H										
eggplant after wheat					S..............................H.......H?							
melon *uri*		S............................H										
melon after wheat					S..............................H.......H?							

Notes: S: sown, planted; T: transplanted; H: harvested.
Nōgyō zue, 1717, by Tsuchiya Matasaburō (1642–1719), manager of a domain owned by the Maeda family in Kaga (Ishikawa) (coll. Nihon nōsho zenshū, vol. 26). Crop times established by the author from the drawings in the *Nōgyō zue*.
Calendar: The months refer to the traditional calendar; approximately one month needs to be added for the Gregorian calendar (the Japanese 3rd month corresponds to the present-day 4th month).
Source: © Verschuer, *Rice, Agriculture, and the Food Supply in Premodern Japan*, Routledge, 2016.

Table A.5 **Calorific values of foods (per 100 grammes)**

Japanese	English	gross (kcal)	Edible part to be deducted
shiinomi*#	castanopsis	256	–30% = 179 net kcal
(oni)gurumi*#	walnut	673	–50% = 337 net kcal
dried kaya*#	*kaya* (torreya)	610	–30% = 427 net kcal
kuri*#	chestnut	156	–30% = 109 net kcal
tochi #	horse chestnut	365	–23% = 281 net kcal
hishi*#	water chestnut	190	
hasunomi*	lotus	121	–35% = 79 net kcal
yuri*	lily (rhizome)	124	
yamanoimo*	yam	118	
satoimo*	taro	60	
warabi	bracken (root)	23	–20% (stem)
warabi	bracken (cooked and dried)	276	
kuzu*	pueraria (rhizome)	16.8	
kuzu*	pueraria (100g of starch)	347	
iwashi	sardines	134	–35%
hamaguri	clams	60	–70%
inoshishi	wild boar (flesh)	149	
matsunomi	pinenuts	634	
goma*#	sesame (dried)	578	
egoma*#	perilla (dried)	544	
asanomi*#	hemp (dried)	463	
kome*#	rice	351 (whole)	356 (white)
soba #	buckwheat	300 (grain)	361 (flour)
kibi*#	common millet	299 (grain)	

Japanese	English		Edible part
awa*#	foxtail millet	307 (whole)	364 (hulled)
hie*#	barnyard millet	311 (grain)	
komugi*#	wheat	333 (grain)	368 (flour)
daizu*#	soybeans	417 (raw)	180 (boiled)
azuki*#	red beans	339 (raw)	143 (boiled)
sasage*#	cowpeas	336 (raw)	145 (boiled)

Notes: # Discovered in Jōmon and/or Yayoi archaeological sites.
* Found in the *Engishiki*.
Sources: Kagaku gijutsuchō Shigen chōsakai ed. (2000).
© Verschuer, *Rice, Agriculture, and the Food Supply in Premodern Japan*, Routledge, 2016.

Bibliography

Primary sources

Aizu nōsho, 1684, Sase Yojiemon (1630–1711), Nihon nōsho zenshū, vol. 19.
Akazome Emon shū, ed. Akazome Emon (active ca. 974–1045), SKT vol. 3.
Akishino gesseishū, 1204, Fujiwara no Yoshitsune, comp., SKT vol. 3.
Azuma kagami, chronicle from 1180–1266, written ca. 1264–1275, and 1288–1306, anon., SZKT vols 32, 33.
Chōshū eisō, Fujiwara no Toshinari (1114–1204), comp., NKBT vol. 80.
Chōshūki, 1105–1136, Minamoto no Morotoki (1088–1136), Zōho shiryō taisei, vols 16, 17.
Chōya gunsai, 1117, Miyoshi no Tameyasu, ed., SZKT vol. 29.
Chūjiruiki, after 1295, Ki no Munenaga, GR vol. 19.
Chūyūki, 1087–1138, Fujiwara no Munetada (1062–1141), Zōho shiryō taisei, vols 9–15.
Dainagon Tsunenobu shū, Minamoto no Tsunenobu (1016–1097), NKBT vol. 80.
Daini Takatō shū, ca. 1012, Fujiwara no Takatō (949–1013), SKT vol. 3.
Daisaiin saki no gyoshū, after 986, Senshi naishinnō, SKT vol. 3.
Denryaku, 1098–1118, Fujiwara no Tadazane (1078–1162), Dainihon kokiroku, 5 vols.
Dōbun tsūkō, 1711–1716, Arai Hakuseki (1657–1725), Arai Hakuseki zenshū, vol. 4.
Ehon tsūhōshi, 1729, Tachibana Morikuni, xylographic edition, private collection.
Eiga monogatari, ca. 1028–1036, attrib. Akazome Emon, completed ca. 1092–1106, NKBT vols 75, 76.
Eishōki, 1105–1129, Fujiwara no Tametaka (1070–1130), Zōho shiryō taisei, vol. 8.
Engishiki, 927, Fujiwara no Tokihira, Fujiwara no Tadahira *et al.*, SZKT vol. 26; see Torao Toshiya.
Enryaku kōtaishiki, 803, Sugano no Mamichi *et al.*, eds, SZKT vol. 26.
Fuboku wakashō, ca. 1310, Fujiwara (Katsumata) no Nagakiyo, comp., SKT vol. 2.
Fudoki, Fudoki itsubun, regional monographs, early-mid 8th century, anon., NKBT vol. 2.
Fukushima-ken kannai gunson nōgu no zu, 1872, anon., in Fukushima kenritsu hakubutsukan chōsa hōkoku 25, 1993, pp. 55–72.
Fusō ryakki, Kōen (?–1169), SZKT vol. 12.
Gen dainagon ke utaawase, 1041, Minamoto no Morofusa (1008–1077), comp., SKT vol. 5.
Genji monogatari, early 11th century, Murasaki Shikibu, NKBT vols 14–18.
Genkō shakusho, 1322, Kokan Shiren (1278–1346), SZKT vol. 31.
Gishiki, Shingishiki, ca. 873–877, attrib. Murakami Tennō, GR vol. 6.

Gōke shidai, ca. 1111, Ōe no Masafusa (1041–1111), Kaitei zōho kojitsu sōsho, vol. 2.

Gosen wakashū, commissioned in 951, Kiyohara no Motosuke *et al.*, comp., SKT vol. 1.

Goshūi wakashū, 1086, Fujiwara no Michitoshi (1047–1099), comp., SKT vol. 1.

Gyokuyō, 1164–1203, Kujō (Fujiwara no) Kanezane (1149–1207), 3 vols, Kokusho kankōkai, 1965.

Gyokuyō wakashū, ca. 1312, Fujiwara (Kyōgoku) Tamekane, comp., SKT vol. 1.

Heian ibun, Takeuchi Rizō, ed., 15 vols, Tōkyōdō shuppan, 1947–1970.

Honchō gatsuryō, 930–946, Koremune no Kinkata, GR vol. 6.

Honchō kōsōden, 1702, Mangen Shiban (1626–1710), ed., BZ vols 102, 103.

Honchō shojaku mokuroku, 1277–1294, anon., GR vol. 28.

Honchō shokkan, printed in 1697, Hitomi Hitsudai, Heibonsha, coll. Tōyō bunko, 5 vols, 1976–1981, 1993–1999.

Horikawa-in hyakushu, ca. 1105–1106, Ōe no Masafusa *et al.*, SKT vol. 4.

Hyakurenshō, ca. 1259–1274, anon., SZKT vol. 11.

Hyakushō denki, 1681–1683, anon., Nihon nōsho zenshū, vols 16, 17.

Hyōta jiruishō, before 1305, anon., ZGR vol. 30 ge.

Ippen shōnin eden, Ippen hijiri'e, 1299, Shōkai ed., Nihon no emaki, vol. 20.

Iroha jiruishō, (a) 1144–1165, 2 books, Tachibana no Tadakane, ed., (b) 1177–1181, 3 books, anon., Kazama shobō, 1964, 1987; (c) ca. 1200, 10 books, anon., Kazama shobō, 1988.

Ishiyamadera engi, 1324–1326, Gōshu and Takashina Takakane, completed at a later date, Nihon no emaki, vol. 16.

Jikata hanreiroku, ca. 1794, Ōishi Hisataka, Nihon shiryō sensho, vol. 1.

Jikata no kikigaki, Saizōki, 1688–1704, Ōhata Saizō, Nihon nōsho zenshū, vol. 28.

Jindai no maki moshiokusa, Tamaki Masahide (1670–1736), Meiji edition.

Jōgan gishiki, ca. 871–872, anon., Shintei zōho kojitsu sōsho, vol. 31.

Kaden, Fujiwara no Nakamaro (706–764) and Enkei (active ca. 753–758), Nara ibun, vol. 3.

Kagerō nikki, ca. 971, mother of Fujiwara no Michitsuna, NKBT vol. 20.

Kaidōki, ca. 1223(?), anon., GR vol. 18.

Kakyō hyōshiki, 772–784, Fujiwara no Hamanari (723–790), ed., Nara ibun, vol. 3.

Kamakura ibun, Takeuchi Rizō, ed., 51 vols, Tōkyōdō shuppan, 1971–1991.

Kanke bunsō, 900, Sugawara no Michizane (845–903), NKBT vol. 72.

Kasuga gongen genki'e, after 1307, Takatsukasa Mototada, Takashina Takakane *et al.*, Zoku Nihon no emaki, vols 13, 14.

Keiun hōin shū, Keiun (active between 1315 and 1369), SKT vol. 4.

Kenkyū sannen Kōtaijingū nenjūgyōji, 1192(?), Arakida no Tadanaka *et al.*, ZGR vol. 1 jō.

Kenmu nenjūgyōji, Godaigo Tennō (1288–1339), GR vol. 6.

Kintō shū, Fujiwara no Kintō (966–1041), SKT vol. 3.

Kin'yō wakashū, 1124–1127, Minamoto no Toshiyori (1055–1129), comp., SKT vol. 1.

Kodai kayōshū, Tsuchihashi Yutaka *et al.*, eds, Iwanami, 1957, NKBT vol. 3.

Kogo shūi, 807, Inbe no Hironari, Iwanami, 1991 (1st printing 1985).

Kojidan, 1212–1215, Minamoto no Akikane (1160–1215), ed., SZKT vol. 18.

Kojiki, 712, Ō no Yasumaro; see *Kojiki Norito*.

Kojiki Norito, Kurano Kenji *et al.*, eds, Iwanami, 1958, NKBT vol. 1.

Kojikiden, 1798, Motoori Norinaga (1730–1801), Motoori Norinaga zenshū, vols 9–12.

Kojiruien, 1879–1914, Monbushō *et al.*, 51 vols, Yoshikawa kōbunkan, 1978–1985.

Kojiruien Sangyōbu 1, 1908, Yoshikawa kōbunkan, 1984.

Kōjishō, 11th–12th century, attrib. Koremune no Toshimichi, ZGR vol. 30 ge.

Kōka shunjū, 1707, Tsuchiya Matasaburō (1642–1719), Nihon nōsho zenshū, vol. 4.

Kokawadera engi emaki, 12th century, anon., Nihon no emaki, vol. 5.

Kokin waka rokujō, late 10th century, anon., SKT vol. 2.

Kokin wakashū, ca. 913, Ki no Tsurayuki *et al.*, comp., SKT vol. 1.

Kokon chomonjū, 1254, Tachibana no Narisue, NKBT vol. 84.

Kokutai no hongi kaisetsu taisei, 1940, Magota Hideharu, Hara Fusataka, Daimeidō.

Kokutai no hongi to nōdō, 1942, ed. Dainippon nōdō kyōdōtai, Taibunkan.

Kokutai shikan, 1929, Takeda Kentarō, Shunjūsha.

Kōninshiki, ca. 803, Sugano no Mamichi *et al.*, eds, SZKT vol. 26.

Konjaku monogatari, after 1120, anon., NKBT vols 22–26.

Kōtai jingū gishikichō, 804, Arakida Kiminari *et al.*, eds, GR vol. 1.

Makura no sōshi, ca. 1000, Sei Shōnagon, NKBT vol. 19.

Man'yōshū, ca. 759, Ōtomo no Yakamochi *et al.*, eds, NKBT vols 4–7.

Michinari shū, Minamoto no Michinari (?–1036), SKT vol. 3.

Midō kanpakuki, 995–1021, Fujiwara no Michinaga (966–1027), Dainihon kokiroku, 3 vols.

Minamoto Shitagō shū, Minamoto no Shitagō (911–983), SKT vol. 3.

Mitsune shū, Ōshikōchi no Mitsune (?–?), SKT vol. 3.

Moromitsu nenjūgyōji, Nakahara no Moromitsu (1206–1265), ZGR vol. 10 jō.

Moromoto nenjūgyōji, Nakahara no Moromoto (1109–1175), ZGR vol. 10 jō.

Morotō nenjūgyōji, Nakahara no Morotō (1070–1130), ZGR vol. 10 jō.

Mumyōshō, Kamo no Chōmei (1155–1216), GR vol. 16.

Myōgoki, 1275, Kyōson, Benseisha, 1983.

Nagayoshi shū, Fujiwara no Nagayoshi (949–1009), SKT vol. 3.

Nara ibun, Takeuchi Rizō, ed., 3 vols, Tōkyōdō shuppan, 1962 (1st printing 1943).

Nenjūgyōji, undated, anon., handed down by the Kamo family, ZGR vol. 10 jō.

Nenjūgyōji goshōjimon, 885, Fujiwara no Mototsune (836–891), ZGR vol. 10 jō.

Nenjūgyōji hishō, after 1195, anon., GR vol. 6.

Nenjūgyōji shō, undated, anon., ZGR vol. 10 jō.

Nihon eitaikura, 1688, Ihara Saikaku (1642–1693), SNKBZ, vol. 68.

Nihon jōmin seikatsu ebiki, see Shibusawa Keizō.

Nihon kiryaku, 11th century, anon., SZKT vols 10, 11.

Nihon kōki, 840, Fujiwara no Fuyutsugu *et al.*, SZKT vol. 3.

Nihon Montoku tennō jitsuroku, 879, Fujiwara no Mototsune *et al.*, SZKT vol. 3.

Nihon ryōiki, 810–824, Keikai (?–?), NKBT vol. 70.

Nihon sandai jitsuroku, 901, Fujiwara no Tokihira *et al.*, SZKT vol. 4.

Nihon shoki, 720, Toneri shinnō *et al.*, SZKT vol. 1.

Nihon shoki den, Suzuki Shigetane, Ban Nobutomo, Hirata Atsutane, Suzuki Shigetane zenshū, vols 1–9, 13.

Nihon shoki tsūshō, 1751, Tanikawa Kotosuga (1709–1776), Nihon seishin bunka bunken, vol. 15.

Nihongi shiki, 721–965, Ō no Yasumaro *et al.*, SZKT vol. 8.

Nihonkoku genzaisho mokuroku, Fujiwara no Sukeyo (847–898), ZGR vol. 30 ge.

Nihonshi jiten, Nagahara Keiji *et al.*, eds, Iwanami shoten, 1999.

Nōgu benriron, 1822, Ōkura Nagatsune (1768–?), Edo kagaku koten sōsho, vol. 4, Kōwa shuppan, 1977.

Nōgu zoroe, 1865, Ōtsubo Futaichi, Nihon nōsho zenshū, vol. 24.

Nōgyō zensho, 1697, Miyazaki Yasusada (1623–1697), Iwanami, 1988 (1st printing 1936).

Nōgyō zue, 1717, Tsuchiya Matasaburō (1642–1719), Nihon nōsho zenshū, vol. 26.

Ōkagami, before 1134, anon., NKBT vol. 21.

Ononomiya nenjūgyōji, before 1046, Fujiwara no Sanesuke (957–1046), GR vol. 6.

Owari no gebumi, *Owari no kuni gunji hyakushōra gebumi*, 988, anon., Heian ibun vol. 2, no. 339.

Qiming yaoshu, 6th century, Jia Sixie, Sanyōsha, 1984 (1st printing 1969).

Rakuyō dengakuki, 1096(?), Ōe no Masafusa (1041–1111), GR vol. 19.

Ritsu, 718, Fujiwara no Fuhito *et al.*, SZKT vol. 22.

Ritsuryō, *Taihō ritsuryō*, 701, Fujiwara no Fuhito; *Yōrō ritsuryō*, ca. 718, Yamato no Nagaoka, Nihon shisō taikei, vol. 3.

Rōnō ruigo, 1722, Suyama Donō, Nihon nōsho zenshū, vol. 32.

Ruiju fusenshō, 737–1093, compiler anon., SZKT vol. 27.

Ruiju kokushi, 892, attrib. Sugawara no Michizane (845–903), SZKT vols 5, 6.

Ruiju myōgishō, ca. 1100, re-edited early 12th century, anon., Kazama shobō, 1986.

Ruiju sandai kyaku, early 11th century, anon., SZKT vol. 25.

Ruiju zatsuyōshō, after 1146, anon., GR vol. 26.

Ryōjin hishō, Goshirakawa-in (1127–1192) comp., NKBT vol. 73.

Ryō no gige, 833, Kiyohara no Natsuno *et al.*, SZKT vol. 22.

Ryō no shūge, 859–877, Koremune no Naomoto, SZKT vols 23, 24.

Saigūki, *Seikyūki*, 957–964, Minamoto no Takaakira (914–982), Shintei zōho kojitsu sōsho, vols 6, 7.

Saigūki, *Seikyūki*, 957–964, Minamoto no Takaakira (914–982), Shintei zōho kojitsu sōsho, vols 6, 7.

Saizōki, see *Jikata no kikigaki*.

Sanbokuki kashū, 1128, Minamoto no Toshiyori, comp., SKT vol. 3.

Sangō shiiki, 797, Kūkai (774–835), NKBT vol. 71.

Sankaiki, 1151–1185, Fujiwara no Tadachika (1131–1195), Zōho shiryō taisei, vols 26–28.

Sarashina nikki, ca. 1060, daughter of Sugawara no Takasue, NKBT vol. 20.

Satsujōshū, 1454, Iino Takatsura, ZGR vol. 30 ge.

Seiji yōryaku, 1002, Koremune no Tadasuke, ed., SZKT vol. 28.

Seikei zusetsu, 1806, Sō Han (1758–1834), Shirao Kunihashira, ed., Kokutai shuppansha, 1933.

Seiryōki, ca. 1628, Doi Suiya (1546–1629), Nihon nōsho zenshū, vol. 10.

Senchū wamyō ruijushō, 1827, Kariya Ekisai (1775–1835), Zenkoku shobō, 1943.

Sendai kuji hongi, 807–901, anon., SZKT vol. 7.

Senmen Hokekyō(sasshi), mid-12th century, anon., Kokushi daijiten, vol. 8.

Shaku Nihongi, ca. 1302, Urabe Kanekata, SZKT vol. 8.

Shasekishū, 1279–1283, Mujū Dōgyō, Iwanami, 1997 (1st printing 1943).

Shigisan engi emaki, ca. 1155–1180, anon., Nihon no emaki, vol. 4.

Shingishiki, see *Gishiki*.

Shinkokin wakashū, 1210, Fujiwara no Teika *et al.*, comp., SKT vol. 8.

Shinpen Musashi fudoki kō, 1828, Mamiya Kotonobu, Dainihon chishi taikei, vol. 18.

Shinpen tsuika, after 1444, anon., Nihon chūsei hōsei shiryōshū, vol. 1.

Shinsarugakuki, 1053–1064, attrib. Fujiwara no Akihira (989–1066), Heibonsha, coll. Tōyō bunko, 1983.

Shinsen jikyō, 898–901, Shōjū (?–?), Kyōto: Rinsen shoten (Tenjibon of 1124), 1991 (1st printing 1958).

Shinsen shōjiroku, 815, Manda shinnō *et al.*, Ōkura seishinbunka kenkyūjo, 1936.

Shinsen waka rokujō, mid-13th century, Kinugasa Ieyoshi, SKT vol. 2.

Shinyaku Kegonkyō ongishiki, 8th century(?), anon., Nara ibun vol. 3.

Shinyō wakashū, 1381, Munenaga shinnō, comp., SKT vol. 1.

Shisseishoshō, after 1121, anon., ZGR vol. 10 jō.

Shoku gosen wakashū, 1251, Fujiwara no Tameie, comp., SKT vol. 1.

Shoku goshūi wakashū, 1326, Nijō no Tamesada *et al.*, comp., SKT vol. 1.

Shoku kokin wakashū, 1265, Fujiwara no Tameie *et al.*, comp., SKT vol. 1.

Shoku Nihongi, 797, Sugano no Mamichi *et al.*, SZKT vol. 2.

Shoku Nihon kōki, 869, Fujiwara no Yoshifusa *et al.*, SZKT vol. 3.

Shōsōin monjo, 1901–, Shiryō hensanjo ed., DNK vols 1–25.

Shōyūki, 982–1032, Fujiwara no Sanesuke (957–1046), Zōho shiryō taisei, bekkan, 3 vols.

Shūgaishō, before 1293, Tōin Kinkata and Tōin Sanehiro, Shintei zōho kojitsu sōsho, vol. 22.

Shūgyokushū, 1346, Jien (1155–1225); Son'en shinnō (1298–1356), comp., SKT vol. 3.

Shūi wakashū, commissioned ca. 1005 by Kazan-in, SKT vol. 1.

Sōkonshū, Shōtetsu (1381–1459), SKT vol. 8.

Sone no Yoshitada shū, see *Yoshitada shū*.

Taiheiki, 14th/15th century, anon., NKBT vols 34–36.

Taihō ritsuryō, see *Ritsuryō*.

Tamemasa senshu, 1415, Reizei Tamemasa, SKT vol. 4.

Tameshige shū, 1380, Nijō Tameshige, SKT vol. 7.

Tametada(kyō) senshu, 1415, Reizei Tametada (1361–1417), SKT vol. 7.

Tango no kami ke hyakushu, Fujiwara no Tametada (?–1136) *et al.*, comp., SKT vol. 4.

Teikin ōrai, late 14th century, anon., Heibonsha, coll. Tōyō bunko, 1987 (1st printing 1973).

Teiō hennenki, 1364–1380(?), anon., SZKT vol. 12.

Tōdaiji monjo, Shiryō hensanjo ed., 1901–, DNK Iewake series 18.

Tōdaiji yōroku, 1106–1134, Kangon *et al.*, eds, Zenkoku shobō, 1944.

Toyukegū gishikichō, 804, Watarai no Satsukimaro, GR vol. 1.

Tsuika hō, 13th–14th century, anon., Chūsei hōsei shiryōshū, vol. 1.

Udaijin Kanezane utaawase, 1175, Fujiwara no Kanezane *et al.*, SKT vol. 5.

Uji shūi monogatari, 1212–1221, anon., NKBT vol. 27.

Utsuho monogatari, before 972, anon., NKBT vols 10–12.

Wajiga, 1688, Kaibara Yoshifuru, Ekiken zenshū, vol. 7.

Wakansansaizue, 1712, Terashima Ryōan, Heibonsha, coll. Tōyō bunko, 18 vols, 1994.

Wakun no shiori, 1777, Tanikawa Kotosuga (1709–1776), 3 vols, Meiji edition 32 (1899).

Wamyō ruijushō, 931–938, Minamoto no Shitagō (911–983), Kazama shobō, 1977.

Yamashiro no kuni Kadono-gun handen zu, 9th century, anon., in Shiryō hensanjo ed., Nihon shōen ezu shūei, vol. 2, Kinki 1.

Yamato kōsaku eshō, before 1703, Ishikawa Tomonobu, Nihon fūzoku zue, vol. 5.

Yamato no kuni Sōnoshimo-gun keihoku handen zu, 743–811, anon., in Shiryō hensanjo ed., Nihon shōen ezu shūei, vol. 2, Kinki 1.

Yoshinobu shū, Ōnakatomi no Yoshinobu (921–991), SKT vol. 7.

Yoshitada shū, late 10th century, Sone no Yoshitada (?–ca.1003), NKBT vol. 80.

Yukimune shū, Minamoto no Yukimune (1064–1143), SKT vol. 7.

Zander Handwörterbuch der Pflanzennamen, Dictionary of plant names, Dictionnaire des noms de plantes, 16th edition, W. Erhardt, E. Götz, N. Bödeker, S. Seybold, eds, Stuttgart: Ulmer, 2000 (1st edition 1927).

Secondary sources

Abe Takeshi, *Man'yōbito no seikatsu*, Tōkyōdō shuppan, 1995.

Abe Takeshi, *Gekokujō no shakai*, Tōkyōdō shuppan, 1998.

Abe Yoshio, *Terres à riz en Asie: Essai de typologie*, Paris: Masson, 1995.

Abe Yoshio, *Le 'décorticage' du riz: Typologie, répartition géographique et histoire des instruments à monder le riz*, Paris: Maison des Sciences de l'Homme, 2007.

Abe Yoshio, Cozette Griffin-Kremer, Perrine Mane, François Sigaut, Éric Trombert, Charlotte von Verschuer, eds, *Agricultural Technology: A French–English–Chinese–Japanese Glossary (Grains and Horticulture) Draft Version 2013* [an internet database].

Abel, Wilhelm, *Geschichte der deutschen Landwirtschaft vom frühen Mittelalter bis zum 19. Jahrhundert*, Stuttgart, 1978 (1st printing 1962).

Akimichi Tomoya, *Nawabari no bunkashi*, Shōgakukan, 1995.

Akiyama Takashi *et al.*, eds, *Zuroku nōmin seikatsushi jiten*, Kashiwa shobō, 1991 (1st edition 1979).

Amino Yoshihiko, *Nihon chūsei no minshū zō*, Iwanami, 1980.

Amino Yoshihiko, *Chūsei saikō rettō no chiiki to shakai*, Nihon editā sukūru, 1986.

Amino Yoshihiko, *Nihonron no shiza*, Shōgakukan, 1990.

Amino Yoshihiko *et al.*, eds, *Kōza Nihon shōenshi*, vol. 1, Yoshikawa kōbunkan, 1991 (1st printing 1989).

Amino Yoshihiko, *Nihon chūsei tochi seidoshi no kenkyū*, Hanawa shobō, 1991.

Amino Yoshihiko, "Chūsei minshū seikatsu no ichi sokumen", in Kinoshita Tadashi *et al.*, eds, *Seisan gijutsu to busshitsu bunka*, Yoshikawa kōbunkan, 1993.

Amino Yoshihiko, "Inasaku ichigenron no kuppuku", in Tanigawa Kenichi *et al.*, eds, *Fūdo to bunka*, "Nihon minzoku bunka taikei", vol. 1, Shōgakukan, 1995 (1st printing 1986), pp. 60–73.

Amino Yoshihiko, Ishii Susumu, *Kome, hyakushō, tennō*, Daiwa shobō, 2000.

Andō Hiromichi, "Suiden chūshin shikan hihan no kōzai", *Kokuritsu rekishi minzoku hakubutsukan kenkyū hōkoku* 185, Feb. 2014, pp. 405–448.

Aoba Takashi, *Yasai no Nihonshi*, Yasaka shobō, 1991.

Araki Toshio, "Heian jidai no ochibo hiroi kankō to inekari rōdō", in Takeuchi Rizō, ed., *Kodai tennō sei to shakai kōzō*, Azekura shobō, 1980, pp. 257–276.

Araki Toshio, "Kodai kokka to minkan saishi", *Rekishigaku kenkyū* 560 (7), 1986, pp. 45–51.

Araki Toshio, "Kodai no matsuri to asobi", in Nihon sonrakushi kōza henshū iinkai ed., *Seikatsu 1 genshi, kodai, chūsei*, Yūzankaku, 1991, pp. 136–151.

Arizono Shōichirō, *Zairai nōkō no chiiki kenkyū*, Kokin shoin, 1997.

Arizono Shōichirō *Higanbana ga Nihon ni kita michi*, Ōtsu: Kaiseisha, 1998.

Asahi Graph ed., "Sannai-Maruyama iseki", special issue, *Asahi Graph* 25, Oct. 1994.

Ashikaga Kenryō, *Keikan kara rekishi o yomu*, NHK Ningen daigaku, 1997.

Barnes, Gina L., *Protohistoric Yamato: Archaeology of the First Japanese State*, Ann Arbor, MI: University of Michigan Press, 1988.

Barnes, Gina L., "Landscape and Subsistence in Japanese History", in Peter Martini, Ward Chesworth, eds, *Landscapes and Societies*, Doordrecht: Springer Verlag, 2010, pp. 321–340.

Barnes, Gina L., "Japan's Natural Setting", in Karl F. Friday, ed., *Japan Emerging: Premodern History to 1850*, Boulder, CO: Westview Press, 2012, pp. 3–15.

Barnes, Gina L., *Archaeology of East Asia: The Rise of Civilization in China, Korea and Japan*, London: Oxbow Books, 2015.

Barrau, Jacques, "Culture itinérante, culture sur brûlis, culture nomade, écobuage ou essartage?: un problème de terminologie agraire", *Études rurales* 45, 1972, pp. 99–103.

Batten, Bruce L., Philip C. Brown, eds, *Environment and Society in the Japanese Islands: From Prehistory to the Present*, Corvallis, OR; Oregon State University Press, 2015.

Berger, Günther, Métailié, Georges, Watabe Takeshi, "Une chinoiserie insolite: étude d'un papier peint chinois", *Arts Asiatiques* 51, 1996, pp. 96–116.

Berque, Augustin, "La montagne et l'écoumène", *L'espace géographique* 2, 1980, pp. 151–162.

Berque, Augustin, ed., *Dictionnaire de la Civilisation Japonaise*, Paris: Hazan, 1994.

Berthier-Caillet, Laurence, *Fêtes et rites des 4 saisons au Japon*, Paris: Publications Orientalistes de France, 1981.

Braudel, Fernand, *Civilisation matérielle, économie et capitalisme, XVe–XVIIIe siècle*, vol. 1. *Les structures du quotidien*, Paris: Armand Colin, 1979.

Bray, Francesca, *Agriculture*, in Joseph Needham, ed. *Science and Civilisation in China*, Cambridge University Press, 1984, Reprint Taipei: Caves Books, n.d.

Bray, Francesca, *The Rice Economies: Technology and Development in Asian Societies*, Stanford, CA: University of California Press, 1986.

Brown, Philip C., *Central Authority and Local Autonomy in the Formation of Early Modern Japan: Case of Kaga Domain*, Stanford, CA: Stanford University Press, 1994.

Brown, Philip C., *Cultivating Commons: Joint Ownership of Arable Land in Early Modern Japan*, Honolulu, HI: University of Hawai□i Press, 2011.

Brunhes-Delamarre, Mariel J., *La vie agricole et pastorale dans le monde*, Grenoble: Glénat, 1999.

Burte, Jean-Noël, ed., *Le Bon Jardinier*, 3 vols, Paris: La Maison Rustique, 1992.

Cartier, Michel, "La marginalisation des animaux en Chine", *Anthropozoologica* 18, second semester 1993, pp. 7–15.

Cauvin, Jacques, *Naissance des divinités, naissance de l'agriculture*, Paris: CNRS, 1997.

Chiba Tokuji, "Yamashita Tadajirō ke sho zōjiki ni tsuite", *Aichi daigaku sōgō kyōdo kenkyūjo kiyō* 31, 1986, pp. 9–20.

Chiba Tokuji, "Yama no minzoku", in Ōbayashi Taryō, ed., *Sanmin to ama*, Shōgakukan, 1995 (1st printing 1983), pp. 173–218.

Chiku, Katei, Wan Min-i, *Bukkōgaku dai shizengo no tehiki*, Chinese translation by Tansei sōgō kenkyūjo, Tanseisha, 1988.

Cobbi, Jane, *Le végétal dans la vie japonaise*, Paris: Publications Orientalistes de France, n.d. (1978).

Cobbi, Jane, 'La vie alimentaire des Japonais, son évolution récente', doctoral thesis, n.d. (1980).

Comet, Georges, *Le paysan et son outil: Essai d'histoire technique des céréales (France, VIIIe–XVe siècles)*, Rome: École Française de Rome, 1992.

Condominas, Georges, *Nous avons mangé la forêt*, Paris: Flammarion, 1982 (1st edition Mercure de France, 1957).

Condominas, Georges, "Aspects écologiques d'un espace social restreint en Asie du Sud-Est. Les Mnong Gar et leur environnement", *Études rurales* 89, 90, 91, 1983, pp. 11–76.

Daniels, Christian, Watabe Takeshi, eds, *Unnan no seikatsu to gijutsu*, Keiyūsha, 1994.

Debaine-Francfort, Corinne, Idriss, Abdurassul, Wang Binghua, "Agriculture irriguée et art bouddhique ancien au cœur du Taklamakan (Karadong, Xinjiang, IIe–IVe siècles)", *Arts Asiatiques* 49, 1994, pp. 34–52.

Dictionnaire des sources du Japon classique/Dictionary of Sources of Classical Japan, ed. Joan Piggott, Ivo Smits, Ineke van Put, Michel Vieillard-Baron, Charlotte von Verschuer, Paris: Collège de France, Institut des Hautes Études Japonaises, 2006.

Duby, Georges, *L'économie rurale et la vie des campagnes dans l'occident médiéval*, 2 vols, Paris: Flammarion, 1977 (1st edition 1962).

Ducourtieux, Olivier, *Du riz et des arbres: L'interdiction de l'agriculture d'abattis-brûlis, une constante politique au Laos*, IRD and Karthala, 2009.

Endō Motoo, *Orimono no Nihonshi*, NHK Books, 1978 (1st edition 1971).

Farris, W. Wayne, *Population, Disease, and Land in Early Japan, 645–900*, Cambridge, MA: Harvard University Press, 1985.

Farris, W. Wayne, *Sacred Texts and Buried Treasures: Issues in the Historical Archaeology of Ancient Japan*, Honolulu, HI: University of Hawai□i Press, 1998.

Farris, W. Wayne, "Pieces in a Puzzle: Changing Approaches to the Shōsōin Documents", *Monumenta Nipponica* 62 (4), 2007, pp. 397–453.

Farris, W. Wayne, *Daily Life and Demographics in Ancient Japan*, Ann Arbor, MI: University of Michigan Press, 2009.

Farris, W. Wayne, *Japan to 1600: A Social and Economic History*, Honolulu, HI: University of Hawai□i Press, 2009.

Ferdière, Alain, *Les campagnes en Gaule romane*, vol. 2, *Les techniques et la production rurales en Gaule (52 av. J.-C.-486 ap. J.-C.)*, Paris: Errance, 1988.

Friday, Karl F., ed., *Japan Emerging: Premodern History to 1850*, Boulder, CO: Westview Press, 2012.

Fujii Kazutsugu, *Shoki shōenshi no kenkyū*, Hanawa shobō, 1986.

Fujii Kazutsugu, *Kodai Nihon no shiki goyomi*, Chūō kōronsha, 1997.

Fujimori Eiichi, *Jōmon nōkō*, Gakuseisha, 1970.

Fujio Shin'ichirō, *Shin Yayoi jidai, gohyakunen hayakatta suiden inasaku*, Yoshikawa kōbunkan, 2011.

Fujio Shin'ichirō, ed., "Nōkō bunka fukugō to Yayoi bunka", special issue, *Kokuritsu rekishi minzoku hakubutsukan kenkyū hōkoku* 185, Feb. 2014.

Fujiwara Hiroshi, *Inasaku no kigen o saguru*, Iwanami, 1998.

Fukuda Toyohiko, "Bunken shiryō yori mita kodai no seitetsu", in Tōkyō kōgyō daigaku Seitetsushi kenkyūkai ed., *Kodai Nihon no tetsu to shakai*, Heibonsha, 1985 (1st printing 1982), pp. 163–190.

Fukui Katsuyoshi, *Yakihata no mura*, Asahi shinbunsha, 1974.

Fukui Katsuyoshi, "Yakihata nōkō no fuhensei to shinka", in Ōbayashi Taryō, ed., *Sanmin to ama*, Shōgakukan, 1995 (1st printing 1983), pp. 235–274.

Furuhashi Nobutaka, *Kotoba no kodai seikatsushi*, Kawade shobō shinsha, 1989.

Furushima Toshio, *Nihon nōgyō gijutsushi*, "Furushima Toshio chosakushū", vol. 6, Tōkyō daigaku, 1983 (2nd edition 1975; 1st edition 1947) (2nd printing of 1st edition 1949).

Gourou, Pierre, *Riz et civilisation*, Paris: Fayard, 1984.

Guilaine, Jean, ed., *Pour une archéologie agraire*, Paris: Armand Colin, 1991.

Guilaine, Jean, "Civilisation de l'Europe au Néolithique et à l'Age du Bronze", *Annuaire du Collège de France*, 1996–1997, pp. 697–719.

Guilaine, Jean, "Les premiers paysans du monde: où? quand? comment? pourquoi?", Collège de France Seminar, 1999–2000.

Guilaine, Jean, ed., *Les premiers paysans du monde, naissance des agricultures*, Paris: Errance, 2001.

Guilaine, Jean, *Les racines de la Méditerranée et de l'Europe*, Paris: Collège de France, Fayard, 2008.

Gunma-ken kyōiku iinkai ed., *Hidaka iseki*, 1982.

Guyonvarch, Olivier, "Les outils agricoles dans la Chine du XIVe siècle d'après le Nongshu de Wang Zhen", *Études Chinoises* 12 (2), 1993, pp. 9–49.

Haga Noboru, Ishikawa Hiroko, *Kome, mugi, zakkoku, mame*, "Zenshū Nihon no shoku bunka", vol. 3, Yūzankaku, 1998.

Haga Noboru, Ishikawa Hiroko, *Shoku seikatsu to shokumotsushi*, "Zenshū Nihon no shoku bunka", vol. 2, Yūzankaku, 1999.

Hagiwara Hidesaburō, *Ine to tori to taiyō no michi*, Taishūkan, 1996.

Harada Nobuo, *Rekishi no naka no kome to niku*, Heibonsha, 1996 (1st printing 1993).

Harada Nobuo, *Kome o eranda Nihon no rekishi*, Bungei shunju, 2006.

Harada Nobuo, *Chūsei no mura no katachi to kurashi*, Kadokawa sensho, 2008.

Harada Nobuo, *Nihonjin wa nani o tabete kita ka*, Kadokawa gakugei shuppan, 2010.

Harada Nobuo, "Laos hokubu sonraku no keikan to nōmin", *Kikan Tōhokugaku* 29, Oct. 2011, pp. 250–274.

Harada Nobuo, ed., *Chiiki kaihatsu to sonraku keikan no rekishiteki tenkai*, Kyōto: Shibunkaku shuppan, 2011.

Harada Nobuo, Kurata Takashi, eds, *Yakihata no kankyōgaku – ima yakihata to wa*, Kyōto: Shibunkaku shuppan, 2011.

Harashima Reiji, "Hasseiki no inasaku ni kansuru ni-san no mondai", *Rekishi hyōron* 148, Dec. 1962, pp. 24–39.

Hashimoto Seiji, "Nihon no nōkō bunka to imo saibai – nanpō no shiten kara", in Kansai daigaku Tōzaigakujutsu kenkyūjo ed., *Kansai daigaku Tōzaigakujutsu kenkyūjo sōritsu gojusshūnen kinen ronbunshū*, Ōsaka, 2001, pp. 313–326.

Hatai Hiromu, *Ritsuryō shōen to nōmin no kenkyū*, Yoshikawa kōbunkan, 1981.

Hattori Hideo, "Kataarashi no gogi to nimōsaku no kigen", *Rekishi o yomihiraku*, Seishi shuppan, 2003.

Haudricourt, André-Georges, Brunhes-Delamarre, Mariel, *L'homme et la charrue*, Paris: La Manufacture, 1986 (1st edition Gallimard, 1955).

Haudricourt, André-Georges, Brunhes-Delamarre, Mariel, *La technologie science humaine*, Paris: Maison des Sciences de l'Homme, 1987.

Hayama Teisaku, ed., *Seisan no gijutsu*, "Nihon no kinsei", vol. 4, Chūō kōronsha, 1992.

Heian bungaku rindokukai ed., *Nagayoshishū chūshaku*, Hanawa shobō, 1999.

Heian jidaishi jiten, Tsunoda Bunei *et al.*, eds, 3 vols, Kadokawa shoten, 1994.

Hérail, Francine, *Notes journalières de Fujiwara no Michinaga: Traduction du Midō kanpakuki*, 3 vols, Genève, Paris: Droz, 1987, 1988, 1991.

Hérail, Francine, *La cour et l'administration du Japon à l'époque de Heian*, Genève, Paris: Droz, 2006.

Hidaka Masaharu, *Kodai Hyūga no kuni*, NHK Books, 1993.

Hirakawa Minami, "Shin hakken no tane fuda to kodai no inasaku", *Kokushigaku* 169, Oct. 1999, pp. 1–56.

Hirata Yoshinobu, Misaki Hisashi, *Waka shokubutsu hyōgen jiten*, Tōkyōdō, 1998 (1st printing 1994).

Hirono Takashi, *Shoku no Man'yōshū*, Chūō kōronsha, 1998.

Hirose Kazuo, *Jōmon kara Yayoi e no shin rekishi zō*, Kadokawa shoten, 1997.

Hirose Tadahiko, *Koten bungaku to yasai*, Ōsaka: Tōhō shuppan, 1988.

Horio Hisashi, "Nōgu kara kikaika no taidō e", in Oka Mitsuo, Numata Jirō, Horio Hisashi, eds, *Inasaku no gijutsu to riron*, Heibonsha, 1990, pp. 213–259.

Hotate Michihisa, "Beitō nengu no shūnō to ina nio, hakari sadame", in Kamakura ibun kenkyūkai ed., *Kamakura jidai no shakai to bunka*, Tōkyōdō, 1999, pp. 108–130.

Hotate Michihisa, "Waka shiryō to suiden inasaku shakai", *Rekishigaku o mitsume naosu*, Azekura shobō, 2004, pp. 199–280.

Hudson, Mark J., "Rice, Bronze, and Chieftains: An Archaeology of Yayoi Ritual", *Japanese Journal of Religious Studies* 19 (2, 3), 1992, pp. 139–189.

Hyōgo-ken kyōiku iinaki ed., *Yamagaki iseki*, 1990.

Ichikawa Takeo, *Fūdo no naka no ishokuju*, Tōkyō shoseki, 1990 (1st printing 1978).

Ichikawa Takeo, Saitō Isamu, *Saikō Nihon no shinrin bunka*, NHK Books, 1992 (1st printing 1985).

Ichikawa Takeo, Yamamoto Shōzō, Saitō Isao, eds, *Nihon no buna-tai bunka*, Asakura shoten, 1994 (1st printing 1984).

Ichikawa Takeo, *Heike no tani Shinetsu no hikyō Akiyama-gō*, Nagano: Reibunsha, 1998 (1st printing 1961).

Ide Itaru, "Kodai no chimei to jōdaigo", *Gekkan gengo* 5 (7), 1976, pp. 10–17.

Ienaga Saburō, *Jōdai yamatoe nenpyō*, Meichō kankōkai, 1998 (1st edition 1942).

Iida Tasuhiko, "Shiiba no yakihata to shoku bunka", in Harada Nobuo *et al.*, eds, *Yakihata no kankyōgaku*, Kyōto: Shibunkaku shuppan, 2011.

Ikata Sadaaki, "Nihon kodai yakihata kō", *Nōgyō to keizai* 7 (6), 1940, pp. 82–90.

Ikata Sadaaki, *Nōgu no rekishi*, Shibundō, 1965.

Ikata Sadaaki, *Nihon kodai kokumotsushi no kenkyū*, Yoshikawa kōbunkan, 1985 (1st printing 1977).

Ikebe Wataru, "Kodai no awa kō", *Seijō tanki daigaku kiyō* 19, Dec. 1987.

Ikebe Wataru, *Wamyō ruijushō gun-gō-ri-ekimei kōshō*, Yoshikawa kōbunkan, 1988 (1st printing 1981).

Ikeya Kazunobu, Shirōzu Satoshi, eds, *Yama to mori no kankyōshi*, vol. 5 of Yumoto Takazu, ed., *Shiirizu Nihon rettō no sanman gosen nen hito to shizen no kankyōshi*, Bun'ichi sōgō shuppan, 2011.

Imamura Keiji, *Prehistoric Japan: New Perspectives on Insular East Asia*, Honolulu, HI: University of Hawai□i Press, 1996.

Inoue Makoto, *Yakihata to nettai rin Karimantan no dentōteki yakihata shisutemu no henyō*, Kōbundō, 1995.

Inoue Mitsusada, *Nihon kokka no kigen*, Iwanami, 1960.

Inoue Takuya, "Akiyama-gō ni okeru sansai kinoko riyō no hensen to saishū katsudō", in Ikeya Kazunobu, Shirōzu Satoshi, eds, *Yama to mori no kankyōshi*, Bun'ichi sōgō shuppan, 2011.

Isaji Yasunari, "Ritsuryōseika no domō kōshin ni kansuru oboegaki", *Historia* 147, June 1995, pp. 1–23.

Ishige Naomichi, "Kome shoku minzoku hikaku kara mita Nihonjin no seikatsu", in Nakabachi Masami, ed., *Seikatsugaku no hōhō*, Domesu, 1986, pp. 10–26.

Ishino Hironobu, Iwasaki Takuya, Kawakami Kunihiko, Shiraishi Taichirō, eds, *Seisan to ryūtsū 1*, "Kofun jidai no kenkyū", vol. 4, Yūzankaku, 1991.

Isogai Fujio, "Kodai chūsei ni okeru zakkoku no kyūkōteki sakutsuke ni tsuite, suiden nimōsaku tenkai no rekishiteki zentei to shite", *Tōkyō gakugei daigaku fuzoku kōtōgakkō kenkyū kiyō* 26, 1988, pp. 107–130.

Isogai Fujio, *Chūsei no nōgyō to kikō – suiden nimōsaku no tenkai*, Yoshikawa kōbunkan, 2002.

Itō Toshikazu, "Kodai chūsei no nobatake ni kansuru rekishi chirigakuteki kenkyū", *Nihon joshi daigaku daigakuin Bungaku kenkyūka kiyō* 1, March 1995, pp. 1–20.

Itō Toshikazu, "Heian Kamakura jidai no yamabata (yakihata) ni kansuru rekishi chirigakuteki kenkyū", *Nihon joshi daigaku kiyō Bungakubu* 45, March 1996, pp. 79–95.

Itō Toshikazu, "Kodai chūsei no hatasaku to hatake seido ni kansuru kisoteki kenkyū", *Jōrisei kenkyū* 12, Dec. 1996, pp. 1–15.

Itō Toshikazu, "Chūsei kōki ni okeru Tōdaiji-ryō Yamato no kuni Kawakami-shō no yakihata keiei to cha no saibai", *Nihon joshi daigaku kiyō Bungakubu* 48, March 1999, pp. 29–47.

Itō Toshikazu, "Kii no kuni no yamabata (yakihata) ni kansuru rekishi chirigakuteki kenkyū", *Shikyō* 41, Sept. 2000, pp. 1–24.

Itō Toshikazu, "Oka no seigyō", in Uehara Mahito *et al.*, eds, *Rettō no Nihonshi 2 Kurashi to seigyō*, Iwanami, 2005.

Iwanami Nihonshi jiten, see *Nihonshi jiten*.

Iyanaga Teizō, *Nihon kodai shakai keizaishi no kenkyū*, Iwanami, 1980.

Janata, Alfred, Kreiner, Joseph, Pauer, Erich, "Materialien zu Kuwa (Erdhacke) und Suki (Spaten)", *Archiv für Völkerkunde* 23, 1969, pp. 101–160.

Janata, Alfred Kreiner, Pauer, "Zur Geschichte des Pfluges (*Karasuki*)", *Archiv für Völkerkunde* 24, 1970, pp. 107–164.

Janata, Alfred Kreiner, Pauer, "Geräte der Tagoshirae Feldbestellung und des Taue Reisauspflanzens", *Archiv für Völkerkunde* 25, 1971, pp. 67–126.

Janata, Alfred Kreiner, Pauer, Klaus Müller, "Bewässerung (*mizu-hiki*) und Bewässerungsgeräte. Bodenbaugeräte Japans IV", *Archiv für Völkerkunde* 26, 1972, pp. 59–118.

Janata, Alfred Kreiner, Pauer, "Geräte für das Jäten und für die Schädlingsbekämpfung (Bodenbaugeräte Japans V)", *Archiv für Völkerkunde* 27, 1973, pp. 15–68.

Janata, Alfred Kreiner, Pauer, "Reiserte (*inekari*), Erntemesser und Sichel (*kama*). Bodenbaugeräte Japans VI", *Archiv für Völkerkunde* 30, 1976, pp. 31–100.

Jidaibetsu kokugo daijiten, Jōdaihen, Jōdaigo jiten henshū iinkai, ed., Sanshōdō, 1967.

Kadokawa shoten ed., *Shinpen kokka taikan*, CD-ROM, n.d. [1998?]

Kagaku gijutsuchō Shigen chōsakai ed., *Shinpen shokuhin seibun hyō*, Hitotsubashi shuppan, 2000.

Kagawa Mitsuō, "Yakihata to suiden", Shiomi Hiroshi sensei taikan kinen jigyōkai ed., *Kōko ronshū*, Hiroshima, 1993, pp. 309–317.

Kajiwara iseki chōsakai ed., *Kajiwara minami iseki hakkutsu chōsa hōkokusho*, Ōsaka-fu Takatsuki, 1988.

Kakubayashi Fumio, *Nihon shoki Jindai no maki zenchūshaku*, Hanawa shobō, 1999.

Kameda Takashi, "Kodai no kannō seisaku to sono seikaku", in Iyanaga Teizō, ed., *Nihon keizaishi taikei*, vol. 1, Tōkyō daigaku, 1965.

Kameda Takashi, *Nihon kodai yōsuishi no kenkyū*, Yoshikawa kōbunkan, 1973.

Kanai Tenbi, *Shitsugen saishi*, Hōsei daigaku, 1993 (1st printing 1977).

Kasarhérou, Emmanuel *et al.*, *Guide des plantes du chemin Kanak*, Nouméa, 1989.

Katsuki Setsuko, Katsuki Yōichirō, *Mura no kajiya*, Heibonsha, 1986.

Kawane Yoshiyasu, "Nimōsaku no kigen ni tsuite", *Nihonshi kenkyū* 77, March 1965, pp. 74–80.

Kawasaki-shi shimin myūjiamu ed., *Yayoi no shokuten: Himikotachi no tabemono*, exhibition catalogue, 1995.

Kelly, William, *Irrigation Management in Japan: A Critical Review of Japanese Social Science Research*, Ithaca, NY: Cornell University Press, 1982.

Kidder, J. Edward, "The Sannai-Maruyama Site: New Views on the Jōmon Period", *Southeast Review of Asian Studies* 20, 1998, pp. 29–52.

Kierstead, Thomas, *The Geography of Power in Medieval Japan*, Princeton, NJ: Princeton University Press, 1992.

Kimura Shigemitsu, *Nihon kodai chūsei hatasakushi no kenkyū*, Azekura shobō, 1992.

Kimura Shigemitsu, "Chūsei nōmin no shiki", in Toda Yoshimi, ed., *Chūsei no seikatsu kūkan*, Yūhikaku, 1993, pp. 75–128.

Kimura Shigemitsu, *Hatake to Nihonjin*, Chūō kōronsha, 1996.

Kimura Shigemitsu, "Kaihatsu to jōrisei kōchi, kōya o chūshin ni", *Jōrisei kenkyū* 12, 1996a, pp. 16–24.

Kimura Shigemitsu, *Zakkoku: Hatasaku nōkōron no chihei*, Aoki shoten, 2003.

Kimura Shigemitsu, *Zakkoku: Funshoku bunkaron no kanōsei*, Aoki shoten, 2006.

Kimura Shigemitsu, "Rekishi no naka no hatakechi to suiden (ge): hatasaku to inasaku no seisanryokuteki tokuchō", *Rekishi chiri kyōiku*, June 2008.

Kimura Shigemitsu, ed., *Nihon nōgyōshi*, Yoshikawa kōbunkan, 2010.

Kimura Yōjirō, ed., *Zusetsu sōboku meii jiten*, Kashiwa shobō, 1991.

Kinda Akihiro, Ishigami Eiichi, Kamada Motokazu, Sakaehara Towao, eds, *Nihon kodai shōen zu*, Tōkyō daigaku, 1996.

Kinda Akihiro, ed., *A Landscape History of Japan*, Kyōto: Kyōto University Press, 2010.

Kinoshita Tadashi, *Nihon nōkō gijutsu no kigen to dentō*, Yūzankaku, 1985.

Kinoshita Tadashi, Amino Yoshihiko, Kamino Yoshiharu, eds, *Seisan gijutsu to busshitsu bunka*, Yoshikawa kōbunkan, 1993.

Kishi Toshio, "Suki-kuwa setsu no dai-ni an", *Nihon kodai bunbutsu no kenkyū*, Hanawa shobō, 1988, pp. 178–183.

Kitamura Yasuhiro, "Jiden no seiritsu – Yamato no kuni Gufukuji o rei ni shite", *Shigaku zasshi* 121 (3), March 2012, pp. 38–61.

Kitō Hiroshi, *Nihon nisennen no jinkōshi*, PHP, 1983.

Kitō Hiroshi, "Edo jidai no komeshoku", in Haga Noboru, Ishikawa Hiroko, *Kome, mugi, zakkoku, mame*, "Zenshū Nihon no shoku bunka", vol. 3, Yūzankaku, 1998, pp. 47–58.

Kobayashi Yukio, *Kodai no gijutsu*, Hanawa shobō, 1982 (1st printing 1962).

Kodama Kōta, Itō Yoshiichi, *Kantō no fūdo to rekishi*, Yamakawa shuppan, 1978.

Kōgetsu Keigo, *Hōken jidai zenki no sangyō keizai*, Chūō kōronsha, 1950.

Kōgetsu Keigo, *Chūsei kangaishi no kenkyū*, Yoshikawa kōbunkan, 1983 (1st edition 1943).

Kojima Noriyuki, ed., *Nihon shoki*, "Shinpen Nihon koten bungaku zenshū", vol. 2, Shōgakukan, 1994.

Kokubu Naoichi, *Kita no michi, minami no michi*, Daiichi shobō, 1992.

Kokugakuin daigaku Nihon bunka kenkyūjo ed., *Kōhon Nihon shoki*, vol. 3, Kadokawa shoten, 1989.

Kokuritsu kokugo kenkyūjo ed., *Nihon gengo chizu*, 6 vols, Ōkurashō Insatsukyoku, 1966–1975.

Kokuritsu rekishi minzoku hakubutsukan ed., *Shōen ezu to sono sekai*, exhibition catalogue, 1993.

Komeie Taisaku, *Chū-kinsei sanson no keikan to kōzō*, Azekura shobō, 2002.

Komeie Taisaku, "Kinsei Dewa no kuni ni okeru yakihata no kenchi, keiei, nōhō", *Rekishi chirigaku* 47 (2), March 2005, pp. 1–23.

Komeie Taisaku, "Yakihata ni yoru sanchi shokusei no riyō to kaihatsu", in Miyamoto Shinji, Nonaka Kenichi, eds, *Shizen to ningen no kankyōshi*, Ōtsū: Kaiseisha, 2014, pp. 213–236.

Kōmoto Masayuki, "Yayoi to Jōmon", *Shūkan Asahi hyakka Nihon no rekishi* 39, 4 Jan. 1987, pp. 179–186.

Kōmoto Masayuki, "Yayoi jidai no shokuryō jijō", in Sahara Makoto, Tsude Hiroshi, eds, *Kankyō to shokuryō seisan*, "Kodaishi ronten" 1, Shōgakukan, 2000, pp. 167–182.

Kondō Hideo, *Shikoku tabemono minzokugaku*, Matsuyama: Atlas shuppan, 1999.

Kōno Michiaki, "Kamakura kaiga ni miru karasuki", *Kinki mingu* 9, Oct. 1985, pp. 1–12.

Kōno Michiaki, *Nihon nōkōgushi no kisoteki kenkyū*, Ōsaka: Izumi shoin, 1994.

Kōno Michiaki, "Nihon ni okeru rikō kokunai hassei setsu no sai kentō", *Bukkyō daigaku sōgō kenkyūjo kiyō* 1, March 1994, pp. 216–244.

Kōno Michiaki, "Gyūba kō no dōnyū to tenkai", *Shizuokaken maizō bunkazai chōsa kenkyūjo*, 1994, pp. 77–84.

Kōno Michiaki, "Kinsei nōgyō to chōshōri, jō, chū, ge 1", *Shōkei ronsō* 30 (1), 1994, pp. 85–126, 30 (3), Jan. 1995, pp. 135–212, 31 (3), March 1996, pp. 1–64.

Kōno Michiaki, "Ryūkotsusha no Chūgoku de no hattatsu to Nihon e no denrai ni tsuite", *Bukkyō daigaku sōgō kenkyūjo kiyō* 2 bessatsu, March 1995, pp. 52–95.

Kōno Michiaki, "Ine no kake-boshi no kigen ni tsuite no kisoteki kōsatsu", *Kokuritsu rekishi minzoku hakubutsukan kenkyū hōkoku* 71, March 1997, pp. 729–751.

Kōno Michiaki, "Heian jidai no momisuriusu", in Ōsaka daigaku bungakubu Nihonshi kenkyūshitsu ed., *Kodai chūsei no shakai to kokka*, Ōsaka: Seibundō, 1998, pp. 325–343.

Kōno Michiaki, "Tachibana Morikuni Ehon tsūhōshi no kisoteki kenkyū, jō", *Shōkyō ronsō* 36 (1), May 2000, pp. 1–28.

Kōno Michiaki, "Mingu to iu himoji shiryō no taikeika no tame no zairai nōgu no hikaku chōsa", in Kōno Michiaki, Kanagawa daigaku nijūisseki COE puroguramu, eds, *Jinrui bunka kenkyū no tame no himoji shiryō no taikeika* (Kenkyū seika hōkokusho), Kanagawa daigaku, 2008.

Koyama Shūzō, Matsuyama Toshio, Akimichi Tomoya, Fujino Yoshiko, Sugita Shigeharu, "Hida gofūdoki ni yoru shokuryō shigenteki kenkyū", *Kokuritsu minzokugaku hakubutsukan kenkyū hōkoku* 6 (3), 1981, pp. 363–596.

Koyama Shūzō, Gotō Yoshiko, "Nihonjin no shushoku no rekishi", in Ishige Naomichi, ed., *Ronshū Higashi ajia no shokuji bunka*, Heibonsha, 1985.

Koyama Shūzō, Gotō Yoshiko, "Nihonshokushi komeshoku no seiritsu made", in Ōsaka furitsu Yayoi hakubutsukan, 1999, pp. 80–93.

Koyama Shūzō, "Jōmon zan'ei: yakihata nōkō", in Harada Nobuo *et al.*, eds, *Yakihata no kankyōgaku*, Kyōto: Shibunkaku shuppan, 2011.

Kreiner, Joseph, Janata, Alfred, Pauer, Erich, "Methoden und Geräte des Reistrocknens. Bodenbaugeräte Japans VII", *Archiv für Völkerkunde* 36, 1982, pp. 95–146.

Kubota Kurao, *Zōho kaitei Tetsu no minzokushi*, Yūzankaku, 1991.

Kudō Yuichirō, ed., "Jōmon jidai no hito to shokubutsu no kankeishi", special issue, *Kokuritsu rekishi minzoku hakubutsukan kenkyū hōkoku* 185, July 2014.

Kuraku Yoshiyuki, *Suiden no kōkōgaku*, Tōkyō daigaku, 1991.

Kuroda Hideo, "Chūsei nōgyō gijutsu no yōsō", in Nagahara Keiji *et al.*, eds, *Nōgyō, nōsan kakō*, Nihon hyōronsha, 1987 (1st printing 1983).

Kuroda Hideo, *Nihon chūsei kaihatsushi no kenkyū*, Azekura shobō, 1995 (1st printing 1984).

Kurosaki Tadashi, "Nōgu", in Ishino Hironobu *et al.*, 1991, pp. 69–88.

Lachiver, Marcel, *Dictionnaire du monde rural*, Paris: Fayard, 1997.

Lamouroux, Christian, "Crise politique et développement rizicole en Chine; la région du Jiang-Huai (VIIIe–Xe siècles)", *Bulletin de l'École française d'Extrême-Orient* 82, 1995, pp. 145–184.

Laurent, Éric, "Nom d'un mushi, Nomenclature populaire des insectes dans le Japon rural", *Cipango* 8, autumn 1999, pp. 65–118.

Leroi-Gourhan, André, *Milieu et technique*, Paris: Albin Michel, 1973 (1st printing 1945).

Li Liu, Xingcan Chen, *The Archaeology of China: From the Late Paleolithic to the Early Bronze Age*, Cambridge: Cambridge University Press, 2012.

Mae Hisao, *Dōgu kojiki*, Tōkyō bijutsu, 1983.

Makino Tomitarō, *Shin Nihon shokubutsu zukan*, Hokuryūkan, 1974.

Mane, Perrine, *Calendriers et techniques agricoles (France–Italie, XIIe–XIIIe siècle)*, Paris: Le Sycomore, 1983.

Mane, Perrine, "L'outil agricole dans l'iconographie médiévale", in *L'outillage agricole médiéval et moderne et son histoire, 23ème Colloque de Flaran*, G. Comet (ed.), Toulouse: Presses Universitaires du Mirail, 2003, pp. 245–263.

Mane, Perrine, *La vie dans les campagnes au Moyen Âge d'après les calendriers*, Paris: La Martinière, 2004.

Mane, Perrine, *Le travail à la campagne au Moyen Âge. Étude iconographique*, Paris: Picard, 2006.

Masuda Shōko, *Awa to hie no shoku bunka*, Sammii shoten, 1995 (1st printing 1990).

Masuda Shōko, *Zakkoku no shakaishi*, Yoshikawa kōbunkan, 2001.

Matsui Akira, "Kachiku to maki, uma no seisan", in Ishino Hironobu *et al.*, 1991, pp. 105–120.

Matsumura Keiji, "Kodai inekura o meguru shomondai", *Bunkazai ronsō* 18, March 1983.

Matsuo Hikaru, "Bunken shiryō ni miru kodai no inasaku", in Takemitsu Makoto, Yamagishi Ryōji, eds, *Kodai Nihon no inasaku*, Yūzankaku, 1994.

Matsuyama Toshio, "Inewara to shokubutsu sen'i", in Sasaki, Matsuyama, 1988, pp. 289–306.

Matsuyama Toshio, *Ki no mi*, Hōsei daigaku, 1990 (1st printing 1982).

Mayuzumi Hiromichi, *Kodai Nihonjin no nazo*, Daiwa shobō, 1985.

Mazoyer, Marcel, Roudart, Laurence, *Histoire des agricultures du monde*, Paris: Seuil, 1997.

Menzies, Nicholas, "Three Hundred Years of Taungya: A Sustainable System of Forestry in South China", *Human Ecology* 16 (4), 1988, pp. 361–376.

Métailié, Georges, Stäuble Tercier, Nicole, *Au jardin potager chinois*, Vevey: Musée de l'Alimentation, 1997.

Métailié, Georges, Stäuble Tercier, Nicole, "Noms de plantes asiatiques dans les langues européennes: Essai en forme de vade-mecum", in Viviane Alleton,

Michael Lackner, eds, *De l'un au multiple: Traductions du chinois vers les langues européennes*, Paris: Maison des Sciences de l'Homme, 1999, pp. 275–291.

Métailié, Georges, Stäuble Tercier, Nicole, Fèvre, Francine, *Dictionnaire RICCI des plantes de Chine: chinois-français-latin-anglais, Li shi han fa la ying zhongguo zhiwu mingcheng cidian*, Paris: Association Ricci–Éditions du Cerf, 2005.

Mikami Yoshitaka, "Suiko nōgyō keiei to chiiki shakai", *Rekishigaku kenkyū* 781, Oct. 2003 (zōkan gō), pp. 39–48.

Mikkyō jiten, Sawa Ryūken ed., Kyōto: Hōzōkan, 1975.

Minamikawa Takashi, "Jidai kubun ron to rekishigaku kenkyū no genzai", *Shigaku zasshi* 121 (3), 2012, pp. 35–38.

Minzoku bunka eizō kenkyūjo ed., *Narada no seikatsu to shizen to no tsunagari, yakihata o chūshin ni*, Yamanashi-ken: Hayakawa-chō kyōiku iinkai, 1987.

Mishina Akahide, "Tenson kōrin shinwa no shomondai", *Kenkoku shinwa no shomondai*, Heibonsha, 1971.

Miwa Shigeo, *Usu*, Hōsei daigaku, 1991 (1st printing 1978).

Miyahara Takeo, *Nihon kodai no kokka to nōmin*, Hōsei daigaku, 1976 (1st printing 1973).

Miyamoto Kazuo, "Jōmon nōkō to Jōmon shakai", in Sahara and Tsude, 2000, pp. 115–138.

Miyamoto Tsuneichi, *Nihon bunka no keisei*, vol. 1, Chikuma gakugei, 1994.

Mizoguchi Tsunetoshi, "Shifting cultivation and land tenure in Shirakawa-go: changes from the 1690s to the 1880s". Paper presented at the Workshop on Population Change and Socioeconomic Development in the Nobi Region, Stanford University, 15 March 1987.

Mori Kōichi, ed., *Man'yōshū no kōkogaku*, Chikuma shobō, 1985 (1st printing 1984).

Mori Kōichi, ed., *Jōmon Yayoi no seikatsu*, 1986.

Morita Kikuo, *Nihon kodai no ōken to sanya kakai*, Yoshikawa kōbunkan, 2009.

Morita Tei, *Nihon kodai no kōchi to nōmin*, Daiichi shobō, 1986.

Murai Yasuhiko, *Heian kizoku no sekai*, Tokuma shoten, 1986 (1st printing 1968).

Murakami Shigeyoshi, *Tennō to saishi*, Iwanami, 1977.

Murao Jirō, *Ritsuryō zaiseishi no kenkyū*, Yoshikawa kōbunkan, 1978 (1st printing 1961).

Musashino bijutsu daigaku Minzoku shiryōkan ed., *Kurashi no zōkei: wara, tsuru, kusa*, exhibition catalogue, 1994.

Nakanishi Ryōtarō, "Meiji makki Ibaraki-ken ka chōson no shokumotsu shōhi ryō", in Haga and Ishikawa, 1998, pp. 59–90.

Nakanishi Susumu, ed. annot., *Man'yōshū*, 4 vols, Kōdansha, 1978–1983.

Nakanishi Susumu, *Man'yōshū jiten*, Kōdansha, 1993 (1st edition 1985).

Nakazawa Michihiko, "Jōmon nōkōron o megutte", in Shitara Hiromi et al., eds, *Shokuryō no kakutoku to seisan*, vol. 5 of *Yayoi jidai no kōkogaku*, Dōseisha, 2009.

Naoki Kōjirō, ed., *Shōsōin monjo sakuin*, Heibonsha, 1982 (1st printing 1981).

Naoki Kōjirō, *Nara jidai no shomondai*, Hanawa shobō, 1989 (1st printing 1968).

Naora Nobuo, *Nihon kodai nōgyō hattatsushi*, Saera shobō, 1956.

Nara kokuritsu bunkazai kenkyūjo ed., *Heijōkyū hakkutsu chōsa shutsudo mokkan gaihō*, vols 20–34, 1989–1998.

Nasu Hiroo, "Zassō kara mita Jōmon jidai banki kara Yayoi jidai kōki ni okeru ine to zassō no saibai keitai", *Kokuritsu rekishi minzoku hakubutsukan kenkyū hōkoku* 187, July 2014, pp. 95–110.

Negita Yoshio, "Inasaku no hajimari", in Sahara and Tsude, 2000, pp. 139–166.

Nespoulous, Laurent, "L'apparition des premières sociétés agricoles dans l'archipel japonais", in Jean-Paul Demoule, Pierre Souyri, eds, *Archéologie et patrimoine au Japon*, Paris: Maison des Sciences de l'Homme, 2008, pp. 25–60.

Nespoulous, Laurent, "Cadres chrono-culturels et dynamiques protohistoriques", in J.-P. Demoule, P. Souyri, eds, *Archéologie et patrimoine au Japon*, Paris: Maison des Sciences de l'Homme, 2008, pp. 19–24.

Niels, Daniel, Abe Kōichi, eds, *Humanity and Nature in the Japanese Archipelago*, Kyōto: Research Institute for Humanity and Nature, 2015.

Nihon gakushiin ed., *Meiji zen Nihon nōgyō gijutsushi*, Inoue shoten, 1980 (1st edition 1944).

Nihon shokubutsu hōgen shūsei, Yasaka shobō comp., Yasaka shobō, 2001.

Nishimiya Kazutami, ed. annot., *Kojiki*, "Shinchō Nihon koten shūsei", Shinchōsha, 1991 (1st printing 1979).

Nishitani Masaru, "Suiden to yakihata", *Kokuritsu rekishi minzoku hakubutsukan kenkyū hōkoku* 153, Dec. 2009, pp. 87–114.

Nishizaki Tōru, ed., *Nihon kojisho o manabu hito no tame ni*, Sekai shisōsha, 1995.

Nissen Jaubert, Anne, "Les finages et leurs rendements: l'exemple danois", in L. Feller, P. Mane, F. Piponnier, eds, *Le village médiéval et son environnement*, Paris: Publications de la Sorbonne, 1998, pp. 551–570.

Nōkō bunka kenkyū shinkōkai ed., *Suiden ikō shūsei*, Kyōto, 1988.

Nomoto Kanichi, "Man'yōshū to yamada no minzoku", in Mori Kōichi, ed., *Man'yōshū no kōkogaku*, Chikuma shobō, 1984a.

Nomoto Kanichi, *Inasaku minzoku bunkaron*, Yūzankaku, 1993.

Nomoto Kanichi, *Yakihata minzoku bunkaron*, Yūzankaku, 1995 (1st printing 1984).

Nomoto Kanichi, "Yakihata bunka no keisei", in Ōbayashi Taryō, ed., *Sanjin no seigyō*, Chūō kōronsha, 1996, pp. 139–208.

Nomoto Kanichi, "Shiiba no shigen seigyō minzoku", *Museum Kyūshū* 60, May 1998, pp. 13–17.

Notō Takeshi, "Hatasaku nōkō", in Ishino Hironobu *et al.*, eds, *Seisan to ryūtsū 1*, "Kofun jidai no kenkyū", vol. 4, Yūzankaku, 1991, pp. 89–104.

Nunome Junrō, *Kinu to nuno no kōkogaku*, Yūzankaku, 1988.

Obata-Reiman, Etsuko, *Nihonjin no tsukutta kanji*, Nanundō, 1990.

Ōbayashi Taryō, *Insaku no shinwa*, Kōbundō, 1973.

Ōbayashi Taryō, *Higashi ajia no ōken shinwa: Nihon, Chōsen, Ryūkyū*, Kōbundō, 1984.

Ōbayashi Taryō *et al.*, "Tōnan ajia Oceania ni okeru sho minzoku bunka no data base no sakusei to bunseki", *Kokuritsu minzokugaku hakubutsukan kenkyū hōkoku bessatsu* 11, 1990.

Ōbayashi Taryō, "Seikatsu yōshiki to shite no yakihata kōsaku", in Ōbayashi Taryō, ed., *Sanmin to ama*, Shōgakukan, 1995 (1st printing 1983), pp. 219–234.

Ōga Ikuo, "Kinseiki no kiroku ni miru Shiibayama", *Museum Kyūshū* 60, May 1998, pp. 36–40.

Ōgata Tōru, "Kaya ni tsuite, sono jujutsuteki kōyō o megutte", *Nihon kenkyū* 18, Sept. 1998, pp. 151–175.

Ogawa Naoyuki, "Hatasaku to tsumida", in Kinoshita Tadashi, ed., *Shitsuden nōkō*, Iwasaki bijutsusha, 1988, pp. 89–111.

Ogawa Naoyuki, "Yakihata to tsumida", in Kinoshita Tadashi *et al.*, eds, 1993, pp. 142–175.

Ogawa Naoyuki, *Tsumida inasaku no minzokugakuteki kenkyū*, Iwata shoin, 1995.

Ohnuki-Tierney, Emiko, *Rice as Self*, Princeton, NJ: Princeton University Press, 1993.

Ōishi Naomasa, "Jūichi-jūni seiki no tochi no shurui to sono riyōkōei no jōtai", "Jūichi-jūni seiki no sakumotsu to saibai hō", in Toyoda Takeshi, ed., *Sangyōshi*, Yamakawa shuppan, 1964, pp. 172–182, 189–198.

Oka Mitsuo, Iinuma Jirō, Horio Hisashi, eds, *Inasaku no gijutsu to riron*, Heibonsha, 1990.

Okada Seishi, *Kodai ōken no saishi to shinwa*, Hanawa shobō, 1970.

Okimori Takuya, Yajima Izumi, Satō Makoto, Hirasawa Ryūsuke, *Kakyō hyōshiki chūshaku to kenkyū*, Ōfūsha, 1993.

Okimori Takuya, Satō Makoto, Yajima Izumi, *Tōshi Kaden Kamatari, Jōei, Muchimaro den chūshaku to kenkyū*, Yoshikawa kōbunkan, 1999.

Ōkuma Kiichirō, Inui Katsumi, *Jōdai setsuwa jiten*, Yūzankaku, 1993.

Ono Takeo, *Nihon nōgyō kigenron*, Nihon hyōronsha, 1942.

Orikuchi Shinobu, "Niiname to azuma uta", in Niiname no kenkyūkai ed., *Niiname no kenkyū*, vol. 1, 1953.

Ōsaka furitsu Yayoi bunka hakubutsukan ed., *Himiko no shokutaku*, Yoshikawa kōbunkan, 1999.

Ōsawa Masaaki, *Chin Fu nōsho no kenkyū*, Nōsan gyoson bunka kyōkai, 1993.

Ōsawa Masaaki *Tō Sō henkakuki nōgyō shakaishi kenkyū*, Kyūko shoin, 1996.

Ōta Akira, *Seishi kakei daijiten*, 3 vols, Kadokawa shoten, 1963.

Otomasu Shigetaka, *Yayoi nōgyō to mainō shūzoku*, Rokukō shuppan, 1992.

Ōtsuka Hatsushige *et al.*, eds, *Nihon kodai iseki jiten*, Yoshikawa kōbunkan, 1995.

Ōtsuka Hatsushige *et al.*, eds, *Kōkogaku ni yoru Nihon rekishi 2, Sangyō 1*, Yūzankaku, 1996.

Pauer, Erich, *Technik-Wirtschaft-Gesellschaft: der Einfluss wirtschaftlicher und gesellschaftlicher Veränderungen auf die Entwicklung der landwirtschaftlichen Geräte in der vorindustriellen Epoche Japans ab dem 17. Jahrhundert*, Vienna: Institut für Japanologie der Universität, 1973.

Pauer, Erich, *Papers on the History of Industry and Technology of Japan 1: From the Ritsuryō System to the Early Meiji Period*, Marburg: Förderverein der Marburger Japan-Reihe, 1995.

Pearson, Richard J. *et al.*, eds, *Windows on the Japanese Past: Studies in Archaeology and Prehistory*, Ann Arbor, MI: University of Michigan Press, 1986.

Pearson, Richard J. *et al.*, *Ancient Japan*, New York: George Braziller, 1992.

Pelletier, Philippe, "Livre deuxième, le Japon", pp. 224–430, in *Géographie Universelle: Chine, Japon, Corée*, Roger Brunet, ed., Paris: Belin/Reclus, 1994.

Péronny, Claude, *Les plantes du Man'yōshū*, Paris: Maisonneuve et Larose, 1993.

Pigeot, Jacqueline, *Manuel de japonais classique*, Paris: Asiathèque, 1998.

Pirazzoli-t'Serstevens, Michèle, "Pour une archéologie des échanges. Apports étrangers en Chine: transmission, réception, assimilation", *Arts Asiatiques* 49, 1994, pp. 21–33.

Reizei Tamehito, Kōno Michiaki, Iwasaki Takehito, *Mizuho no kuni: Nihon shiki kōsaku zu no sekai*, Kyōto: Tankōsha, 1996.

Rotermund, Hartmut, ed., *Religions, croyances et traditions populaires du Japon*, Paris: Maisonneuve et Larose, 2000.

Ruellan, Francis, *La production du riz au Japon*, Paris: Larose, 1938.

Saeki Arikiyo, *Nihon kodai shizoku jiten*, Yūzankaku, 1994.

Sahara Makoto, *Nihon bunka o horu*, NHK ningen daigaku, 1992.

Sahara Makoto, *Nihonjin no tanjō*, Shōgakukan, 1994 (1st printing 1987).

Sahara Makoto, "Kome to Nihon bunka", *Kokuritsu rekishi minzoku hakubutsukan kenkyū hōkoku* 60, March 1995, pp. 107–132.

Sahara Makoto, *Shoku no kōkogaku*, Tōkyō daigaku, 1996.

Sahara Makoto, Tsude Hiroshi, eds, *Kankyō to shokuryō seisan*, "Kodaishi ronten" 1, Shōgakukan, 2000.

Saigō Nobutsuna, *Kojiki no sekai*, Iwanami, 1984 (1st printing 1967).

Saitō Masami, Shiiba Kuniko, *Obāsan no shokubutsu zukan*, Fukuoka: Ashi shobō, 1995.

Sakamoto Shintarō, "Engishiki kara mita shokoku no sanbutsu hyō", *Waseda shōgaku* 281, Dec. 1979, pp. 655–706, *Waseda shōgaku* 284, July 1980, pp. 357–388.

Sakamoto Shōzō, "Edo jidai o kinsei to iu koto", *Nihon rekishi* 769, June 2012, pp. 105–114.

Sakurai Junya, "Kaizuka to karorii keisan", in Ōtsuka Hatsushige *et al.*, eds, *Kōkogaku ni yoru Nihon rekishi* 2, Sangyō 1, Yūzankaku, 1996, pp. 174–185.

Sakurai Mitsuru, ed., *Man'yōshū no minzokugakuteki kenkyū*, Ōfūsha, 1995.

Sano Masami, "Man'yōshū hatako kō, tago no ura no imi", *Chūō daigaku kokubun* 6, March 1963.

Sasaki Akira, Shiiba Kuniko, *Obāsan no yamazato nikki*, Fukuoka: Ashi shobō, 1998.

Sasaki Kōmei, *Nihon no yakihata, sono chiikiteki hikaku kenkyū*, Kokin shoin, 1972.

Sasaki Kōmei, *Shōyū jurin bunka no michi*, NHK Books, 1984 (1st printing 1982).

Sasaki Kōmei, Matsuyama Toshio, eds, *Hatasaku bunka no tanjō*, Nihon hōsō shuppankai, 1988.

Sasaki Kōmei, *Inasaku izen*, NHK Books, 1989 (1st printing 1971).

Sasaki Kōmei, *Nihonshi no tanjō*, Shūeisha, 1991.

Sasaki Kōmei, *Nihon bunka no kiso o saguru*, NHK Books, 1994 (1st printing 1993).

Sasaki Kōmei, *Nihon bunka no tajū kōzō*, Shōgakukan, 1998 (1st printing 1997).

Sasaki Takeo, "Aizu nōsho no nōgu o megutte", in Kinoshita Tadashi *et al.*, eds, *Seisan gijutsu to busshitsu bunka*, Yoshikawa kōbunkan, 1993, pp. 52–76.

Satō Masatoshi, "Kodai tennō no shokuji to nie", *Heian jidai no tennō to kanryōsei*, Tōkyō daigaku shuppankai, 2008.

Satō Toshio, "Sado no yakimaki to soba", *Kōshiji* (Niigata-ken minzoku gakkai ed.) 309, Nov. 1993, pp. 1–11.

Satō Tsuneo, Tokunaga Mitsutoshi, Etō Akihiko, *E nōsho* 1, "Nihon nōsho zenshū", vol. 71, Nōsan gyoson bunka kyōkai, 1996.

Satō Yōichirō, *DNA ga kataru inasaku bunmei*, NHK Books, 1996.

Satō Yōichirō, *Jōmon nōkō no sekai*, PHP, 2000.

Satō Yōichirō, "DNA kara mita ine no michi", in NHK Special Nihonjin Project ed., *Nihonjin harukana tabi 4: Ine, shirarezaru ichimannen no tabi*, NHK, 2001, pp. 118–135.

Satō Yōichirō, *Ine no Nihonshi*, Kadokawa shoten, 2002.

Sawada Gōichi, *Narachō jidai minsei keizai sūteki kenkyū*, Hanawa shobō, 1978 (2nd printing of 1st edition 1972) (1st edition 1927).

Seki Kazuhiko, *Kodai nōmin Oshiha o tazunete: Nara jidai tōkokujin no kurashi to shakai*, Chūō kōronsha, 1998.

Sekine Shinryū, *Narachō shoku seikatsu no kenkyū*, Yoshikawa kōbunkan, 1989 (1st printing 1969).

Sekine Shinryū, *Shōsōin monjo jikō sakuin*, Yoshikawa kōbunkan, 2001.

Senda Minoru, *Takachiho no gensō*, PHP, 1999.

Settsu-shi kyōiku iinkai ed., *Settsu-shi no mingu to kurashi*, Settsu, 1993.

Shibusawa Keizō, "Engishiki nai suisan shinsen ni kansuru kōsatsu jakkan", *Saigyo dōzatsukō*, Okushoin, 1954, pp. 241–307.

Shibusawa Keizō, Kanagawa daigaku Nihon jōmin bunka kenkyūjo ed., *Shinpan Emakimono ni yoru: Nihon jōmin seikatsu ebiki*, 5 vols, Heibonsha, 1988 (1st printing 1984).

Shikanai Hirotane, "Kujōkebon Engishiki oboegaki", *Shoryōbu kiyō* 52, 2000, pp. 1–22.

Shikinaisha kenkyūkai ed., *Shikinaisha chōsa hōkoku*, 24 vols, Ise: Kōgakukan daigaku shuppanbu, 1977–1990.

Shiomi Hiroshi, "Jōmon jidai no shokuyō shokubutsu", in Saitō Tadashi, ed., *Nihon kōkogaku ronshū*, vol. 2, *Shūraku to ishokuju*, Yoshikawa kōbunkan, 1986, pp. 430–465.

Shiraishi Akiomi, *Hatasaku no minzoku*, Yūzankaku, 1988.

Shiraishi Akiomi, "Nōkō no seikatsu", in Sakurai Mitsuru, ed., *Man'yōshū no minzokugaku*, Ōfūsha, 1993, pp. 77–90.

Shiraishi Akiomi, *Ine to mugi no minzoku*, Yūzankaku, 1994.

Shirōzu Satoshi, *Shirarezaru Nihon – sanson no kataru rekishi sekai –*, Nihon Hōsō shuppankai, 2005.

Shiryō hensanjo ed., *Nihon shōen ezu shūei*, vol. 2 Kinki 1, 1992; vol. 1 jō Higashi Nihon 1, 1995; vol. 1 ge Higashi Nihon 2, 1996; vol. 3 Kinki 2, 1998; vol. 5 Kinki 3, 1999; vol. 4 jō Nishi Nihon 1, 2001; vol. 4 ge, Nishi Nihon 2, 2002; Tōkyō daigaku Shiryō hensanjo, 1988–2002.

Shitara Hiromi, "Nōkō bunka fukugō to Yayoi bunka", *Kokuritsu rekishi minzoku hakubutsukan kenkyū hōkoku* 185, Feb. 2014, pp. 449–469.

Shizuoka-ken maizō bunkazai chōsa kenkyūjo ed., *Kodai ni okeru nōgu no hensen: inasaku gijutsu o nōgu kara miru*, conference proceedings, 3 vols, Shizuoka, 1994.

Shōgakukan ed., *Shokuzai zuten*, 1996 (1st printing 1995).

Sieffert, René, *Man'yōshū*, vols 1, 2, 3, Paris: Publications Orientalistes de France/ Unesco, 1997, 1998, 2001.

Sigaut, François, *L'agriculture et le feu*, Paris: Mouton, 1975.

Sigaut, François, "Quelques notions de base en matière de travail du sol dans les anciennes agricultures européennes", *Journal d'Agriculture Traditionnelle et de Botanique Appliquée*, 24 (2/3), 1977, pp. 139–169.

Sigaut, François, "Identification des techniques de récolte des graines alimentaires", *Journal d'Agriculture Traditionnelle et de Botanique Appliquée*, 25 (3), 1978, pp. 145–161.

Sigaut, François, "Identification des techniques de conservation et de stockage des grains", in Marceau Gast, François Sigaut, eds, *Les techniques de conservation des grains à long terme*, Paris: CNRS, 1981, pp. 156–180.

Sigaut, François, "Essai d'identification des instruments à bras de travail du sol", *Cahiers Orstom*, Série Sciences Humaines, 20 (3, 4), 1984, pp. 359–374.

Sigaut, François, "A method for identifying grain storage techniques and its application for European agricultural history", *Tools and Tillage*, 6 (1) (1988), pp. 3–33.

Sigaut, François, "Storage and threshing in pre-industrial Europe: Additional notes", *Tools and Tillage*, 6 (2), (1989), pp. 119–124.

Sigaut, François, "Pour une méthode d'identification des systèmes et des techniques d'égrenage", in *Le traitement des récoltes: un regard sur la diversité du néolithique au présent: XXIIIe rencontres internationales d'archéologie et d'histoire d'Antibes*, P. C. Anderson, L. S. Cummings, T. K. Schippers, B. Simonel (eds), Antibes: Éditions APDCA, 2003, pp. 515–521.

Sigaut, François, *La troublante histoire de la jachère: Pratiques des cultivateurs, concepts de lettrés et enjeux sociaux*, Versailles: Éditions Quae, 2008.

Smits, Ivo, *The Pursuit of Loneliness: Chinese and Japanese Nature Poetry in Medieval Japan, ca. 1050–1150*, Stuttgart: Franz Steiner, 1995.

Smolarz, Bruno, "La riziculture au Japon", doctoral thesis, Paris: Inalco, 1997.

Sonoda Kōyū, *Nihon kodai zaiseishi no kenkyū*, Hanawa shobō, 1981.

Steensberg, Axel, *Draved*, Copenhagen: National Museum of Denmark, 1979.

Steensberg, Axel, *Fire-Clearance Husbandry: Traditional Techniques Throughout the World*, Hering, Denmark: Poul Kristensen, 1993.

Sutō Isao, *Tagayasu*, "Shashin de miru Nihon seikatsu zuhiki", vol. 1, Kōbundō, 1994 (1st edition 1989).

Tachibana Reikichi, *Hakusanroku no yakihata nōkō*, Hakusuisha, 1995.

Takagi Tokuo, "Kinokawa ryūiki shōen ni okeru konsaku to detsukuri", *Nihon chūsei chiiki kankyōshi no kenkyū*, Azekura shobō, 2008.

Takagi Tokuo, "Chūsei wa seisanryoku ga kōjō shita jidai dewa nakatta no desu ka", *Nihon rekishi* 764, Jan. 2012, pp. 30–33.

Takahashi Kazuki, "Hatakeda o tsūjite miru jūgo seiki no Kinai kinkoku ni okeru nōgyō seisan", in Nakajima Keiichi, ed., *Chūsei o owaraseta 'seisan kakumei' – ryōsanka gijutsu no hirogari to eikyō –*, Kiban kenkyū B Kenkyū seika hōkokusho, 2015.

Takahashi Kazuo, "Kodai no seitetsu", in Amakasu Ken et al., eds, *Saikō to yakin*, "Nihon gijutsu no shakaishi", vol. 5, Nihon hyōronsha, 1989 (1st printing 1983), pp. 7–28.

Takai Yoshihiro, "Engishiki naizenshiki kōsaku enpu jō chūshaku", *Kenkyū kiyō* (Gunma-ken maizō bunkazai chōsa jigyōdan ed.) 15, March 1998, pp. 101–117.

Takase Katsunori, "Yayoi jidai no zakkoku saibai to kinomi no hyōka", in Shitara Hiromi et al., eds, *Shokuryō no kakutoku to seisan*, vol. 5 of *Yayoi jidai no kōkogaku*, Dōseisha, 2009.

Takashige Susumu, *Kodai chūsei no kōchi to sonraku*, Daimeidō, 1994 (1st printing 1975).

Takaya Yoshikazu, *Kome o dō toraeru ka*, NHK Books, 1990.

Takei Noriko, "Kodai ni okeru sōkō zuinō gyōmu no jittai", *Kokuritsu rekishi minzoku hakubutsukan kenkyū hōkoku* 194, March 2015, pp. 101–126.

Takemoto Toyoshige, "Jitō to chūsei sonraku Bitchū no kuni Niimi-shō", in Ishii Susumu, ed., *Chūsei no sonraku to gendai*, Yoshikawa kōbunkan, 1991, pp. 255–350.

Takemoto Toyoshige, Ishii Susumu, "Niimi-shō o kataru", in Ishii Susumu, ed., *Chūsei no sonraku to gendai*, Yoshikawa kōbunkan, 1991, pp. 351–370.

Takigawa Masajirō, "Tengai kō", *Nihon shakai keizaishi ronkō*, Nikkō shoin, 1939, pp. 194–222.

Takigawa Masajirō, *Ritsuryō jidai no nōmin seikatsu*, Tōkō shoin, 1943; Meichō fukyūkai, 1988 (1st printing 1969).

Tamada Yoshihide, ed., *Rettō bunka no hajimari*, vol. 1 of *Shiseki de yomu Nihon no rekishi*, Yoshikawa kōbunkan, 2009.

Tamura Noriyoshi, "Shizen kankyō to chūsei shakai", in Ōtsu Tōru, ed., *Iwanami kōza Nihon rekishi*, vol. 9, Chūsei 4, Iwanami, 2015.

Tanaka Toyoharu, "Yakihata, maki, makibata to Nihon hatasaku nōgyō tenkai mondai", *Rekishi chirigaku kiyō* 23, 1981, pp. 85–106.

Tanigawa Kenichi, Amino Yoshihiko, Ōbayashi Taryō, Miyata Noboru, Mori Kōichi, eds, *Nihon zō o toinaosu*, Shōgakukan, 1994 (1st printing 1993).

Tanigawa Kenichi *et al.*, eds, *Fūdo bunka*, "Nihon minzoku bunka taikei", vol. 1, Shōgakukan, 1995 (1st printing 1986).

Taranczewski, Detlev, *Lokale Grundherrschaft und Ackerbau in der Kamakura Zeit*, "Bonner Zeitschrift für Japanologie", vol. 10, Japanologisches Seminar der Universität Bonn, 1988.

Tateno Kazumi, "Sonraku no saijiki", in Nihon sonrakushi kōza henshū iinkai ed., *Seikatsu 1 genshi, kodai, chūsei*, Yūzankaku, 1991, pp. 169–189.

Tatsumi Kazuhiro, "Zōhō mokkan ni miru shokubutsusei shokumotsu to sono sanchi", in Saitō Tadashi, ed., *Nihon kōkogaku ronshū 2, Shūraku to ishokuju*, Yoshikawa kōbunkan, 1986, pp. 466–475.

Terasawa Kaoru, Terasawa Tomoko, "Yayoi jidai shokubutsushitsu shokuryō no kisoteki kenkyū", *Kashihara kōkogaku kenkyūjo kiyō (Kōkogaku ronshū)* 5, March 1981, pp. 1–130.

Terasawa Kaoru, "Inasaku gijutsu to Yayoi no nōkō", in Mori Kōichi, ed., *Jōmon Yayoi no seikatsu*, Chūō kōronsha, 1986, pp. 291–350.

Terasawa Kaoru, "Hatasakumotsu", *Kikan kōkogaku* 14, Feb. 1989, pp. 23–31.

Terasawa Kaoru, "Shūkaku to chozō", in Ishino Hironobu *et al.*, eds, *Seisan to ryūtsū 1*, "Kofun jidai no kenkyū", vol. 4, Yūzankaku, 1991, pp. 50–68.

Terasawa Kaoru, "Yayoi jidai no shokuryō jijō", in Ōsaka bunka sentā ed., *Genshi kodai no komezukuri*, 1992, pp. 16–19.

Terasawa Kaoru, "Yayoi jidai no hatasakumotsu", in Haga Noboru, Ishikawa Hiroko, *Shoku seikatsu to shokumotsushi*, "Zenshū Nihon no shoku bunka", vol. 2, Yūzankaku, 1999, pp. 33–48.

Terasawa Kaoru, *Ōken tanjō*, Kōdansha, 2000.

Thiede, Ulrike, *Auf Haustierspuren zu den Ursprüngen der Japaner. Vor- und frühgeschichtliche Haustierhaltung in Japan*, Munich: Iudicium, 1998.

Toda Yoshimi, *Nihon ryōshusei seiritsushi no kenkyū*, Iwanami, 1995 (1st printing 1967).

Toda Yoshimi, *Shoki chūsei shakaishi no kenkyū*, Tōkyō daigaku, 1999 (1st printing 1991).

Torao Toshiya, "Kōden o meguru futatsu no mondai", in Takeuchi Rizō hakase kanreki kinenkai ed., *Takeuchi Rizō hakase kanreki kinen Ritsuryō kokka to kizoku shakai*, Yoshikawa kōbunkan, 1969, pp. 265–292.

Torao Toshiya, trans., annot., *Engishiki*, 2 vols, Shūeisha, 2000, 2007.

Totman, Conrad, *The Green Archipelago: Forestry in Preindustrial Japan*, Stanford, CA: University of California Press, 1989.

Totman, Conrad, *Nihonjin wa dono yō ni mori o tsukutte kita ka*, Tsukiji shokan, 1998.

Totman, Conrad, *A History of Japan*, Malden, MA: Blackwell, 2000.

Toyoda Takeshi, ed., *Sangyōshi*, Yamakawa shuppan, 1964.

Tozawa Mitsunori, "Jōmon nōkō", in Katō Shunpei *et al.*, eds, *Jōmon bunka no kenkyū 2, Seigyō*, Yūzankaku, 1988 (1st printing 1983), pp. 254–266.

Trombert, Éric, *Le crédit à Dunhuang. Vie matérielle et société en Chine médiévale*, Paris: Collège de France, 1995.

Trombert, Éric, "Un moulin chinois du XIIe siècle", *Arts Asiatiques* 51, 1996, pp. 81–95.

Tsuboi Hirofumi, *Imo to Nihonjin*, Miraisha, 1981 (1st printing 1979).

Tsuboi Hirofumi, "Nōkō minzoku no nigensei", in Mori Kōichi, ed., *Jōmon Yayoi no seikatsu*, "Nihon no kodai", vol. 4, Chūō kōronsha, 1986, pp. 351–374.

Tsuboi Hirofumi, "Inasaku bunka no tagensei", in Tanigawa Kenichi, ed., *Fūdo to bunka*, "Nihon minzoku bunka taikei", vol. 1, Shōgakukan, 1995 (1st printing 1986), pp. 294–342.

Tsuda Sōkichi, *Nihon koten no kenkyū*, "Tsuda Sōkichi zenshū", vol. 1, Iwanami, 1963 (1st edition 1948).

Tsude Hiroshi, *Kofun jidai no ō to minshū*, Kōdansha, 1989.

Tsukamoto Manabu, "Yōsui fushin", in Nagahara Keiji *et al.*, eds, *Doboku*, "Kōza Nihon gijutsu no shakaishi", vol. 6, Nihon hyōronsha, 1988 (1st printing 1984).

Tsukuba Hisaharu, *Nihon no nōsho*, Chūō kōronsha, 1987.

Ueyama Shunpei, ed., *Shōyō jurin bunka*, Chūō kōronsha, 1985 (1st printing 1969).

Umehara Takeshi, Yasuda Yoshinori, eds, *Jōmon bunmei no hakken, kyōi no Sannai-Maruyama iseki*, PHP, 1996 (1st printing 1995).

Verschuer, Charlotte von, *Les relations officielles du Japon avec la Chine aux VIIIe et IXe siècles*, Genève, Paris: Droz, 1985.

Verschuer, Charlotte von, "L'habitat rural du Japon ancien", *Archéologie médiévale* 23, 1993, pp. 1–55.

Verschuer, Charlotte von, "L'autre agriculture: les cultures sur brûlis dans le Japon ancien", *Journal d'Agriculture Traditionnelle et de Botanique Appliquée*, 37 (2), 1995, pp. 129–164.

Verschuer, Charlotte von, "Jōjin découvre la ville de Hangzhou en 1072", in J. Pigeot, H. Rotermund, eds, *Le vase de béryl*, Arles: Philippe Picquier, 1997, pp. 353–362.

Verschuer, Charlotte von, "De la houe à la sandale chevaline dans le Japon ancien", in L. Feller, P. Mane, F. Piponnier, eds, *Le village médiéval et son environnement*, Paris: Publications de la Sorbonne, 1998, pp. 427–443.

Verschuer, Charlotte von (Veashua), "Nihon kodai ni okeru yakihata to kaikon kankei no kokuji ni tsuite", *Tōkyō daigaku Shiryō hensanjo kenkyū kiyō* 13, March 2003, pp. 1–7.

Verschuer, Charlotte von, "Les épis 'de riz' du jardin céleste: deux lectures d'un même texte", in J. Kyburz, F. Macé, C. von Verschuer, eds, *Éloge des sources: Reflets du Japon ancien et moderne*, Arles: Philippe Picquier, 2004, pp. 17–42.

Verschuer, Charlotte von, *Across the Perilous Sea: Japanese Trade with China and Korea from the Seventh to Sixteenth Centuries*, translated from French by Kristen Lee Hunter, Ithaca, NY: Cornell University Press, 2006.

Verschuer, Charlotte von (Veashua), "Nihon kodai ni okeru gokoku to nenjūgyōji", *Shigaku zasshi* 118 (1), 2009a, pp. 37–59.

Verschuer, Charlotte von (Veashua), "Kodai no Nihonjin wa kome o dore gurai tabete ita ka", *Hikaku Nihongaku kyōiku kenkyū sentā nenpō* (Ochanomizu joshi daigaku) 5, 2009b, pp. 53–62.

Verschuer, Charlotte von, "Demographic Estimates and the Issue of Staple Food in Early Japan", *Monumenta Nipponica* 64 (2), 2009c, pp. 337–362.

Verschuer, Charlotte von (Veashua), "Kiki no jindai maki ni okeru inaho setsu no seiritsu katei", in Suzuki Yasutami, ed., *Nihon kodai no ōken to higashi ajia*, Tōkyō: Yoshikawa kōbunkan, 2012, pp. 165–182.

Verschuer, Charlotte von, *Le commerce entre le Japon, la Chine et la Corée à l'époque médiévale, VIIe–XVIe siècle*, Paris: Publications de la Sorbonne, 2014.

Vieillard-Baron, Michel, *Fujiwara no Teika (1162–1241) et la notion d'excellence en poésie: Théorie et pratique de la composition dans le Japon classique*, Paris: Collège de France, 2001.

Wada Sei, Ishihara Michihiro, ed. annot., *Gishi Wajinden, Gokanjo Yamatoden, Sōsho Wakokuden, Suisho Wakokuden*, Iwanami, 1976 (1st edition 1951).

Wappenschmidt, Friederike, "Die Konferenz der Pflanzen: Gartenpanoramen der Antiken- und Chinamode", *Pantheon*, 57, 1999, pp. 109–115.

Watabe Tadayo *et al.*, eds, *Ine no Ajiashi 3: Ajia no naka no Nihon inasaku bunka*, Shōgakukan, 1987.

Watabe Tadayo *et al.*, *Ine no taichi*, Shōgakukan, 1993.

Watabe Takeshi, "Chūgoku nōsho kōshokuzu no ryūden to sono eikyō ni tsuite", *Tōkai daigaku kiyō*, 46, 1986, pp. 1–36.

Watabe Takeshi, "Tanyū shukuzu naka no kōshokuzu to Kōyasan Henshōson-in shozō no shokuzu ni tsuite", *Tōkai daigaku kiyō* 50, 1988, pp. 19–59.

Watabe Takeshi, *Gazō ga kataru Chūgoku no kodai*, Heibonsha, 1991.

Watabe Takeshi, "Chūgoku nōsho ni mirareru shizenkan to gijutsu shisō", photocopy, University of Tōkyō, 1996–1997.

Watanabe Akihiro, "Nifuda mokkan kara mita Nara jidai no kakuchi no toku-sanbutsu", in Ōtsuka Hatsushige *et al.*, eds, *Kōkogaku ni yoru Nihon rekishi 2, Sangyō 1*, Yūzankaku, 1996, pp. 67–70.

Watanabe Makoto, *Jōmon jidai no shokubutsu shoku*, Yūzankaku, 1984 (1st edition 1975).

Watanabe Makoto, "Yokozuchi no kōko mingugakuteki kenkyū", *Kōkogaku zasshi* 70 (3), 1985, pp. 52–93.

Watanabe Makoto, "Kenkarui", *Kikan kōkogaku* 14, Feb. 1986, pp. 32–35.

Watanabe Makoto, Esaka Teruya, *Kodaijin no kurashi o saguru*, Fukutake shoten, 1990 (1st edition 1986).

Watanabe Makoto, "Jōmon jidai no keizai kiban", in Ōtsuka Hatsushige *et al.*, eds, *Kōkogaku ni yoru Nihon rekishi 2, Sangyō 1*, Yūzankaku, 1996, pp. 27–38.

Watanabe Makoto, "Jōmon jidai no tabemono", in Haga Noboru, Ishikawa Hiroko, *Shoku seikatsu to shokumotsushi*, "Zenshū Nihon no shoku bunka", vol. 2, Yūzankaku, 1999, pp. 15–32.

Wright, Michael, *The Complete Handbook of Garden Plants*, London: Michael Joseph/ Rainbird, 1984.

Yamada Masahisa, "Inasaku gijutsu", in Ishino Hironobu *et al.*, eds, *Seisan to ryūtsū 1*, "Kofun jidai no kenkyū", vol. 4, Yūzankaku, 1991, pp. 36–49.

Yamaguchi Hirofumi, "Sakumotsu to zassō no kita michi", in Takahara Hikaru, Murakami Noriaki, *Kankyōshi o toraeru gihō*, vol. 6 of Yumoto Takakazu, ed., *Shiirizu Nihon rettō no sanman gosen nen hito to shizen no kankyōshi*, Bun'ichi sōgō shuppan, 2011.

Yamaguchi Sadao, "Yakihata no shomondai", *Minkan denshō* 2 (5), 1937, p. 41.

Yamaguchi Sadao, "Yakihata no chiriteki bunpu, sono ta", *Chirigaku hyōron* 14 (1), 1938, pp. 1–23.

Yamaguchi Yaichirō, *Tōhoku no muramura*, Kōshunkaku shobō, 1943.

Yamaguchi Yaichirō, *Tōhoku chihō no yakihata kankō*, "Yamaguchi Yaichirō zenshū", vol. 3, Sekai shuppan, 1972 (1st edition 1944).

Yamamoto Kiyoshi, ed., *Izumo no kuni fudoki no maki*, "Fudoki no kōkogaku", Dōseisha, 1995.

Yamanaka Yutaka, *Heianchō no nenjūgyōji*, Hanawa shobō, 1994 (1st printing 1972).

Yanagita Kunio, Minzokugaku kenkyūjo ed., *Sōgō Nihon minzokugoi*, 5 vols, Heibonsha, 1956.

Yanagita Kunio, *Chimei no kenkyū*, "Yanagita Kunio zenshū", vol. 20, Chikuma shobō, 1990 (1st edition 1936).

Yasuda Yoshinori, *Sekaishi no naka no Jōmon bunka*, Yūzankaku, 1987.

Yin Shaoting, *Yunnan wenzhi wenhua*, 2 vols, Yunnan jiaoyu chubanshe, 1993 (1st printing 1991).

Yin Shaoting, *Yunnan daogenghuo zhongzhi*, Yunnan renmin chubanshe, 1994.

Yin Shaoting, Shirasaka Shigeru, Lin Hong, *Unnan no yakihata*, Nōrin tōkei kyōkai, 2000.

Yoshida Atsuhiko, *Hōjō to fushi no shinwa*, Seidosha, 1990.

Yoshida Shigeki, "Kodai no chimei kara mita no to hara no shomondai", *Higashi ajia no kodai bunka* 16, summer 1978, pp. 170–181.

Yoshida Shigeki, *Kompakuto-ban Nihon chimei jiten*, Shinjinbutsu ōraisha, 1994 (1st printing 1991).

Yoshida Shūji, *Nihon kodai shakai hensei no kenkyū*, Hanawa shobō, 2010.

Yoshida Takashi, *Ritsuryō kokka to kodai shakai*, Iwanami, 1983.

Yoshida Toshihiro, "Chūsei nōgyō no tenkai to shuson-ka genshō", Taranchevski, Detlev *et al.*, eds, *Chūsei seiō to Nihon: tashu to bunken no jidai*, Yoshikawa kōbunkan, 2009.

Yoshizaki Shōichi, "Kodai zakkoku no kenshutsu", *Kōkogaku jānaru* 355, Dec. 1992.

Yun So Sot, *Kankoku no shoku bunkashi*, Domesu, 1995.

Zhongyao dacidian, Jiangsu Xinyixueyuan ed., 3 vols, Shanghai: Renmin chubanshe, 1977.

Index